The Use of Synthetic Environments for Corrosion Testing

P. E. Francis and T. S. Lee, editors

Special Technical Publication (STP) 970

1916 Race Street, Philadelphia, PA 19103

ASTM Publication Code Number (PCN) 04-970000-27
ISBN 0-8031-0977-6

NOTE

The Society is not responsible, as a body,
for the statements and opinions
advanced in this publication.

Peer Review Policy

Each paper published in this volume was evaluated by three peer reviewers. The authors addressed all of the reviewers' comments to the satisfaction of both the technical editor(s) and the ASTM Committee on Publications.

The quality of the papers in this publication reflects not only the obvious efforts of the authors and the technical editor(s), but also the work of these peer reviewers. The ASTM Committee on Publications acknowledges with appreciation their dedication and contribution of time and effort on behalf of ASTM.

Printed in Baltimore, MD
March 1988

Foreword

The symposium on the Use of Synthetic Environments for Corrosion Testing was presented at the National Physical Laboratory, Teddington, Middlesex, England on 10–12 Feb. 1986. The symposium was sponsored by ASTM Committee G-1 on Corrosion, Deterioration, and Degradation of Materials and the National Physical Laboratory. P. E. Francis, National Physical Laboratory, and T. S. Lee, National Association of Corrosion Engineers, served as chairmen of the symposium and are editors of the resulting publication.

Contents

Introduction

The most reliable indication of the corrosion performance of a material is its service history. However, there are many occasions when this is not available or it may be necessary to ascertain the influence of minor changes in the specification of a material or in the composition of the environment on the material's corrosion performance. The use of field trials can provide useful information, but such tests require a reasonable exposure time, and the experimental variables may not be precisely controllable, so that some speculation may be necessary on the interpretation of the outcome of the trials.

In most cases laboratory test methods can provide valuable information on the probable service performance and the influence of deviations from normal operating conditions. Furthermore, laboratory tests will allow a ranking of materials and enable the engineer to make a sensible selection of materials for service in a particular environment.

To avoid misleading results, standard procedures are increasingly used to specify the corrosion performance of metallic engineering materials. Many of these procedures rely on simulated natural environments and the specification of these synthetic environments is probably the most critical aspect of corrosion performance testing. In some instances the synthetic environment is chosen so that the corrosion process is accelerated. Other synthetic environments are formulated to simulate the natural environment as closely as possible, and sometimes a synthetic environment is only used because the natural environment is not readily accessible, for example, corrosion tests in body fluids and in seawater when natural seawater is not readily available.

The need for careful assessment of the effectiveness of the chemical composition of the synthetic environment in providing meaningful data is obvious. Some synthetic environments are specific to one industrial application, for example, steel in concrete, while others may relate to a range of industrial applications, for example, tests for stress corrosion cracking of metallic materials. Thus ASTM Committee G-1 on Corrosion, Deterioration, and Degradation of Materials in the United States jointly with the National Physical Laboratory in the UK sponsored a conference on the use of synthetic environments for corrosion testing that took place at the National Physical Laboratory. The aim of the conference was to provide a forum for an exchange of information on the effectiveness of synthetic environments in providing a prediction of long-term material performance.

The papers from this conference form the basis of this special technical publication (STP). They cover environments for tests relating to atmospheric corrosion, the food industry, body fluids, microbial corrosion, environmental fracture, potable waters, seawater, steel in concrete, flue gases, and so forth. The synthetic environments have been incorporated in test procedures involving exposure tests, electrochemical tests, erosion corrosion tests, and fracture tests.

Thus the 22 papers included in this STP provide the reader, with a very wide range of experience in the formulation and testing of synthetic environments and will familiarize them with the limitations and advantages associated with synthetic environments. For the reader unfamiliar to a particular application this STP will provide an excellent starting point.

P. E. Francis

National Physical Laboratory
Teddington, Middlesex, England; symposium chairman and editor.

T. S. Lee

National Association of Corrosion Engineers
Houston, TX 77218; symposium chairman and editor.

Gerhardus H. Koch,[1] J. Michael Spangler,[2] and Neil G. Thompson[3]

Corrosion Studies in Complex Environments

REFERENCE: Koch, G. H., Spangler, J. M., and Thompson, N. G., **"Corrosion Studies in Complex Environments,"** *The Use of Synthetic Environments for Corrosion Testing, ASTM STP 970,* P. E. Francis and T. S. Lee, Eds., American Society for Testing and Materials, Philadelphia, 1988, pp. 3–17.

ABSTRACT: Corrosion studies have typically been performed in relatively simple environments where variations in corrosion behavior are determined as functions of one variable at a time. However, the desire to more closely simulate actual service environments requires more complex test environments. To study the effects of environmental variables on the corrosion behavior of materials, a statistical approach is highly desirable. In this paper, a sequential experimental design approach for complex experimental environments will be described. Examples are given where this approach is used to study the effect of solution variables in complex environments on the electrochemical parameters for different alloys.

KEY WORDS: corrosion studies, experimental design, complex environments, screening experiments, main effects, interactions, regression analysis

Synthetic environments which have been used to simulate actual service environments in laboratory corrosion studies are usually relatively simple. An example is the 3.5% aqueous sodium chloride (NaCl) solution, which for years served to simulate seawater. In the last decade, researchers realized that actual environments are far more complex and that several of the trace elements present in actual environments can have considerable effects on the corrosion performance of materials. Complex laboratory environments are being used more frequently to simulate actual service conditions. However, in these complex environments, the effect on the corrosion performance of materials of changing one variable at a time is typically studied. This methodology not only may result in large numbers of corrosion experiments, but also, the observed effect may be due to synergistic effects between other constituents in the test environments, which go undetected.

In this paper, a methodology is developed to handle the effects of a large number of variables on the corrosion behavior of materials and to better simulate fluctuations of chemical species within these complex environments. By using an experimental design, a mathematical expression can be developed that relates a dependent variable or response, such as corrosion rate, to a number of independent solution variables or factors, such as species concentration, temperature, and pH. These expressions generally include main effect terms (linear), interactions (cross products), and quadratic terms. Higher-order terms, such as cubic and three-way interactions, could also be evaluated depending on their significance or prior physical evidence for their existence.

[1]Section Manager, Battelle Columbus Division, Columbus, OH 43201.
[2]Principal research scientist, Battelle Columbus Division, Columbus, OH 43201.
[3]President, Cortest Columbus, Columbus, OH 43085.

3

In this paper, a sequential and cost-effective approach for developing expressions that describe the corrosion behavior of materials in complex environments will be developed. Following a general outline of this experimental approach, some examples will be presented where experimental designs have been used successfully to describe corrosion of alloys in complex laboratory environments which closely simulated actual field environments.

Experimental Design

Screening Experiments

During the initial investigation of a complex environment, it is likely that a large number of variables could have a significant effect on the corrosion performance of a material. With the aid of fractional factorial designs, it is possible to simultaneously evaluate a large number of variables with relatively few experiments. Subsequent regression analyses will then provide good estimates of the significance of each environmental variable for each measured corrosion parameter.

Initial screening can be performed by using a two-level saturated fractional factorial design of resolution III where k variables are evaluated in $k + 1$ experiments [1,2]. For example, up to 15 variables may be evaluated with 16 experiments or up to 31 variables with 32 experiments. An example of a screening test is shown in Table 1 where 15 solution variables are tested in 16 experiments. The − and + symbols in the figure refer to low and high values of the various solution variables, e.g., species concentration in the test solution, or temperature. At least four midpoints with average values of the variables should be selected to provide an estimate of experimental error.

The resolution III design provides estimates of main effect terms influenced by one another. However, two variable (and higher) interactions are linked with the main effect terms. Therefore, a significant main effect cannot be differentiated from certain significant two variable interactions. Recent experience with complex chemical environments has confirmed the presence of numerous two-factor interactions. With large systems, the possible number of two-factor interactions increases rapidly. For example, with 15 solution variables (factors) there are 105 possible two-factor interactions; perhaps ten or more of these may be significant. If these interaction effects are confounded with main effects, then it is possible that nonsignificant variables will be carried to the next experimental design phase. The additional variables will result in artificially large designs for the next phase with a substantial increase in the number of required experiments.

To estimate the main effects of solution variables free and clear of other main effects and any interactions, a design of resolution IV is recommended (see Table 1). These resolution IV designs require approximately twice the number of experiments as a saturated (resolution III) design, but provide clear estimates of each of the variables and estimates of blocks of interactions between two variables. However, with a resolution IV design, three- and higher factor interactions are confounding the main effect terms. The higher-order interaction terms are generally small compared with main effects and two-factor interactions, so that a good estimate of the main effects can generally be obtained.

The screening approach described above can utilize any of various fractional factorial designs as taught by Box and Hunter [1,2], and numerous other authors [3-6]. These designs are useful in a variety of experimental environments and are a first choice. However, several limitations may reduce their usefulness under certain conditions. Two such conditions are given below:

(1) Certain combinations of test variables or factor combinations are impossible to achieve. For example, at pH 7 or above, it is generally impossible to achieve high levels of Fe in solution. These constraints do not necessarily reduce the need to evaluate each variable over the intended range, but only to avoid specific combinations of variables.

TABLE 1—*Partial factorial design used for the screening experiments.*

DESIGN OF RESOLUTION III

Variable	Al	Ca	Cr	Cu	Fe	Mo	N	P	Si	Br	I	H₂S	O₂	SO₂	pH
1	−	−	+	+	+	+	+	+	−	−	−	−	−	+	−
2	−	+	+	+	+	−	−	−	−	+	+	+	−	−	−
3	+	−	−	−	+	+	−	+	−	+	+	−	+	−	−
4	+	+	−	−	+	−	+	−	−	−	−	+	+	+	−
5	−	−	+	−	−	−	+	+	+	+	−	+	+	−	−
6	−	+	+	−	−	+	−	−	+	−	+	−	−	+	−
7	+	−	−	+	−	−	−	+	+	−	+	+	−	+	−
8	+	+	−	+	−	+	+	−	+	+	−	−	−	−	−
9	−	−	−	+	−	+	+	−	−	−	+	+	+	−	+
10	+	+	−	+	−	−	−	+	−	+	−	−	+	+	+
11	+	−	+	−	−	+	−	−	−	+	−	+	−	+	+
12	−	+	+	−	−	−	+	+	−	−	+	−	−	−	+
13	−	−	−	−	+	−	+	−	+	+	+	−	−	+	+
14	+	+	−	−	+	+	−	+	+	−	−	+	−	−	+
15	+	−	+	+	+	−	−	−	+	−	−	−	+	−	+
16	+	+	+	+	+	+	+	+	+	+	+	+	+	+	+

FOLD OVER DESIGN TO PRODUCE A RESOLUTION IV DESIGN

Variable	Al	Ca	Cr	Cu	Fe	Mo	N	P	Si	Br	I	H₂S	O₂	SO₂	pH
1	+	+	−	−	−	−	−	−	+	+	+	+	+	−	+
2	+	−	−	−	−	+	+	+	+	−	−	−	+	+	+
3	−	+	+	+	−	−	+	−	+	−	−	+	−	+	+
4	−	−	+	+	−	+	−	+	+	+	+	−	−	−	+
5	+	+	−	+	+	+	−	−	−	−	+	−	−	+	+
6	+	−	−	+	+	−	+	+	−	+	−	+	+	−	+
7	−	+	+	−	+	+	+	−	−	+	−	−	+	−	+
8	−	−	+	−	+	−	−	+	−	−	+	+	+	+	+
9	+	+	+	−	+	−	−	+	+	+	−	−	−	+	−
10	−	−	+	−	+	+	+	−	+	−	+	+	−	−	−
11	−	+	−	+	+	−	+	+	+	−	+	−	+	−	−
12	+	−	−	+	+	+	−	−	+	+	−	+	+	+	−
13	+	+	+	+	−	+	−	+	−	−	−	+	+	−	−
14	+	−	+	+	−	−	+	−	−	+	+	−	+	+	−
15	−	+	−	−	−	+	+	+	−	+	+	+	−	+	−
16	−	−	−	−	−	−	−	−	−	−	−	−	−	−	−

(2) A significant quantity of previously generated hard data exists for the particular response, e.g., corrosion rate, and test variables. but the previous experimental plan does not fit the requirement of the proposed new design. In this case, it may be desirable to retain the original database and to build a new design around the previous work. This could provide a full evaluation of each of the proposed variables with significantly fewer additional experiments.

If either or both of the above conditions exists, then a customized screening design could be developed with the aid of an advanced computer algorithm. Recent work with such a computer program has shown that high quality designs can be produced by augmenting previous work while avoiding impossible factor combinations [7]. The computer algorithm used in this program is called COED® (Computer Optimized Experimental Design).[4] The COED program is an interactive computer program that selects an optimal subset of experiments to be carried out from the total number of possible experiments. The selection process is based on determinant optimality theory; that is, it determines the experiments which minimize the integrated error of prediction over all experimental design points. COED can be used to generate an entire experimental design or to augment an existing experimental design.

At the completion of the experiments based on the screening designs, the test variables are analyzed with respect to the corrosion parameters studied. The purpose of the analysis is

(1) to determine which variables have a significant effect on a corrosion parameter of a particular material at a given probability;

(2) to establish the magnitude and direction of the effect; and if possible, and

(3) to determine the significance of groups of two-factor interactions.

Statistical analytical tools include forward regression with extended analysis of variance. Typically, the statistical package calculates two useful statistics for each of the corrosion parameters, the F ratio and the adjusted R^2 [8,10]. The F ratio is the ratio of two variances; i.e., the sum of squares explained by each factor when entered in the equation, divided by the residual mean square (error). In general, when the calculated F ratio for a variable is large, it indicates that a large amount of experimental variation is explained by this term compared to the error variation. If a calculated F ratio exceeds the appropriate tabulated F statistic, then it can be assumed that the solution variable has a statistically significant effect on a particular corrosion parameter. A 90% probability that a test solution variable is significant is usually acceptable for most experimental work, and implies that a 10% probability of being wrong can be accepted in assuming that the variable has a significant effect.

The adjusted R^2 values indicates the percentage of the observed variation for the corrosion parameters that can be explained by the solution variables. An adjustment is made for the number of experiments and the number of parameters estimated in the regression.

Interactions and Higher Order Terms

Once the main effects of solution variables on the corrosion behavior of a material are determined, an additional level of experimental design can be applied to examine interactions between two variables and higher order effects. At this design level, the following information is needed:

(1) a list of environmental variables that are known by screening tests to have a significant effect on the corrosion performance of materials,

(2) the maximum and minimum levels for each variable included,

(3) a list of interactions between two variables that are suspected to affect the corrosion behavior,

---

[4]COED is a registered trademark of B. F. Goodrich Co.

(4) any combination between variables that must be avoided (constraints), and
(5) prior data that can be incorporated in the design.

With the above information available, the most cost effective approach can be developed for individual cases. A simple case would involve an environment with only a few (three or four) significant variables, no constraints, and no additional considerations. For this case, a Box-Behnken design [8] of 15 or 27 experiments would provide all the information needed to support a full quadratic equation for the three or four variables.

However, the following complexities may be encountered and require a more involved design:

(1) A large number of variables (greater than 10) has been identified as significant.
(2) There are several interactions between solution variables that are suspected to have significant effects of the corrosion behavior.
(3) There are several constraints.
(4) It would be desirable to use data from previous experimental programs.
(5) Data are available from previous screening experiments.

An experimental design either for such a complex environment or to complement existing data, can be created using a sequential approach and an appropriate computer aided design program, such as COED. During the first iteration, a design can be developed to estimate main effect and interaction terms. This can be achieved by the computer program by augmenting the screening design and other forced design points while avoiding constraints. The number of additional experiments designed by the algorithm will be determined by the balance of the forced-in experiments, the complexity introduced by constraints, and the number of new terms that are to be estimated by the design.

After the additional experiments required by the experimental design have been performed, the results are analyzed using an appropriate statistical analysis. The analysis objectives are:

(1) to determine and select the significant two-factor interactions;
(2) to reevaluate the significance of each of the main effect terms;
(3) to estimate the strength of the best model to predict the measured corrosion performance; and
(4) to attempt to estimate the significance of the quadratic terms.

The results of the statistical analysis would to a large extent determine the next steps to be taken. For example, a mathematical expression with main effect and two-factor interaction terms may explain most of the experimental variation in the corrosion parameters measured. However, if these terms do not sufficiently explain the experimental variation of the measured parameter, a second design iteration would be performed to add quadratic terms to the final model. Results from the additional experiments would then be subject to appropriate statistical analysis.

The mathematical expressions describing corrosion parameters as functions of solution variables can be prepared with the aid of various statistics software packages such as SPSS [9], SAS [10] and Minitab [11]. These software packages provide many different multiple regression techniques. Forward step-wise regression is particularly useful. This technique starts with no variables in the model and adds variables one at a time. For each variable not in the model, an F statistic is calculated that reflects the variable's contribution to the model if it were to be included next. The variable which contributes most to the model is added provided that the F statistic is significant at the 90% significance level. Variables that are added to the model do not necessarily stay there. After a variable is added, a backward elimination is performed to delete any variable that does not contribute significantly to the model. This process continues until no variables can be added or deleted. The result of this regression analysis is a model in which the selected corrosion parameters are a function of significant solution variables. Such an empirical model will take on the form of:

$$Y = a_0 + \Sigma\, a_1 x_i + \Sigma\, a_2 x_i^2 + \Sigma\, a_3 x_i x_j \tag{1}$$

where

$\Sigma\, a_1 x_i$ and $\Sigma\, a_2 x_i^2$ represent the main and quadratic effects of solution variables x_i, and $\Sigma\, a_3\, x_i x_j$ represent the interactions.

The main effect of a solution variable on a corrosion parameter represents the net effect of this variable in the presence of all other variables at an average value. Thus, an observed main effect represents the effect of a species by itself on the corrosion of an alloy. If the measured synergistic effects are significant, the variation in corrosion parameter cannot be explained by the main effects only and the interactive terms should be entered into the empirical model. Interactions between x_i and x_j occur when the rate of change of y with increasing x_i is different at low and high values of variable x_j. This is illustrated schematically in Fig. 1.

Complex Simulated Environments

In this section of the paper, examples are discussed in which experimental designs were applied to study the effect of complex environments on the corrosion performance of alloys. The different levels of the experimental design will be discussed.

Example 1, Corrosion of Stainless Steels in Condensates of High-Efficiency Domestic Furnaces [12,13]

There is an increased interest in the use of residential high-efficiency gas furnaces and boilers. For maximum efficiency, residential heating equipment must be designed to operate in the condensing mode. Flue gas condensate is corrosive with respect to stainless steels [12]. The principal variables of the flue gas condensate that may affect the corrosion of stainless steels are pH, chloride, fluoride, nitrite, nitrate, sulfate, and reduced sulfur oxanions (i.e., S^x where $x = -2$ to $+3$). To determine which of these variables had a significant effect on the corrosion of stainless steel, a fractional factorial design was selected for the experimental evaluation. The selected design included 19 experimental conditions of condensate composition. Of these, three test solutions, numbers 17, 18, and 19, were duplicated to test the reproducibility of the experimental results. The test matrix using the seven variables at the various concentrations is given in Table 2. Potentiodynamic anodic polarization experiments were used to determine the corro-

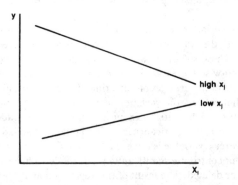

FIG. 1—*Schematic diagram showing interaction between* x_i *and* x_j. *Y is the measured corrosion parameter and* x_i *and* x_j *are the solution variables.*

TABLE 2—*Partial factorial design test matrix to study the effect of solution variables of condensates from high-efficiency domestic furnaces on the corrosion of stainless steels* [13].

Solution No.	pH	Temperature °F[a]	Concentration, log of concentration $(g/m)^3$					
			SO_4^{-2}	Cl^-	F^-	NO_3^-	NO_2^-	$S_2O_3^{-2}$
1	0.4	120	4.7	0	0	4	3.3	3
2	4.0	180	0.7	0	0	1	0.3	3
3	4.0	120	4.7	4	0	1	0.3	0
4	0.4	180	0.7	4	0	4	3.3	0
5	4.0	120	0.7	0	2.7	4	3.3	0
6	0.4	180	4.7	0	2.7	1	0.3	0
7	0.4	120	0.7	4	2.7	1	0.3	3
8	4.0	180	4.7	4	2.7	4	3.3	3
9	4.0	180	0.7	4	2.7	1	0.3	3
10	0.4	120	4.7	4	2.7	4	3.3	0
11	0.4	180	0.7	0	2.7	4	3.3	3
12	4.0	120	4.7	0	2.7	1	0.3	3
13	0.4	180	4.7	4	0	1	0.3	3
14	4.0	120	0.7	4	0	4	3.3	3
15	4.0	180	4.7	0	0	4	3.3	0
16	2.9	120	0.7	0	0	1	0.3	0
17	2.2	150	2.7	2	1.3	2.5	1.8	1.5
18	2.2	150	2.7	2	1.3	2.5	1.8	1.5
19	2.2	150	2.7	2	1.3	2.5	1.8	1.5

[a]$°C = \frac{5}{9} (°F - 32)$.

sion behavior of Type 304L and 316L stainless steels in the simulated flue gas condensate solutions listed in Table 2. The corrosion parameter selected for analysis in this study was the "pitting margin" defined as $\Delta E_{pit} = E_{pit} - E_{pas}$ (Fig. 2).

Following the experimental work, a multiple regression analysis was conducted for the values of ΔE_{pit} and the corresponding environmental variables from the test matrix. The results of this analysis, presented in Table 3, show the probability of significance based on the calculated F ratio and the regression coefficients for those variables with the highest probability of significance. The results of the regression for Type 316L stainless steel indicated that chlorine (Cl) and temperature had the most significant effect on the pitting margin while chlorine and the ratio of nitrate and nitrite (NO_3^-/NO_2^-) significantly affected the pitting margin of Type 304L stainless steel.

Based on the results of these experiments, two solutions were selected to study a range of additional alloys. Solution number 13 was found to be an aggressive pitting solution, whereas solution number 17 was not aggressive.

Example 2, Stress-Corrosion Cracking in Continuous Digesters [14]

Stress-corrosion cracking has recently been observed in the welds and near-weld regions of continuous digesters. Continuous digesters are large carbon steel vessels used in the paper industry to convert wood chips or sawdust into a cellulose pulp by a reaction with a hot solution containing primarily sodium hydroxide (NaOH) and sodium sulfide (Na_2S). Additional chemicals in this solution are elemental sulfur, S^0, sodium carbonate (Na_2CO_3), sodium thiosulfate ($Na_2S_2O_3$), sodium chloride (NaCl) and sodium sulfite (Na_2SO_3). In many metal-environment systems, including carbon steel in caustic solutions [15], stress-corrosion cracking occurs in a narrow region of potentials corresponding to the active-to-passive transition region of anodic polarization curves. Often, various species present in the environment can move the free-corro-

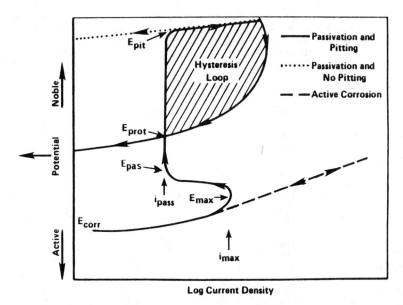

Log Current Density

FIG. 2—*Schematic diagram of typical anodic potentiodynamic polarization curves showing important polarization parameters: E_{cor} = corrosion potential; E_{pit} = potential at which pits initiate on forward scan; E_{prot} = potential at which pits repassivate on reverse scan; i_{max} = current density at active peak; i_{pass} = current density in passive range. $\Delta E_{pit} = E_{pit} - E_{pas}$.*

TABLE 3—*Results of multiple regression analyses for Type 316L and Type 304L test in simulated condensates of high-efficiency domestic furnaces [13].*

	316L, ΔE_{pit}			304L, ΔE_{pit}	
Solution variable	% Probability of significance	Coefficient[a]	Solution variable	% Probability of significance	Coefficient
Cl^-	100	−173	Cl^-	98.0	−127
Temp.	93.7	64	NO_2^-/NO_3^-	89.7	+84
SO_4^{-2}	74.8	...	SO_4^{-2}	50.0	...
S^+	59.1	...	S^+ x	41.5	...
NO_2^-/NO_3^-	56.8	...	Temp.	32.6	...
pH	45.3	...	pH	18.6	...
F^-	35.5	...	F^-	11.8	...
Constant	...	528	Constant	...	464

[a]*A negative sign indicates that an increase in solution variable has a detrimental effect on corrosion behavior, a positive sign indicates a beneficial effect.*

sion potential (E_{cor}) into or out of the cracking range, thereby inducing or eliminating stress-corrosion cracking. An experimental design was set up to determine what chemical variables in the digester liquor would significantly affect E_{cor}, and thus the susceptibility of carbon steel to stress-corrosion cracking.

A Box-Behnken factorial design [8] was employed to make 25 different solutions containing NaOH, Na₂S, Na₂S₂O₃, and S⁰ as variables. The selected ranges of concentrations of these four components were based upon those reported for digesters [15] and are shown in the test matrix

TABLE 4—*Solution compositions of simulated continuous digester liquids based on a Box-Behnken experimental design* [14].

Number	Solution Concentration in g/L Composition			
	NaOH	Na$_2$S	Na$_2$S$_2$O$_3$	S^0
1	80	50	7	1
2	80	10	7	1
3	20	50	7	1
4	20	10	7	1
5	50	30	11	2
6	50	30	11	0
7	50	30	3	2
8	50	30	3	0
9	50	30	7	1
10	80	30	7	2
11	80	30	7	0
12	20	30	7	2
13	20	30	7	0
14	50	50	11	1
15	50	50	3	1
16	50	10	11	1
17	50	10	3	1
18	50	30	7	1
19	80	30	11	1
20	80	30	3	1
21	20	30	11	1
22	20	30	3	1
23	50	50	7	2
24	50	50	7	0
25	50	10	7	2
26	50	10	7	0
27	50	30	7	1

in Table 4. The table indicates that the matrix was designed at three levels for each solution component. Three of the 27 solutions, 9, 18, and 27, were identical, and were intended to test the reproducibility of the data. After measuring E_{cor} for a set period of time in each of the designed test solutions, the results were analyzed using a multiple regression routine. The multiple linear regression analysis of the results indicated that there was a strong correlation between E_{cor} and the Na$_2$S and S^0 contents of the test solutions. The NaOH and Na$_2$S$_2$O$_3$ contents were gave less significant results. The regression equation obtained was:

$$E_{corr} = -659 - NaOH - 15.8\,Na_2S - 24.2\,S^0 + 35.5\,Na_2S_2O_3$$

$$+ 0.129(Na_2S)^2 + 2.63\,Na_2S \cdot S^0 - 2.54(Na_2S_2O_3)^2 \text{ in V SCE} \quad (2)$$

Using this equation, the free-corrosion potential can be predicted for various chemistries, and these agreed with the potentials measured in the liquor near areas where stress-corrosion cracking was observed.

Example 3, Effect of SO$_2$ Scrubber Environments on the Corrosion of Alloys [7,17]

Coal-fired power plants provide an increasingly important source of energy. Sulfur dioxide (SO$_2$) is a major ecological concern resulting from the increased use of coal. Scrubbing of the flue gas generated by combustion, flue gas desulfurization (FGD), is presently the most common method of removing SO_2 emissions. Corrosion of FGD system components is a serious problem in operating these systems. When corrosion was first recognized as a serious problem, Ca, Cl, and Mg concentrations and pH were considered the main environmental variables in FGD systems. However, analyses of these environments [18] indicated that a wide range of elements was present, and that some of these elements could have a significant impact on the corrosion performance of construction materials.

In a recent laboratory study, the individual and synergistic effects of several FGD solution variables on the corrosion behavior of different alloys were studied [7,17]. These alloys were a commercially pure titanium alloy, grade 2, an austenitic stainless steel, Type 317L, and a nickel-base alloy, G3. For the initial screening of the trace elements in FGD environments, 15 variables were selected to determine their effects on the corrosion performance of the alloys studied. A partial factorial design of resolution IV was developed so that significant main effects free and clear of other main effects and two-factor interactions could be obtained (Table 1). The ranges of the different solution variables are listed in Table 5.

It should be noted that several midpoints were included in the design to estimate experimental error. Potentiodynamic polarization experiments were performed in the various test solutions to determine the effect of the solution variables on the following corrosion parameters: the free-corrosion potential, E_{cor}; the corrosion current density, i_{cor}; the passive current density, i_{pas}; and the pitting potential, E_{pit} (Fig. 2). Following the polarization experiments, regression analyses were performed to determine the significance of each solution variable studied.

Based on the results of the screening experiments, solution variables with a high probability of significance (>90 percent) were selected and included in the next level of design. The following solution variables were selected for this main matrix of experiments: Al, Cl, Cr, Cu, F, Fe, I, P, hydrogen sulfide (H_2S), O_2, temperature, and pH for titanium grade 2; Al, Cl, Cr, Cu, F, Fe, Mo, N, sulfur dioxide (SO_2), O_2, temperature, and pH for Type 317L stainless steel; and Al, Br, Cu, Cl, I, N, H_2S, O_2, SO_2, temperature, and pH for alloy G3. The experimental design selected

TABLE 5—*Concentration range of solution variables and the chemical compound used for solution make-up simulating flue gas desulfurization environments* [7,17].

Variable[a]	Concentration range
Al	1–2000 g/m³
Ca	1–1000 g/m³
Cr	1–500 g/m³
Cu	1–1000 g/m³
Fe	1–2000 g/m³
Mo	1–200 g/m³
N	1–200 g/m³
P	1–500 g/m³
Si	1–100 g/m³
Br	1–500 g/m³
I	1–100 g/m³
H₂S	0–5 ml/l
SO₂	0–8%
O₂	0.001–0.3%
pH	1–6
temperature	50–90°C
Cl	1–100 000 g/m³
F	1–10 000 g/m³

[a]Na and SO_4 were used as the balancing cation and anion, respectively.

for these variables was generated by COED, which also allowed for an estimate of 40 two-factor interactions. The screening experiments were forced into the design and additional experiments were added to allow evaluation of the interaction between significant variables. The experimental design called for 117, 127, and 81 experiments for titanium, Type 317L stainless steel and alloy G3, respectively. The results of the screening tests and the additional experiments were combined and analyzed using a forward step-wise regression routine available on "Minitab" [11]. The results of the regression analysis were displayed both in tabular and schematic form so that the effects, both single and synergistic, could be studied. An example of the schematic presentation of the regression results is shown for titanium in Fig. 3. The top part of the diagram indicates the relative magnitude and direction of the significant main effects (>90 percent probability of significance) on the corrosion parameter studied (E_{cor} in this case). When a bar points up, increasing the concentration of the species in the test solution results in an increase in E_{cor}, whereas a bar point down indicates an increase in concentration results in a decrease in E_{cor}. The lower part of the diagram indicates the significant interactions between the two variables.

The results of the study indicated that of the solution variables examined, Al, Cr, Cu, Fe, P, Cl, Fe, O_2, temperature, and pH had significant main effects on the corrosion of titanium; Al, Cl, Cu, Fe, SO_2, temperature, and pH on the corrosion of Type 317L stainless steel; and Al, Cu, Fe, Cl, SO_2, temperature, and pH on the corrosion of Alloy G3. Moreover, several significant interactions between the solution variables were detected. Based on the results of these studies, we developed an understanding of the behavior of alloys in actual SO_2 scrubber environments. Selection of materials for construction of scrubber components could then be made with higher confidence.

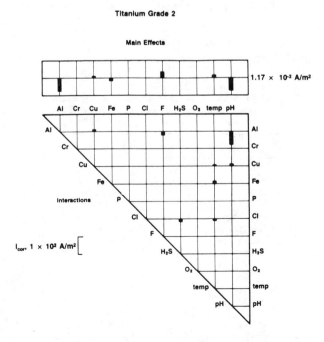

FIG. 3—*Schematic diagram showing relative magnitude and direction of significant main effects and two-factor interactions on* i_{cor} *of titanium grade 2.*

Example 4, Long-term Performance of Materials Used for High-Level Waste Packaging [19]

Studies were performed to investigate the corrosion resistance of container materials for high-level nuclear waste and spent fuel [*19*]. To study the corrosion behavior of carbon steel in high temperature basalt ground water, simulated ground waters were prepared. The effects of the various constituents of the simulated ground waters on the corrosion behavior of 1018 carbon steel were studied using an experimental design similar to that described by SO_2 scrubber environments in Example 3. Again, screening experiments and a computer aided design (COED) were used to determine the significance of main effects and interactions of solution variables on polarization parameters. Table 6 shows the variables that were selected for the potentiodynamic polarization studies. The results of the multiple regression analysis of the polarization parameter and the variables in the simulated ground water solutions were described in predictive equations such as for i_{cor}:

$$\log i_{cor} = -5.54 - 0.87 \text{ pH} - 0.20 \text{ NO}_3 - 0.20 \text{ SiO}_3 + 0.22 \text{ H}_2\text{O}_2 + 0.16 \text{ pH}$$

$$\times \text{ Cl} - 0.28 \text{ Cl} \times \text{CO}_3 + 0.20 \text{ Cl} \times \text{NO}_3 + 0.17 \text{ NO}_3 * \text{CO}_2 - 0.22 \text{ CO}_3 * \text{H}_2\text{O}_2 \quad (3)$$

The results of these studies were used to predict the corrosion performance of carbon steel in basalt ground water. A solution was selected that was extremely aggressive in pitting carbon steel and was used in subsequent work as a standard solution to study the pitting mechanism in carbon steel.

Conclusion

The four examples shown in this paper have indicated that designed experiments are essential in understanding the effects of complex environments on the corrosion behavior of alloys. Depending on the complexity of the environment (i.e., number of solution variables, constraints,

TABLE 6–*High and low concentrations of species selected for electrochemical experiments in synthetic environments as container materials for high-level nuclear waste and spent fuel* [19].

Species	Concentration range
pH	6.0–9.3
Cl^-	100–100 000 g/m³
F^-	10–10 000 g/m³
Fe^{+2}/Fe^{+3}	0.05–100 g/m³
Al^{+3}	0.1–1000 g/m³
CO_3^{-2}/HCO_3^-	0.001–1 M
NO_3^-/NO_2^-	0.1–1000 g/m³ [N]
PO_4^{-3}	0.1–1000 g/m³ [P]
$BO_3^{-3}/B_4O_7^{-2}$	1–1000 g/m³ [B]
SiO_3^{-2}	10–1000 g/m³ [Si]
H_2O_2	0–100 g/m³
ClO_4^-	0–100 g/m³
O_2	0–2% (vapor)
CO	0–1% (vapor)
H_2	1–80% (vapor)

etc.) the design can range from a relatively simple partial factorial design to a complex computer-aided design. The results of such studies can also be used to predict the corrosion performance of alloys as a function of solution composition, which may be very helpful for the selection of materials for various components and structures that are subject to complex environments.

APPENDIX

A Nonrigorous Discussion of Several Statistical Concepts

A *two-level factorial design* includes k factors (independent variables) at two levels for a total of 2^k combinations. If 7 factors are identified for an experimental effort, then there are 2^7 or 128 experimental combinations included in the *full* two-level factorial design. If results were available for each of the 128 trials, then it would be possible to calculate 128 different statistics including:

(1) an average,
(2) seven main effects,
(3) 21 two-factor interactions,
(4) 35 three-factor interactions,
(5) 35 four-factor interactions,
(6) 21 five-factor interactions,
(7) 7 six-factor interactions, and
(8) one seven-factor interaction.

However, it was demonstrated that high level interactions tend to contribute much less to observed variations than do main effects and two- and three-factor interactions [2]. Further, main effects tend to explain more variation than two-factor interactions, which tend to explain more variation than three-factor interactions, etc. For these reasons, when k is large and the objective is to "screen" the k-factors, then a full factorial represents unnecessary capacity when only $k + 1$ effects need to be estimated. In the above example, only eight statistics are needed, an average and seven main effects.

Selection of the appropriate *fraction* of a full factorial design will provide estimates of the desired effects at an equivalent fraction of the experimental effort. However, the "cost" of efficiency is the *confounding* of interaction terms. For example, the effect of x_1x_3 and x_2x_5 may be estimated by the same statistic such that the estimate is the sum of the two two-factor interactions, i.e., their effects cannot be separated. The confounding of four-factor interactions and above is seldom a concern because they are usually not significant. However, the degree of confounding and the confounding pattern of main effects and two- and three-factor interactions should be considered when selecting a partial factorial design. As an aid in describing a design's capability to discriminate between various effects, the term design *resolution* is used. Briefly, a *resolution III* design does not confound main effect terms with one another, but main effects are confounded with two-factor interactions; a *resolution IV* design confounds two-factor interactions with one another, but not with the main effect terms; a *resolution V* design confounds two-factor interactions with three-factor interactions, but not with other two-factor interactions or main effect terms.

After a design is selected, it is recommended that experiments be selected for replication. If the factors are all continuous (such as temperature or species concentration), then the design center point is recommended for replication, and perhaps four or more trials at the mid-level for all factors should be included in the experimental plan. Furthermore, it is important that the complete experimental plan be randomized. The randomization will tend to spread any "bias" or systematic error over all of the factors uniformly; the replication will provide an estimate of the "pure" experimental results during the subsequent analysis.

References

[1] Box, G. E. P. and Hunter, J. S., "The 2^{k-p} Fractional Factorial Design," *Technometrics*, Vol. 3, 1961, pp. 311–351.

[2] Box, G. E. P., Hunter, W. G., and Hunter, J. S., *Statistics for Experimenters*, John Wiley and Sons, New York, 1978.

[3] Plackett, R. L. and Burman, J. P., "The Design of Optimum Multifactorial Experiments," Vol. 33, 1946, pp. 305–325.

[4] Coch, W. G. and Cox, G. M., *"Experimental Designs,"* John Wiley and Sons, New York, 2nd ed., 1957.

[5[Davies, O. L., *"The Design and Analysis of Industrial Experiments,"* Hafner Publishing Co., New York, 2nd ed., 1967.

[6] Hicks, C. R., *"Fundamental Concept in the Design of Experiments,"* Holt, Rinehart, and Winston, New York, 1964.

[7] Koch, G. H., Thompson, N. G., and Spangler, J. M., "The Effects of SO_2 Scrubber Environments on the Corrosion of Alloys," Electric Power Research Institute, EPRI Report CS-4697, Palo Alto, CA, July 1986.

[8] Box, G. E. P. and Behnken, D. W., "Some New Three Level Designs for the Study of Quantitative Variables," *Technometrics*, Vol. 2, 1960, pp. 455–475.

[9] Nie, N. H., Hill, C. H., Jenkins, J. G., Steinbrenner, K., and Bent, D. H., "SPSS, Statistical Package for the Social Sciences," McGraw-Hill Book Co., 1970.

[10] "SAS, Statistical Analysis System," SAS Institute Inc., Cary, NC, 1982.

[11] Ryan, Jr., T. A., Joiner, B. L., and Ryan, B. F., "Minitab," Minitab Project Statistics Department, Pennsylvania State University, University Park, PA, 1985.

[12] Stickford, G. H., et al., "Technology Development for Corrosion-Resistant Heat Exchangers," Gas Research Institute Final Report 8510282, Chicago, IL, Chapter 5.

[13] Hindin, B. and Agrawal, A. K., "Materials Performance of Residential High-Efficiency Condensing Furnaces Using Simulated Environments," *The Use of Synthetic Environments for Corrosion Testing, STP 970* American Society for Testing Materials, Philadelphia, 1986, pp. 274–286 (this publication).

[14] Pednekar, S. P., "A Study of Stress-Corrosion Cracking of Carbon Steel in Simulated Kraft Pulping Liquors," Paper No. 137, *Corrosion 86*, Houston, Texas, March 1986.

[15] Humphries, M. J. and Parkins, R. N., "Stress-Corrosion Cracking of Mild Steels in Sodium Hydroxide Solutions Containing Various Additional Substances," *Corrosion Science*, Vol. 7, 1967, pp. 747–761.

[16] Yeske, R., Final Report to DCRC Project 3544, Institute of Paper Chemistry, Chicago, IL.

[17] Koch, G. H., Thompson, N. G., and Spangler, J. M., "Effect of Solution Species in Wet SO_2 Scrubber Environments on the Corrosion of Alloys," Paper No. 363, *Corrosion 86*, Houston, Texas, March 1986.

[18] Koch, G. H., Thompson, N. G., and Means, J. L., "Effects of Trace Elements in Flue Gas Desulfurization Environments on the Corrosion of Alloys—A Literature Review," Electric Power Research Institute, EPRI Report CS-4374, Palo Alto, CA, Jan. 1986.

[19] Beavers, J. A. and Thompson, N. G., "Long-Term Performance of Materials Used for High-Level Waste Packaging," Compiled by D. Stahl, N. E. Miller, Annual Report, Year Three, April 1984–April 1985, NYREE/CR-3900 BMI-2127, Vol. 4.

DISCUSSION

D. McIntyre[1] *(discusser's question)*—What is the preferred experimental design method when the significant environmental variables are not known?

Koch et al. (authors' response)—A preferred method would be to include all of the suspected environmental variables in the initial experimental design. If the suspected number of variables is large (perhaps >8), a good starting point would be a resolution IV fractional factorial design with replicated center points. After an analysis of the experimental results, a second design and analysis phase would be implemented to refine the model containing only the significant environmental variables (hopefully fewer in number).

J. E. Castle[2] *(discusser's question)*—Could you give an indication of the fraction of the complete factorial design which you have used to illustrate your resolution III and IV experiments.

Koch et al. (authors' response)—Several resolution III designs [*1*] were mentioned including a 2_{III}^{15-11}, 2_{III}^{31-26}, and 2_{III}^{7-4}. The 2_{III}^{15-11} design can be used for up to 15 factors and is a $1/2^{11}$ or $1/2048$ fraction of the 2^{15} full factorial design of 32 768 total combinations. The first 16 trials illustrated in Table 1 represent this design. A second similar fraction is produced by changing each of the signs in the first 16 trials to produce the second 16 trials. The 32 trials together are a fold-over design of resolution IV, i.e., 2_{IV}^{15-10}. A strong "screening" advantage of the resolution III fractional factorials is that if ambiguity exists after analysis of the results, a second folded over fractional design may be implemented to improve design resolution.

[1]Cortest Laboratories, 11115 Mills Rd., Suite 102, Cypress, TX 77429.
[2]University of Surrey, Guildford, Surrey GU2 5XH, United Kingdom.

Gardner S. Haynes[1] and Robert Baboian[1]

Simulating Automotive Exposure for Corrosion Testing of Trim Material

REFERENCE: Haynes, G. S. and Baboian, R., **"Simulating Automotive Exposure for Corrosion Testing of Trim Material,"** *The Use of Synthetic Environments for Corrosion Testing, ASTM STP 970,* P. E. Francis and T. S. Lee, Eds., American Society for Testing and Materials, Philadelphia, 1988, pp. 18–26.

ABSTRACT: To predict the corrosion performance of trim materials on automobiles, it is important that the method of exposure as well as the test environment accurately reflect actual service conditions. The method of exposure must duplicate the various trim configurations. Variables which impact test results include the geometry of the test specimen and area relationships for galvanic effects. Test specimens fulfilling these requirements for trim materials are described. The results of tests in suitable environments for a number of trim systems on various auto-body materials are presented and compared with the service performance of the trim systems.

KEY WORDS: automotive corrosion, road salts, acid precipitation, automotive corrosion testing, automotive trim, stainless steel clad aluminum, field testing, laboratory tests, galvanic corrosion, corrosion of automotive trim

Exterior automotive trim performs a number of important functions in automobiles. These include: (1) shielding the auto-body from mechanical damage (i.e., wheel surrounds and body side moldings), (2) concealing welds and other body panel abutments (window surrounds), (3) corrosion resistance (i.e., of rocker panel moldings), and (4) styling. Corrosion problems associated with exterior automotive trim include: (1) effects of the trim system on the auto-body and (2) corrosion of the trim itself. The problems are caused by the severe corrosivity of the automotive environment as well as the inherent crevice geometry, attachment requirements in the trim/auto-body system, and the susceptibility to paint damage in this area.

Automotive Environment

The corrosivity of the automotive environment has been the subject of numerous investigations [1–8]. These studies have conclusively proven the relationship between the increased use of road salts and automobile corrosion. In addition to road salts, atmospheric pollutants have an important effect on automobile corrosion. These pollutants, sulfur dioxide (SO_2) and nitrogen oxides (NO_x), can react on the metal surface to form corrosive acids, or may convert to sulfuric and/or nitric acid in the atmosphere and be deposited by acid precipitation (rain or snow). The map in Fig. 1 shows the acidity (low pH) of precipitation in the United States. Superimposed on this map are the high automobile corrosion rate areas in the United States. Notice that the Northeast and the Southern coastal areas are the most corrosive to automobiles. These areas have the combined effects of salts (road salts or marine salts) and acid precipitation, which produces a synergestic effect in which the combination is much worse that the sum of the separate effects.

[1]Member of technical staff and head, respectively, Electrochemical and Corrosion Laboratory, Texas Instruments Inc., Attleboro, MA 02703.

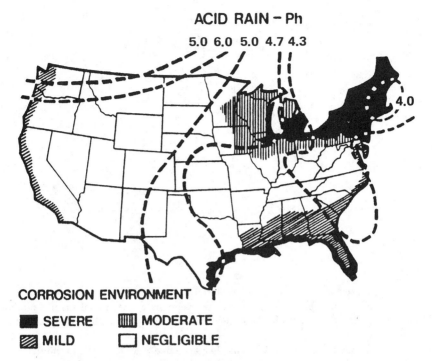

ACID RAIN − Ph

5.0 6.0 5.0 4.7 4.3

4.0

CORROSION ENVIRONMENT

■ SEVERE ▥ MODERATE
▨ MILD ☐ NEGLIGIBLE

FIG. 1—*Map of the United States showing high automobile corrosion rate areas and pH of acid precipitation.*

This complex environment is difficult to duplicate in the laboratory. ASTM standardized accelerated corrosion tests such as the Method of Salt Spray (Fog) Testing (B 117), Method of Acidified Synthetic Sea Water (Fog) Testing (G 43-75), Method for Copper-Accelerated Acetic-Acid Salt Spray (Fog) Testing (CASS Test) (B 368-85), and Practice for Conducting Moist SO_2 Tests (G 87-84) simulate one or more aspects of the automotive environment. However, no single procedure includes all aspects of the automotive environment or adequately predicts the service performance of all materials. For this reason, the automotive industry has resorted to cyclical test procedures involving one or more standardized tests [9-11]. The alternate wet and dry cycles produce more realistic results. However, these procedures do not reproduce all of the mechanisms of corrosion on automobiles. The industry therefore tends to use these tests for screening materials and relies more heavily or proving ground tests and field surveys [12]. The purpose of this work was to evaluate the corrosion of trim materials in field service tests and correlate the results with the service performance of these materials.

Experimental Procedure

Auto trim moldings were mounted on painted auto-body steel panels (8 cm × 5 cm) and held firmly with standard steel auto spring clips. A cross sectional view of the mounting is shown in Fig. 2. Three paint damaged points were made on each panel—one beneath the return flange of the molding, one at the return flange bend, and a third 1 cm above the molding. The damage points were made with an automatic spring-loaded center punch so that both the size and the depth of the points could be controlled. This method simulated the area relationships that occur

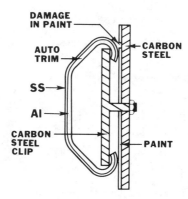

FIG. 2—*Diagrammatic representation of auto trim/auto body assembly.*

in service at paint defects adjacent to trim materials. These test assemblies were then mounted on plexiglass plates for exposure to the tests. After complete assembly of the test plate, each molding and panel unit was checked for electrical continuity between the trim and painted auto-body steel.

Each test plate has a control panel that used an insulated piece of plastic to simulate a molding. The corrosion on this panel represented the normal corrosion of body steel in the environment. Also, a solid stainless steel molding was used in one position, and the excess corrosion of the body steel on this panel represented galvanically accelerated corrosion resulting from the stainless trim. Table 1 lists the various studies performed. With the exception of Kure Beach, NC, the test plates were mounted on the front bumpers of automobiles as shown in Fig. 3. This method of exposure insured that the test plates would be exposed to road splash as well as airborn road salt mist. Exposure was usually from December through April of each year. The average length of each season test was 110 days and the average mileage was between 4000 and 5600 kilometers.

TABLE 1—*Test plate studies.*

DIRECT COMPARISON OF BEHAVIOR OF VARIOUS TRIMS

Stainless steel clad aluminum (SS/Al)
Stainless steel (SS)
Aluminum (Al)

SITE DEPENDENCE

Attleboro area
Detroit area
Pittsburgh area
Stationary racks at Kure Beach, NC, (25 and 250-m lots)

MOUNTING TECHNIQUES

Sealed and unsealed ends on trims
Trim mounting distance on painted body panel
Molding position (horizontal, vertical, etc.)

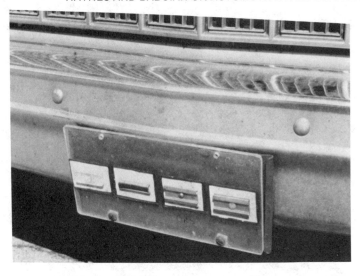

FIG. 3—*Installed field test plate on automobile bumper.*

Results and Discussion

At the time that these studies were conducted, the Pittsburgh area was the most corrosive for field service tests because of the use of cinders as well as deicing salts on the roads. Figures 4 and 5 show the appearance of test plates from the Pittsburgh area. A comparison of the degree of corrosion adjacent to the stainless steel trim and the plastic control show that galvanically accelerated corrosion was severe with stainless steel trim. Both the aluminum and the stainless steel clad aluminum provided galvanic protection to the steel. However, pitting of the aluminum occurred while it provided sacrificial protection. Results in the Detroit area, although not as severe as those in Pittsburgh, were similar. A representative field test plate is shown in Fig. 6. As shown in Fig. 7, results in the Attleboro area were even less severe, although the same mechanisms of corrosion prevailed.

Sealing and not sealing the ends of the trim did not affect the results at any of the test areas. The molding orientation also had little effect on the results and the magnitude of corrosion was independent of the position of the damage sites (above or below trim) as shown in Fig. 5. Studies

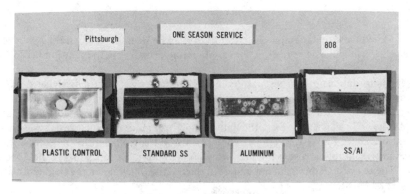

FIG. 4—*Field test plate with side moldings exposed in Pittsburgh area; one season.*

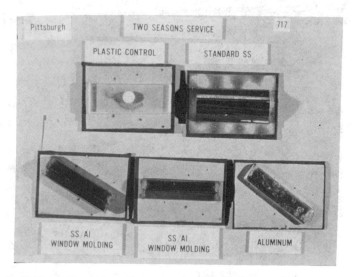

FIG. 5—*Field test plate with window moldings exposed in Pittsburgh area; two seasons.*

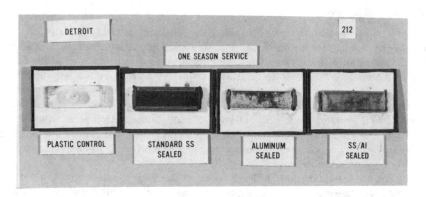

FIG. 6—*Field test plate with side moldings exposed in Detroit area; one season.*

were also conducted to determine the effect of the gap between the trim and the painted auto-body steel by using plastic spacers as shims. Mounting distances in the range of 0.1 to 1.0 mm had no effect on test results.

Attempts to duplicate field test results in the laboratory were not entirely successful. Salt spray and CASS testing of test plates still resulted in galvanically accelerated corrosion adjacent to the stainless steel trim and galvanic protection adjacent to the stainless steel clad aluminum trim. As shown in Fig. 8, however, the nature and appearance of the corrosion was markedly different (compare Fig. 4 through 7). The constant wet environment of the accelerated test resulted in streams of corrosion product, which flowed down the test panel in the direction of gravity. The red rust staining did not adhere in these tests while that on the field test plates was extremely adherent.

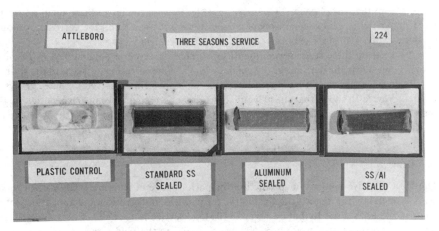

FIG. 7—*Field test plate with side moldings exposed in Attleboro area; three seasons.*

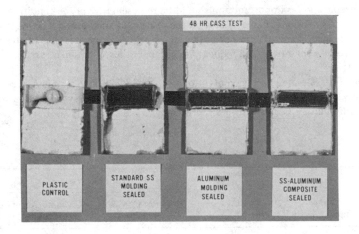

FIG. 9—*Atmospheric testing of various trim materials at Kure Beach 25-m lot; 2.5 months' exposure.*

Results from the Kure Beach, NC, exposures (Fig. 9) were similar to those for the field test plates from vehicles. The mechanisms of corrosion were the same and the corrosion products formed were identical to those observed on plates from vehicles.

The results from these studies were identical to those observed in field surveys in Montreal [13–16]. Table 2 lists the percentage of vehicles with galvanic corrosion adjacent to the stainless steel trim in these surveys. Other trim-related corrosion problems observed in these surveys are also listed. Figure 10 shows that the galvanic corrosion observed on a vehicle is identical to that observed on the field test plates. Also, note the pitting of the aluminum trim that occurred on this vehicle.

Conclusions

The most effective method for predicting the performance of trim materials is field service tests. Although laboratory tests duplicated the mechanisms of corrosion, the appearance of the specimens was not similar to that observed in the field service tests or on vehicles. Results from atmospheric exposure at Kure Beach, NC, were similar to the field service tests, but did not shorten the test duration significantly.

A field service testing technique that duplicates service performance has been described. Specimen preparation is important and should be representative of the conditions of the vehicles. The use of intentional damage sites on the test plates simulates the damage to the paint that can occur during manufacture or use of the vehicle. This insures that the tests reflect the most severe service conditions and reduces the number of test plates necessary to obtain meaningful results.

Galvanically accelerated corrosion of auto-body steel adjacent to stainless steel trim occurs. Stainless steel clad aluminum and aluminum trim provide galvanic protection to auto-body steel. The aluminum pits and corrodes sacrificially to the auto-body steel.

TABLE 2—*Vehicles (percent) with trim related corrosion.*

Vehicle Manufacturer	Trim Material	Type of Corrosion	
		Pitting or Crevice Corrosion	Galvanic Corrosion of Autobody
Audi	SS	100	0
Mercedes	SS	42	17
Jaguar	SS	67	33
Volvo	SS	36	64
Nissan	SS	65	38
Honda	SS	100	6
Mazda	SS	100	41
Subaru	SS	83	17
Toyota	SS	40	60
Buick	SS	25	0
Oldsmobile	SS	78	0
Chrysler	SS	35	42
Chrysler (van)	SS	100	0
Chevrolet	SS	20	20
Chrysler (van)	plastic	...	60
Oldsmobile	Al	92	0
Chrysler	Al	59	0
Chevrolet	Al	80	0
Oldsmobile	SS/Al	0	0
Buick	SS/Al	0	0

FIG. 10—*Galvanic corrosion adjacent to stainless steel trim (door window and bodyside moldings) and pitting of aluminum trim (rear window) in Montreal area; two years' service.*

References

[1] *Automotive Corrosion by Deicing Salts,* Baboian, R., Ed., National Association of Corrosion Engineers, Houston, 1981.

[2] Baboian, R., "Chemistry of the Automotive Environment," in *Designing For Automotive Corrosion Prevention,* SAE Proceedings, Warrendale, PA, 1978, p. 78.

[3] "Vehicle Corrosion Caused By Deicing Salts," *American Public Works Association Special Report No. 34,* Sept. 1970.

[4] "Designing For Automotive Corrosion Prevention," *Society of Automotive Engineers Report,* Warrendale, PA, Nov. 1978, p. 78.

[5] Palmer, J. D., "Corrosive Effects of Deicing Salts on Automobiles," *Materials Performance,* Vol. 10, No. 11, 1971, p. 38.

[6] Wirshing, R. J., "Effect of Deicing Salts on the Corrosion of Automobiles," Bulletin 150, Highway Research Board, 1957.

[7] Fromm, H. J., "The Corrosion of Auto-Body Steel and the Effect of Inhibited Deicing Salt," Report RR-135, Ontario Dept. of Highways, Ontario, November 1967.

[8] Bishop, R. R., "Corrosion of Motor Vehicles by Deicing Salts—Results of a Survey," RRL Report LR 232, Road Research Laboratory, Ministry of Transport, United Kingdom, 1968.

[9] GM Test Method TM 54-21, Environmental Corrosion Cycle Exposure, General Motors Corp., Detroit, 1978.

[10] Wyvill, D., "The Importance of Scab Corrosion Testing," *Metal Finishing,* January 1982.

[11] Hospadaruk, V., Huff, J., Zurilla, R. W., and Greenwood, H. T., "Paint Failure, Steel Surface Quality and Accelerated Corrosion Testing," *Transactions Society of Automotive Engineers,* Vol. 87, Sec. 1, Paper 780186, 1968.

[12] Hook, G., "The Historical Development of a Proving Ground Accelerated Corrosion Test," in *Automotive Corrosion by Deicing Salts*, R. Baboian, Ed., National Association of Corrosion Engineers, Houston, TX, 1981.

[13] R. Baboian, "Causes and Effects of Corrosion Relating to Exterior Trim on Automobiles," Proceedings of the 2nd Automotive Prevention Conference, Society of Automotive Engineers, Warrendale, PA, 1983.

[14] *1983 Montreal Survey on Cosmetic Corrosion in the Trim Area of Automobiles*. TI Technical Report, Texas Instruments Inc., Attleboro, MA, 1983.

[15] *1984 Montreal Survey on Cosmetic Corrosion in the Trim Area of Automobiles*, TI Technical Report, Texas Instruments Inc., Attleboro, MA, 1984.

[16] *1985 Montreal Survey on Cosmetic Corrosion in the Trim Area of Automobiles*, TI Technical Report, Texas Instruments Inc., Attleboro, MA, 1985.

Florian Mansfeld[1] and Samuel L. Jeanjaquet[1]

The Effects of SO₂ Scrubber Chemistry on Corrosion of Structural Materials

REFERENCE: Mansfeld, F. and Jeanjaquet, S. L., **"The Effects of SO₂ Scrubber Chemistry on Corrosion of Structural Materials,"** *The Use of Synthetic Environments for Corrosion Testing, ASTM STP 970,* P. E. Francis and T. S. Lee, Eds., American Society for Testing and Materials, Philadelphia, 1988, pp. 27-57.

ABSTRACT: The effects of 13 trace elements on the corrosion behavior of Ferralium 255, Hastelloy G3, stainless steel (SS) Type 317L, Monit and Ti Grade 2 immersed in an aerated baseline solution containing chlorides and sulfates at pH 1 or 4 in the presence of sulfur dioxide (SO₂) have been studied using electrochemical techniques. In a first series of experiments, a Plackett-Burman test matrix was used to determine which of the trace elements have a significant accelerating or inhibiting effect. Polarization resistance measurements were carried out for electrodes that were totally immersed for the entire six-day test period and for electrodes that were exposed above the solution level, except for 1 h/day, when they were immersed in the test solution. A computerized system was used to perform simultaneous measurements for all five alloys and both types of electrodes using up to ten test electrodes. The results of the electrochemical measurements were confirmed by a correlation analysis using results from a solution analysis and visual observation of the number and dimensions of pits after the test. The concentration dependence of corrosion rates for the five alloys has been studied for those species for which a significant effect was observed in the first test series. For Ferralium 255, Hastelloy G3, SS 317L, and Monit, an accelerating effect of Cu and Fe was observed when a certain concentration threshold was exceeded. For Ti, 1000 ppm fluoride additions produced etching of the surface at pH 1. A comparison with the results of the first test series shows that addition of Al at comparable concentrations can prevent this corrosion damage by complexing the fluoride ions. The effects of single ions added to a modified baseline solution at pH 1 that contained all other trace elements at a median level were further studied by recording potentiodynamic polarization curves after exposure times of 1, 3, and 6 days. For SS 317L, polarization curves are obtained in the baseline solution that are typical for an alloy susceptible to pitting and crevice corrosion. In the presence of 1000 ppm Cu or 2000 ppm Fe, this corrosion damage is increased and the polarization curves are those of highly corroded surfaces. For Ferralium 255, Hastelloy G3, and Monit, which are more corrosion resistant, the anodic polarization curves did not show an active/passive transition, but dissolution at a more or less constant rate. For Ti, the anodic polarization curves showed passive behavior with very low corrosion rates. The susceptibility to localized corrosion was studied further in the same solutions using a multiple crevice device. For SS 317L, severe crevice corrosion was observed in the baseline solution with additions of 1000 ppm Cu or Fe additions exceeding 300 ppm. For the other alloys except Ti, severe damage also occurred for additions of 2000 ppm Fe, but not for 1000 ppm Cu.

KEY WORDS: corrosion rates, scrubber chemistry, polarization resistance, polarization curves, solution chemistry, Ferralium 255, Hastelloy G3, stainless steel Type 317L, Monit, titanium

Unexpected corrosion failures are still quite common despite extensive testing of construction materials by producers and users in environments that simulate the conditions anticipated in practical applications. Unfortunately, the results of field tests often do not correlate well with

[1]Professor and senior research scientist, respectively, Rockwell International Science Center, Thousand Oaks, CA 91360; Prof. Mansfeld's present address: Dept. of Materials Science, University of Southern California, Los Angeles, CA 90089-0241.

results obtained in laboratory tests, which suggests that relatively minor variations in the local chemistry of the corrosive environment may produce pronounced changes in the corrosion behavior of a given material. Also, while a given environment might have a certain aggressivity, which could be characterized based on its chemical composition, its corrosivity can be quite different for different metals and alloys. This project has been carried out to evaluate the influence of the chemistry of a corrosive environment on the corrosion rates and mechanisms of materials of construction in flue gas desulfurization (FGD) plants [1-3].

The materials and chemistry of the environment to be studied in this project were selected based on earlier studies funded by the Electric Power Research Institute (EPRI) [4]. Ferralium alloy 255, Hastelloy alloy G3, Monit, SS 317L, and Ti Grade 2 representing families of alloys of duplex SS, Ni-base alloys, ferritic SS, austenitic SS, and Ti alloys, respectively, were chosen for testing. Thirteen trace metals were selected and added to a baseline solution, which was purged with oxygen and SO_2 and contained magnesium sulfate ($MgSO_4$), sodium sulfate (Na_2SO_4), and sodium chloride (NaCl). A statistical approach using a Plackett-Burman test matrix [8,9] was employed to reduce the number of tests and determine those elements which have a significant effect on corrosion rates. To further reduce the experimental effort, a computerized measurement system was used, which allowed determination of electrochemical parameters for up to ten electrodes at the same time over a six-day test period. The concentration dependence of corrosion rates has been studied further for those species for which a statistical analysis had shown a significant effect. Potentiodynamic polarization curves were obtained under the same conditions to determine the kinetics and mechanisms of the corrosion processes for the different alloy/environment combinations. The susceptibility to crevice corrosion was also studied for these cases.

Experimental Approach

Measurement of Polarization Resistance and Corrosion Potential

The measurements of the polarization resistance, R_p, the electrode capacitance, C, and the corrosion potential, E_{corr}, were carried out for a two-electrode system as a function of time using a computerized system [5,6]. This system allows continuous corrosion monitoring for the five alloys with ten measurements per day of R_p, C, and E_{corr} for each alloy, giving a total of 100 measurements per day (600 per test). R_p and C were calculated from the current-time curves recorded during the application of a potential step of ± 20 mV [5,6].

The measurement system consists of a Hewlett-Packard (HP) Model 9825 desktop computer, which runs the experiment and stores the data on magnetic tape. The HP Model 6904B multiprogrammer (MUX) switches the PAR potentiostat Model 173 sequentially to the ten electrode pairs (five immersed and five wet/dry). The PAR interface Model 276 has an autoranging capability, which insures maximum sensitivity of the current measurement from which R_p is calculated.

At the end of the test, E_{corr}, R_p, and C data were plotted as a function of time and integration of the R_p^{-1}-time curves was carried out using a computer program written for this purpose. The capacitance data have not been analyzed in detail; however, it appears that a sharp increase of C coincides with the initiation of pitting corrosion (see below).

Each cell for the electrochemical measurements contained an electrode pair, which was totally immersed. Each of the two electrodes, in the form of cylinders, was cast in epoxy and polished to expose the circular surface at the end of the cylinder. The two surfaces faced each other; the Vycor tip leading to a calomel reference electrode (SCE) is placed next to the electrode pair. The wet/dry electrode, which is shown in Fig. 1, was constructed similarly to the atmospheric corrosion monitor (ACM) developed earlier [7]. The test cells were thermostatted at 50°C. Sulfur dioxide (SO_2), and for some tests, hydrogen sulfide (H_2S), was bubbled through the cells and regulated by electronic flowmeters.

FIG. 1—*Wet/dry electrode.*

Recording of Polarization Curves

Two different approaches for recording anodic and cathodic polarization curves were used. The initial curves in a baseline solution and in this solution containing leachates of fly ash were obtained with a PAR potentiostat Model 173, a Model 175 function generator, and a Model 376 logarithmic convertor, and recorded on an X-Y recorder. For the later measurements, which were carried out as a function of exposure time over a six-day period, software from PAR ("softcorr" Model 332) was used with a Model 173 potentiostat and a Model 276 interface. An Apple IIe computer was used for recording, display, and analysis of these polarization curves [2,3].

Crevice Corrosion Tests

The effects of selected trace elements on the susceptibility of the five alloys to crevice corrosion was determined using a multiple crevice device as per ASTM Guide for Crevice Corrosion Testing of Iron-Base and Nickel-Base Stainless Alloys in Seawater and Other Chloride-Containing Aqueous Environments G 78-3. Weight-loss data were collected after an immersion time of 6 days, and photographs of the crevice area were taken at different magnifications.

Solution Chemistry

The trace elements studied and their maximum concentrations were selected based on an earlier literature survey by Koch et al. with input from Science Center personnel [4]. SO_2, O_2, and temperature T were held constant, since they have strong effects and/or change the corrosion mechanism if their level was changed. Tests were run at pH 1 or 4.

Two different types of test solutions were used. In one case, a baseline solution contained all trace elements at different combinations of high or low levels in a statistical design to determine main effects (Plackett-Burman tests, Table 1). The total chloride concentration was held at 10 000 ppm, except in the first 16 tests, in which it varied depending on the type and concentration of additive. All solutions contained sulfates and chlorides. Since SO_2 was bubbled continuously through the solution, the sulfate concentration increased somewhat during the test. The

TABLE 1—*Chemical compounds and concentrations used for test solutions.*

Species Added	Compound Added	Concentration as	Low, ppm	High, ppm
F^-	NaF	F	1	1000
Br^-	NaBr	Br	1	500
I^-	NaI	I	1	100
NO_3^-	$NaNO_3$	N	1	200
PO_4^{3+}	$NaH_2PO_4\ H_2O$	P	1	500
Ca^{2+}	$CaCl_2\ 2H_2O$	Ca	1	1000
Cr^{3+}	$CrCl_3\ 6H_2O$	Cr	1	500
Cu^{2+}	$CuCl_2\ 2H_2O$	Cu	1	1000
Mo^{6+}	H_2MoO_4	Mo	1	170
Si^{4+}	$SiCl_4$	Si	1	100
Al^{3+}	$AlCl_3\ 6H_2O$	Al	1	2000
Fe^{3+}	$FeCl_3\ 6H_2O$	Fe	1	2000
S^{2-}	H_2S	H_2S	. . .	5

NOTE: Baseline Solutions (pH 1 and 4): $MgSO_4$ (5000 ppm Mg), Na_2SO_4 (5000 ppm Na), NaCl (10 000 ppm Cl), 8 v/o O_2, 0.3 v/o SO_2, and T = 50°C.

pH decreased by about 0.2 at pH 1 and 2 at pH 4. Not all additions dissolved at pH 4. In the other case, a baseline solution contained all trace elements at a median level except one, the concentration of which was varied to determine the concentration dependence of corrosion rates (Table 2).

Experimental Results and Discussion

Long-Term Polarization Resistance and Corrosion Potential Measurements

In this test series, an attempt was made to determine the main effects of the 13 trace elements on corrosion rates, followed by an evaluation of the concentration dependence of corrosion rates in the presence of those elements for which a significant effect had been observed.

Determination of Main Effects—The main effects of trace elements were determined by a statistical analysis of the R_p data obtained at pH 1 and 4. The validity of the electrochemical data was ascertained by a comparison of these data with the extent of corrosion damage as determined from visual observation, and from the amount of dissolved material at the end of the tests as determined by chemical analysis of the test solution.

Plackett-Burman Tests with Foldover—To determine which of the compounds shown in Table 1 have an accelerating or inhibiting effect on the corrosion behavior of the five alloys, a statistical analysis according to the Plackett-Burman scheme in Table 3 was carried out, which requires 16 tests. The unassigned row Nos. 14 and 15 were used to determine the experimental error. For pH 1, a foldover test series was carried out to eliminate the effects of first-order interactions on the statistical analysis. In this case, all high (+) concentrations were changed to low (−) values and vice versa for each test (Test Nos. 17–32).

The most important results obtained were

(a) the wide variation in corrosion rates for a given alloy due to the presence of different combinations of trace elements, and

(b) the large differences of corrosion rates for the five alloys in a given test solution.

TABLE 2—*Baseline Solution No. 2.*

$MgSO_4$	(5000 ppm Mg)
Na_2SO_4	(5000 ppm Na)
NaCl	(10 000 ppm Cl)[a]
F^-	32 ppm
Br^-	22 ppm
I^-	10 ppm
NO_3^-	14 ppm
PO_4^{3+}	32 ppm
Ca^{2+}	22 ppm
Cu^{2+}	32 ppm
Mo^{6+}	13 ppm
Si^{4+}	10 ppm
Al^{3+}	45 ppm
Fe^{3+}	45 ppm
S^{2-}	2.5 ppm

NOTE: 8 v/o O_2, 0.3 v/o SO_2, and T = 50°C.
[a]Total chloride.

TABLE 3—*Plackett-Burman test matrix variables.*

Test	X1 F	X2 Br	X3 I	X4 NO₃	X5 PO₄	X6 Ca	X7 Cr	X8 Cu	X9 Mo	X10 Si	X11 Al	X12 Fe	X13 H₂S	X14 ÷	X15 ÷
1	+	+	+	+	−	+	−	+	+	−	−	+	−	−	−
2	+	+	+	−	+	−	+	+	−	−	+	−	−	−	+
3	+	+	−	+	−	+	+	−	−	+	−	−	−	+	+
4	+	−	+	−	+	+	−	−	+	−	−	−	+	+	+
5	−	+	−	+	+	−	−	+	−	−	−	+	+	+	+
6	+	−	+	+	−	−	+	−	−	−	+	+	+	+	+
7	−	+	+	−	−	+	−	−	−	+	+	+	+	−	+
8	+	+	−	−	+	−	−	−	+	+	+	+	−	+	−
9	+	−	−	+	−	+	+	−	+	+	+	+	−	+	+
10	−	−	+	−	−	−	+	+	+	+	−	+	−	+	+
11	−	+	−	−	−	+	+	+	+	−	+	−	+	+	−
12	+	−	−	−	+	+	+	+	−	+	−	+	+	−	−
13	−	−	−	+	+	+	+	−	+	−	+	+	−	−	+
14	−	−	+	+	+	+	+	−	+	−	+	+	−	−	+
15	−	+	+	+	+	−	+	−	+	+	+	+	−	+	−
16	−	−	−	−	−	−	−	−	−	−	−	−	−	−	−

NOTE: + = high concentration and − = low concentration.

Examples of experimental data are given in Figs. 2 to 8. Figure 2 shows the time dependence of the corrosion rate (proportional to the reciprocal of the polarization resistance, R_p) for the five immersed alloys in Solution No. 3 (Table 3) at pH 4; Figure 3 shows the corrosion potential E_{corr}. The final R_p values range from about $10^7 \Omega$ for Hastelloy G3 to $10^3 \Omega$ for Ti; in the latter case, activation of the Ti surface occurred in the first day of exposure with a drop of E_{corr} by about 0.5 V and an increase in corrosion rate by a factor of 1000. This result shows that short-term tests of one day or less would miss important features of the corrosion behavior. The wide variations in corrosion behavior are also shown in Figs. 4-5 for Ti in four different solutions at pH 4. The lowest corrosion rate occurs in the baseline solution (Fig. 4). E_{corr} is very noble and similar in value for Solution Nos. 1, 12, and 16 (Fig. 5) despite the differences in corrosion rates.

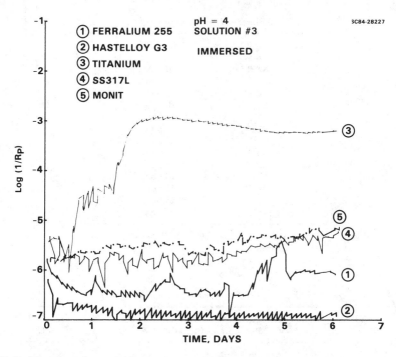

FIG. 2—*Time dependence of $1/R_p$ for the five alloys immersed in Solution No. 3 at initial pH 4.*

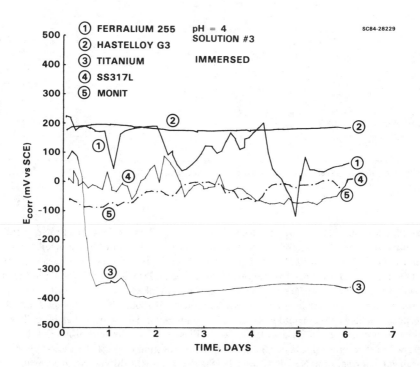

FIG. 3—*Time dependence of E_{corr} for the five alloys immersed in Solution No. 3 at initial pH 4.*

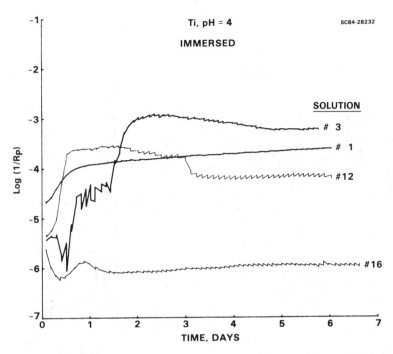

FIG. 4—*Time dependence of $1/R_p$ for Ti, Grade 2 in four different solutions at initial pH 4.*

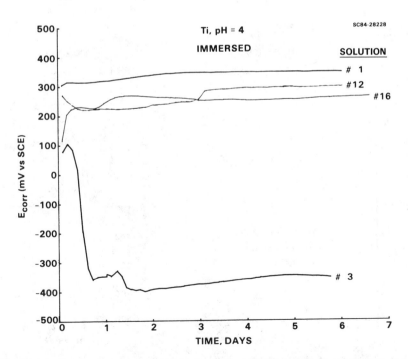

FIG. 5—*Time dependence of E_{corr} for Ti, Grade 2 in four different solutions at initial pH 4.*

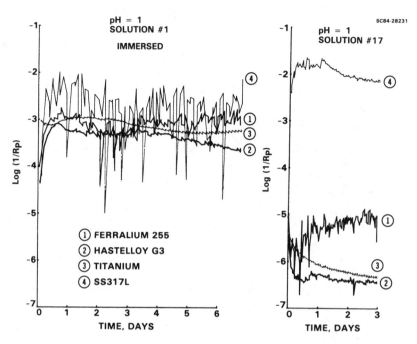

FIG. 6—*Comparison of the time dependence of $1/R_p$ for four alloys in Solution Nos. 1 and 17 at pH 1.*

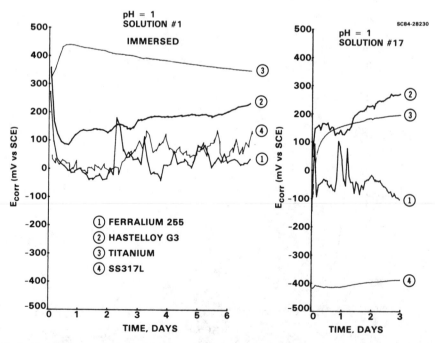

FIG. 7—*Comparison of the time dependence of E_{corr} for four alloys in Solution Nos. 1 and 17 at pH 1.*

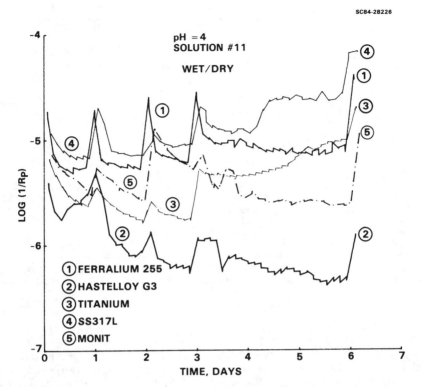

SC84-28226

FIG. 8—*Time dependence of $1/R_p$ for five wet-dry electrodes in Solution No. 1 at initial pH 4.*

The analysis of the results at pH 1 had indicated that interactions of the trace elements could mask the main effects of a single trace element (see below). Therefore, a foldover design was run in which all signs of Table 4 were inverted and the solutions were renumbered Nos. 17-32. The test period in this case was only 3 days. Figures 6 and 7 gives a comparison of the corrosion data for four alloys. Solution No. 17 corresponds to Solution No. 1. The change of the trace element level produced a pronounced decrease in corrosion rates for Ferralium 255, Hastelloy G3, and Ti, but increased the corrosion rate of Type SS 317L, accompanied by a drastic drop of E_{corr}. The large fluctuations of R_p for SS 317L in Fig. 6 usually point to pitting corrosion.

Examples of the corrosion behavior of wet/dry electrodes are given in Fig. 8. The spikes in R_p^{-1}, which occur when the electrodes are dipped into the solution once a day for the first four days of the test, are very obvious, as are the differences in corrosion rates for the five alloys. Similar results were obtained for the other tests at pH 1 and 4.

Table 4 shows the corrosion loss as obtained by integration of the R_p^{-1}-time curves for the 16 tests of Table 3-1 at pH 1. Since Tafel slopes, and therefore factor B, which converts R_p into corrosion current densities i_{corr} ($i_{corr} = BR_p^{-1}$), change with environmental conditions and the nature of the alloy, it is not possible from the present data to calculate exact values of i_{corr} or corrosion rates. Also, since pitting occurs in many cases, only average corrosion rates can be estimated. However, since the corrosion loss changes so drastically in Table 4, one can use $B = 20$ mV as a guideline, for which an integrated corrosion loss of 19 s/Ω would correspond to $i_{corr} = 1$ μA/cm² (for 6 days and a surface area of $A = 0.73$ cm²).

Inspection of the data in Table 4 shows that baseline Solution No. 16 is very corrosive to Ferralium 255 and SS 317L, but does not attack the other three materials. On the other hand,

TABLE 4—*Integrated corrosion loss (s/Ω) for immersed electrodes, pH 1*

Solution Number	Ferralium 255	Hastelloy G3	Ti Grade 2	SS 317L	Monit
1	340	229	448	482	4
2	114	84	3.9	416	101
3	5	0.3	3800	4225	5
4	10	1.3	3.2	15	1.4
5	1580	59	11.7	1195	374
6	708	298	13.0	837	199
7	384	347	8.3	536	189
8	547	104	0.5	175	184
9	537	485	245	1420	691
10	359	290	73	1566	381
11	1773	1376	292	1121	1448
12	1604	380	10.8	2134	600
13	869	236	0.6	1243	290
14	76	62	1.7	193	119
15	3	1.3	1.8	11	6
16	1146	0.8	1.5	2165	1.4

NOTE: 19 s/Ω ≙ 1 μA/cm² for $t = 6d$, $B = 20$ mV, $A = 0.73$ cm². For Ferralium 255, Hastelloy G3, SS 317L, and Monit 1 μA/cm² corresponds to about 10 μm/y or 0.4 mil/y. For Ti, it corresponds to 11.6 μm/y or 0.46 mil/y.

Solution No. 3 is very corrosive to Ti and SS 317L, but benign to the other alloys. In this case, etching of Ti and SS 317L was observed, which most likely was due to the high levels of F⁻, which is present as HF at pH 1, and nitrogen oxide (NO_3^-). The fact that such high corrosion rates do not occur in Solution Nos. 1, 6, and 9, in which F⁻ and NO_3^- are also at a high level, shows that interactions with other trace elements such as Al strongly affect the corrosion behavior. These findings are also confirmed by foldover test results, which are shown in Table 5, for an exposure period of three days. Solution No. 19, which corresponds to No. 3, produced lower corrosion losses for Ti than Solution No. 3, but higher values for Ferralium 255, Hastelloy G3, and Monit. Similar variations of corrosion behavior have also been observed for pH 4 (Table 6).

TABLE 5—*Integrated corrosion loss (s/Ω) for foldover design, pH 1.*

Solution Number	Ferralium 255	Hastelloy G3	Ti Grade 2	SS 317L	Monit
17	1.6	0.1	0.2	3.1	1.6
18	405.3	79.1	1.4	663.1	167.3
19	394.2	97.4	3.0	618.3	43.8
20	222.4	40.0	2.7	478.8	156.4
21	4.0	1.5	0.6	4106	6.1
22	11.2	6.8	12.5	170.7	39.0
23	9.3	5.8	12.9	86.5	41.2
24	28.5	0.4	1.2	218.6	57.5
25	128.3	61.6	0.4	280.2	53.5
26	1.7	0.4	0.3	2537	2.0
27	22.0	10.7	1.0	260.0	39.2
28	2.7	0.7	0.6	7.6	3.6
29	3.7	4.1	226.8	240.8	22.7
30	108.3	18.2	12.5	655.5	70.2
31	190.7	110.7	4.7	400.6	132.6
32	264.7	10.5	3.2	424.5	92.4

TABLE 6—*Integrated Corrosion Loss (s/Ω) for immersed electrodes, pH 4.*

Solution Number	Ferralium 255	Hastelloy G3	Ti Grade 2	SS 317L	Monit
1	100.8	6.96	85.65	799.9	242.5
2	58.95	16.97	0.56	243.7	230.9
3	0.34	0.08	311.3	1.21	0.23
4	0.17	0.12	2.80	2.14	2.44
5	231.9	32.71	0.43	575.0	142.4
6	28.16	0.86	3.35	59.6	89.07
7	0.70	12.65	1.18	100.8	21.61
8	92.73	22.04	0.07	156.6	113.5
9	471.6	92.21	131.0	843.6	774.2
10	457.7	65.16	22.92	386.1	412.5
11	589.8	377.0	16.42	713.5	750.1
12	6.28	27.05	68.21	527.9	345.4
13	0.14	0.07	0.06	1.13	4.87
14	91.57	2.96	0.82	215.1	159.9
15	0.80	0.08	0.24	0.81	2.00
16	1.08	0.76	0.57	2.69	2.81

The integrated corrosion loss for the wet/dry electrodes was, in most cases, much less than for the immersed electrodes. The wide variations shown in Tables 4 to 6 were not observed to the same extent. Considering that the wet/dry electrodes are immersed for only 4 h or 3% of the total test time, corrosion losses are significant in many cases. No detailed analysis of these data has been carried out. It seems that the electrodes never became entirely dry, since they were placed directly above the solution level. Observation of atmospheric corrosion phenomena has shown that the highest corrosion rates occurred when the thin electrolyte film dried out [7].

Post-Mortem Tests—Surface Evaluation and Solution Analysis—After each test, all electrodes were examined under a microscope and pictures were taken at different magnifications to determine the type and extent of corrosion attack. Pit depth and the number of pits were determined. For Solution Nos. 1–16 at pH 1, pitting occurred for SS 317L in all solutions except Nos. 3 and 16, where the surface was etched. Large and deep pits were observed after tests in Solution Nos. 1, 2, 8, 10, and 13. The full extent of pitting could usually be detected only after polishing the surface because of undercutting of the pits. For Ferralium, large and deep pits were found for Solution Nos. 1, 10, 13, 14, and 16. The extent of pitting was less than for SS 317L. In Solution No. 3, severe etching of the Ti surface occurred. Solution No. 3, which contained high levels of fluoride and nitrate at pH 1, also etched the surface of SS 317L. For Hastelloy and Monit, pitting was observed only in a few solutions, such as Nos. 9 and 11.

A ranking of the pitting tendency, as determined by visual examination, produced similar results to the electrochemical data in Table 4. For example, Solution Nos. 9, 11, and 12, which gave the highest corrosion loss data for Monit, were also the ones which produced pitting. A nonparametric statistical analysis was performed to investigate the correlation between the number of pits and the integrated corrosion loss for a given sample. Solution analysis data for dissolved Fe, Ni, and Cr, obtained using an inductively coupled plasma spectrometer, were also compared with integrated corrosion loss. For all cases, a positive correlation was found. For Ferralium 255, there is 99.8% confidence rate for a positive correlation between integrated corrosion loss and solution analysis for iron, and 93.1% confidence rate that a positive correlation exists between the integrated corrosion loss and the observed number of pits.

Statistical Analysis—The results obtained in this test series show that the presence of certain cations and anions at low concentrations can lead to large changes in the corrosion rates for all five alloys investigated. The nature of the corrosion process can also change. For SS 317L, general corrosion was observed in the baseline solution (No. 16 in Table 4), but severe pitting occurred in most solutions containing the additives. Ti grade 2 is passive in most solutions, but conditions can exist (e.g., in Solution No. 3) in which etching occurs and the surface becomes active.

The positive correlation between electrochemical corrosion loss data and solution analysis as well as pitting tendency discussed above confirmed that the electrochemical technique employed for this study produces an accurate measure of the corrosion behavior. Therefore, it was possible to proceed with a statistical analysis of the corrosion data and determine which trace elements have an accelerating or inhibiting effect on the baseline solution.

The linear model

$$Y = \log(\text{COR}) = M + Z_1 A_1 + \ldots + Z_{16} A_{16} + e \tag{1}$$

has been used for the analysis, where COR is the measured corrosion loss, M the mean value, A_i is the effect of the ith factor, and e is a variable representing random experimental error and the effects of second and higher order interactions. The Z_1 takes on the values ± 1 depending on whether the ith factor is at its high ($+1$) or low (-1) level (Table 4).

An example for the analysis according to the linear model in Eq 1 is given in Table 7 for the final 16 tests at pH 1. A decision whether a given trace element has a significant accelerating or inhibiting effect is based in the case of Table 7 on the T-value, which is the ratio of the effect A_i and its standard error. As indicated in Table 7, T-values larger than 4.3 have an effect at the 95% confidence limit. Based on this criterion, Cu and Fe have an accelerating effect in the corrosion rate of Hastelloy, while phosphate (PO_4) and I should act as inhibitors for SS 317L. For Monit, both Cu and Al should accelerate the corrosion rates.

Figure 9 shows the results of the linear model analysis for Ti at pH 4. In this figure, the estimated effects are used instead of T-values. The error bars are explained in this figure. The large error bar is for the 90% multiple confidence interval (c.i.), valid simultaneously for all trace elements. Based on the 90% multiple c.i. for the effects A_i, none of the trace elements has an overwhelming effect by itself, but combinations of several elements lead to the observed wide spread of corrosion rates. For the 90% c.i., one finds that F, Ca, Cr, Cu, Si, and H_2S have an accelerating effect, while PO_4, Fe, and Al inhibit corrosion. It is interesting to note that Al ions complex F^-, which at low pH prevents formation of HF and etching of Ti. This points to an interaction of Al and F.

In carrying out the statistical analysis, one has to consider the possibility that the values of the estimated effects are due to the interactions that are confounded with them and not the main effects themselves. For example, for the test matrix in Table 3, an apparent inhibiting effect of PO_4 could really be an accelerating effect of the interaction of F and NO_3. To eliminate first-order interactions, a foldover design was used at pH 1, as discussed above. Figure 10 gives the results for Ti at pH 1 for the original 16 tests (left), the second 16 tests (right), and all 32 tests (middle). Interactions are evident when the main effect has different signs for the two parts of the test. Such cases are Cr and NO_3 in Fig. 7. For PO_4, the magnitude of the estimated effect is quite different for the two test series. Figures 11–18 show the estimated effects for the other alloys at pH 1 (Figs. 11–14) and pH 4 (Figs. 15–18). The data for Monit (Fig. 14) show that first-order interactions are indicated NO_3 and H_2S. Table 8 summarizes the results of the statistical analysis for all five alloys at both pH values based on the 90% c.i. for all 32 tests at pH 1 and 16 tests at pH 4. Cu and Fe have an accelerating effect in most cases, which is probably due to their effect as oxidizers that polarize the alloys to their pitting potential, or increase corrosion rates by anodic polarization in the case of active dissolution. The effect of Al at pH 1 has already been mentioned; it complexes F^- and prevents formation of HF, which etches Ti. The inhibiting effect of PO_4 has been reported for stainless steels, but not for Ti.

More species that could have an effect on corrosion behavior are listed for pH 4 than for pH 1. This could be because the first-order interactions have not been eliminated for pH 4.

TABLE 7—Estimated effects based on linear models analysis (first 16 tests at pH 1).

	Ferralium 255		Hastelloy G3		Ti		SS 317L		Monit	
	Effect	T-Value	Effect	T-Value	Effect	T-Value	Effect	T-Value	Effect	T-Value
Mean	2.360	20.2	1.733	24.39	1.145	6.3	2.722	33.8	1.793	13.8
1 (F)	-0.133	-1.1	-0.029	-0.4	0.300	1.7[a]	-0.004	-0.1	-0.151	-1.4
2 (Br)	-0.168	-1.4	-0.065	-0.9	0.259	1.4	-0.089	-1.1	-0.061	-0.6
3 (I)	-0.368	-3.2[b]	-0.010	-0.1	-0.129	-0.7	-0.391	-4.9[c]	-0.260	-2.4[a]
4 (NO$_3$)	-0.196	-1.7[a]	-0.102	-1.4	0.238	1.3	0.013	0.2	-0.064	-0.6
5 (PO$_4$)	-0.173	-1.5	-0.147	-2.1[a]	-0.750	-4.1[b]	-0.364	-4.5[c]	0.030	0.3
6 (Ca)	-0.099	-0.9	0.064	0.9	0.250	1.4	0.024	0.3	-0.080	-0.8
7 (Cr)	-0.106	-0.9	0.085	1.2	0.190	1.1	0.137	1.7[a]	0.240	2.2[a]
8 (Cu)	0.316	2.7[a]	0.606	8.5[c]	0.392	2.2[a]	0.201	2.5[a]	0.471	4.4[c]
9 (Mo)	-0.082	-0.7	0.175	2.5[a]	0.056	0.3	-0.268	-3.3[b]	-0.010	0.0
10 (Si)	-0.251	-2.2[a]	-0.024	-0.3	0.101	0.6	-0.022	-0.3	0.181	1.7[a]
11 (Al)	0.260	2.3[a]	0.635	8.9[b]	-0.247	-1.4	0.040	0.5	.568	5.3[c]
12 (Fe)	0.466	4.0[b]	0.589	8.3[c]	-0.129	-0.7	0.187	2.3[a]	0.374	3.5[b]
13 (H$_2$S)	0.052	0.5	0.194	2.6[a]	0.074	0.4	-0.163	-2.0[a]	0.251	2.4[a]
Standard deviation	0.47		0.28		0.73		0.32		0.43	

[a] 75% (1.6).
[b] 90% (2.9).
[c] 95% (4.3).

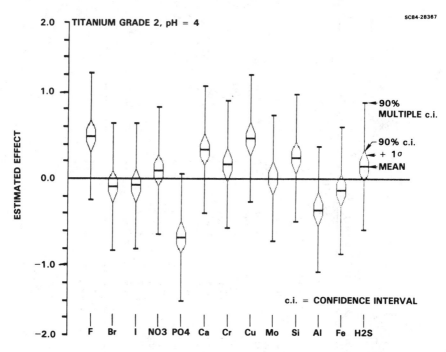

FIG. 9—*Estimated effects of trace elements for Ti, Grade 2 at initial pH 4.*

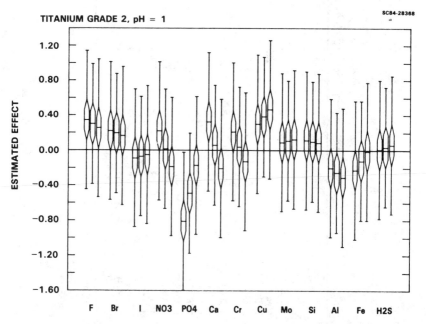

FIG. 10—*Estimated effects of trace elements for Ti, Grade 2 at pH 1.*

FERRALIUM 255

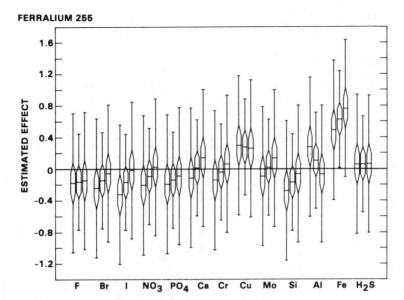

FIG. 11—*Estimated effects of trace elements at pH 1 in Ferralium 255.*

HASTELLOY G3

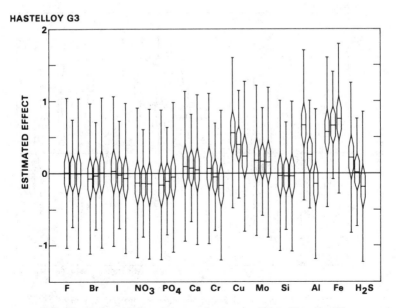

FIG. 12—*Estimated effects of trace elements at pH 1 in Hastelloy G3.*

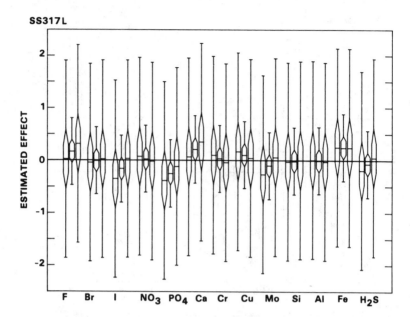

FIG. 13—*Estimated effects of trace elements at pH 1 in SS 317L.*

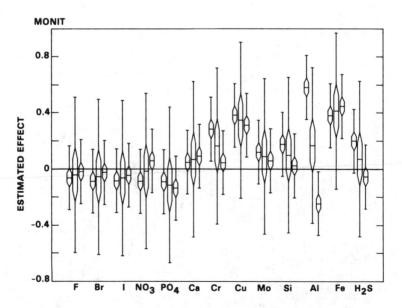

FIG. 14—*Estimated effects of trace elements at pH 1 in Monit.*

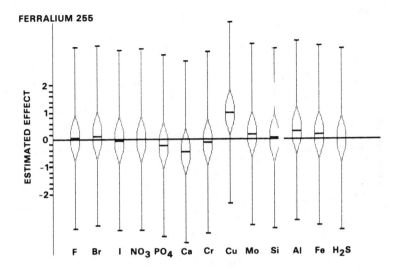

FIG. 15—*Estimated effects of trace elements at pH 4 in Ferralium 255.*

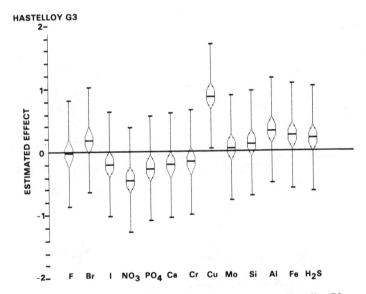

FIG. 16—*Estimated effects of trace elements at pH 4 in Hastelloy G3.*

Concentration Dependence for Single Species

To verify the findings of the previous test series and to determine the concentration dependence of R_p and E_{corr} for the species that had significant accelerating or inhibiting effects, a new test series was initiated. Since corrosion rates are affected by the interaction of many of the trace elements, the concentration dependence of one trace element was evaluated in the presence of all others, held at a constant concentration. For this purpose, a new baseline solution was estab-

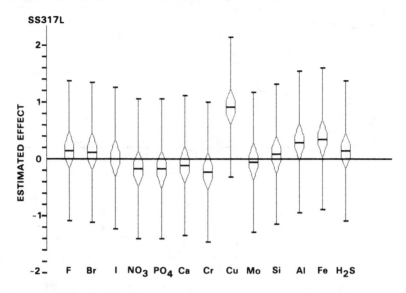

FIG. 17—*Estimated effects of trace elements at pH 4 in SS 317L.*

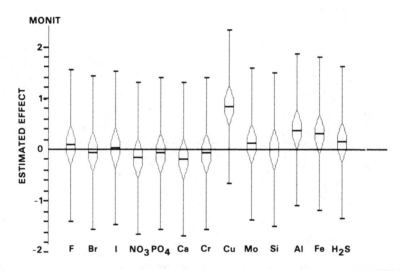

FIG. 18—*Estimated effects of trace elements at pH 4 in Monit.*

lished that contains not only chloride, sulfate, O_2, and SO_2 as before, but also all other elements at a median level except the one to be studied. This value c_m is calculated as

$$\frac{2 \log c_m}{\log c_{hi}} - 1 = 0 \qquad (2)$$

where c_m is the median level (Table 2), and c_{hi} is the high level in the previous test series (Table 1). Taking Ca as an example, for which $c_{hi} = 1000$ ppm, leads to $^2/_3 \log c_m - 1 = 0$ or $c_m =$

TABLE 8—*Significant trace metals.*

Alloy	pH 1		pH 4	
	Accelerates	Inhibits	Accelerates	Inhibits
Ferralium 255	Cu,Fe	I,Si	Cu	. . .
Hastelloy G3	Cu,Fe,Al	. . .	Br,Cu,Al,Fe,H$_2$S	I,NO$_3$,PO$_4$,CA
Ti grade 2	F,Br,Cu	PO$_4$,Al	F,Ca,Cr,Cu,Si,H$_2$S	PO$_4$,Al,Fe
SS 317L	Fe,Ca	PO$_4$	Cu,Al,Fe	. . .
Monit	Cr,Cu,Al,Fe	. . .	Cu,Al	. . .

31.6 ppm Ca. The concentrations for the new baseline solution have been listed in Table 2. The concentration dependence of R_p and E_{corr} was studied for selected species for which a significant effect had been predicted in the first test series (Table 8). Each of these species was added singularly to the baseline solution which contained all the other species at the concentrations shown in Table 2.

The experimental procedure for this test series was the same as that for the previous test series. Only immersed electrodes were studied. All tests were conducted at pH 1, T = 50°C for a six-day period.

For Ferralium 255 and SS 317L, the new baseline solution leads to much lower corrosion rates than the old baseline solution, which contained all trace elements at a low level. This suggests that the inhibiting properties of the added species are stronger than the accelerating effects, which is somewhat surprising, since the previous analysis (Table 8) had identified mainly accelerating species such as Cu and Fe. Figures 19-21 shows the time dependence of R_p, E_{corr}, and the electrode capacitance C_p for Ferralium 255 as a function of Cu concentration. A strong concentration dependence of R_p is observed (Fig. 19). After an initial decrease of corrosion rates

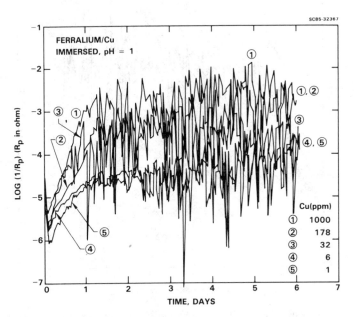

FIG. 19—*Time dependence of $1/R_p$ for Ferralium 255 as a function of Cu concentration.*

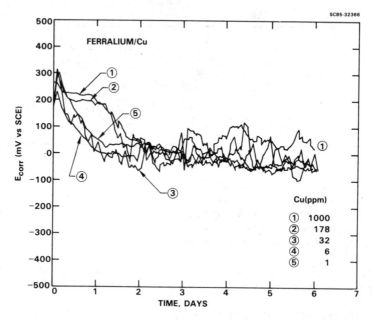

FIG. 20—*Time dependence of* E_{corr} *for Ferralium 255 as a function of Cu concentration.*

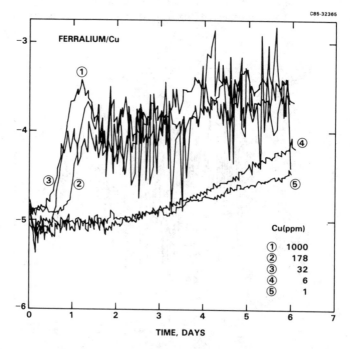

FIG. 21—*Time dependence of* C_p *for Ferralium 255 as a function of Cu concentration.*

for several hours, which is accompanied by a shift of E_{corr} to more noble values (Fig. 20), corrosion rates increase drastically. For 1000 ppm, the increase is about a factor of 1000 in a little more than a day. Corrosion rates continue to increase for the remainder of the test. During the initial time period of sharply increasing corrosion rates, E_{corr} drops to values around 0 mV vs SCE more or less independent of Cu concentration. It will be noted that the R_p-time trace is fairly smooth during the initial rise of corrosion rates, but becomes very noisy from then on. This strong scatter of R_p is considered to be caused by localized corrosion, which does not occur equally on the electrode pair used for these measurements. Visual observation of the electrode pair used for this test showed that a few pits and a large crevice had developed on one electrode, while the other one did not show significant localized corrosion.

Similar changes with time as for R_p occurred for the capacitance C_p (Fig. 21), in which, for higher concentrations, a strong increase is found, which parallels that of the inverse polarization resistance. It is interesting that this strong increase and large scatter of C_p do not occur for 6 and 1 ppm Cu, in which localized corrosion was not observed. Therefore, it can be concluded that the measurement of C_p gives useful information concerning the occurrence of localized corrosion. Similar results have been obtained in the presence of Cu and Fe for all the other alloys except Ti. In general, an incubation period for localized corrosion is observed for Ferralium 255, Hastelloy G3, and Monit, with a strong dependence of corrosion rates on Cu or Fe concentration and exposure time. These results again show that for the system evaluated here, it is imperative to determine the corrosion behavior over extended time periods to obtain meaningful results.

The results obtained in this test series are summarized in Figs. 22-26, which are log-log plots of the integrated corrosion loss (in s/Ω) as a function of concentration of the added species. For each alloy, the value of r that corresponds to a corrosion rate of 100 μm/y (except for Ti, where 10 μm/y is used) is indicated in Fig. 22. These results are compared with the corrosion loss

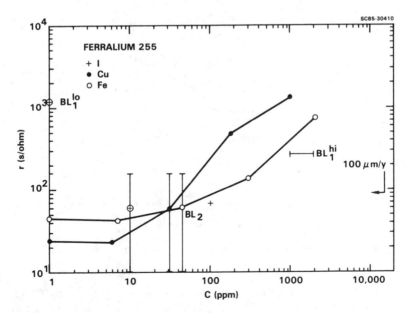

FIG. 22—*Dependence of corrosion loss* r *on concentration of various trace elements at pH 1 in Ferralium 255.*

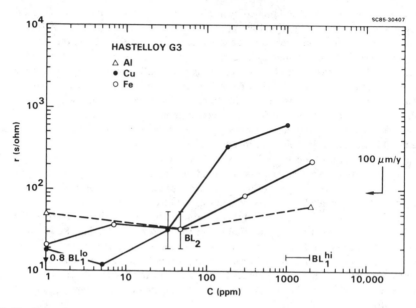

FIG. 23—*Dependence of corrosion loss* r *on concentration of various trace elements at pH 1 in Hastelloy G3.*

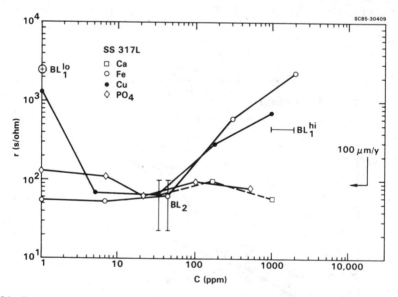

FIG. 24—*Dependence of corrosion loss* r *on concentration of various trace elements at pH 1 in SS 317L.*

observed in the baseline solution (BL_2) used in this test series and the baseline solution used in the first test series with BL_1^{lo} as the solution for Test No. 16 in Table 3, in which all trace elements were at a low level, and BL_1^{hi} referring to Test No. 32, in which all trace elements were at a high level. For BL_2, a scatter band is shown for the results of three tests. The scatter is fairly large in some cases because of the occurrence of localized corrosion.

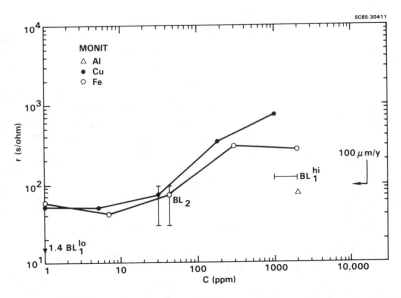

FIG. 25—*Dependence of corrosion loss* r *on concentration of various trace elements at pH 1 in Monit.*

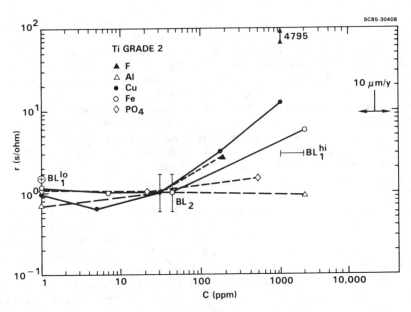

FIG. 26—*Dependence of corrosion loss* r *on concentration of various trace elements at pH 1 in Ti, Grade 2.*

For Ferralium 255 (Fig. 22), corrosion rates in the new baseline solution were between 10 and 200 μm/y, which is lower than the previous baseline solution (BL$_1$, low and high). Both Cu and Fe have an accelerating effect at higher concentrations, and corrosion rates exceed the upper level of the scatter bar for BL$_2$ at 70 ppm Cu and 240 ppm Fe. Addition of iodide at the high level of 100 ppm did not change corrosion rates as compared to 10 ppm in the baseline solution.

For Hastelloy G3 (Fig. 23), a similar concentration dependence of corrosion rates was observed. Cu above 50 ppm and Fe above 110 ppm increase corrosion rates above those for the baseline solution. Additions of Al at 2000 ppm raised corrosion rates slightly above the level for BL_2. Average corrosion rates for BL_2 were about 30 μm/y. Contrary to the case of Ferralium 255, corrosion rates were lower for BL_1 at both low and high levels.

For SS 317L (Fig. 24), the effects of Cu, Fe, Ca, and PO_4 have been evaluated. As for Ferralium, corrosion rates were lower for BL_2 than for BL_1 low and high. For Cu additions, a minimum for corrosion rates is observed, which is caused by a change of the corrosion mechanism at the lowest Cu concentration. When only 1 ppm Cu is present, E_{corr} is about -400 mV as compared to about 0 mV at the other concentrations, and the alloy corrodes in the active state with H_2 evolution. The surface turns black as observed in the tests in the presence of fly ash leachates. For Cu concentrations exceeding 6 ppm and for all Fe concentrations, the alloy is polarized into the passive region and suffers from pitting and crevice corrosion. Phosphate has a slightly inhibiting effect, reducing corrosion rates slightly at concentrations above 10 ppm. Additions of Ca did not produce any changes. The upper level of BL_2 at about 100 μm/y is exceeded for about 60 ppm Cu or Fe.

For Monit (Fig. 25), Cu and Fe produced the accelerating effects observed for the other alloys, each with a threshold value of about 60 ppm. Additions of 2000 ppm Al did not change the corrosion rate observed at the baseline value of 45 ppm.

For Ti Grade 2, corrosion rates were very low in the baseline solution with an average of about 1 μm/y (Fig. 26). Cu and Fe additions increased corrosion rates by about a factor of 10 for 1000 ppm Cu and a factor of about 5 at 2000 ppm Fe. A very interesting effect was noted for fluoride additions. At a concentration of 178 ppm, corrosion rates increased only by a factor of three, but when 1000 ppm were added, corrosion rates increased by a factor of 500 to about 5 mm/y. The surface was etched by the HF formed at pH 1. Apparently, 45 ppm Al in the baseline solution is enough to complex most of the F and prevent formation of HF at the 178 ppm level, but not enough for 2000 ppm F. Additions of 2000 ppm Al did not change corrosion rates as compared to 45 ppm, since the baseline solution contains only 32 ppm F and corrosion rates are already very low. An increase of the PO_4 level also did not change these low corrosion rates.

The results of the electrochemical tests have been confirmed by the investigation of the number and dimensions of the pits and crevices formed in the different tests and for Hastelloy G3 by solution analysis using atomic absorption [3]. High concentrations of Cu and Fe produced a large number of large pits for SS 317L and crevice corrosion for Ferralium 255, Monit, and Hastelloy G3. Additions of Al, PO_4, Ca, or I did not produce any significant changes based on this visual observation, except for the occurrence of a large number of small pits on Monit at 2000 ppm Al.

The results of the solution analysis for all tests with Hastelloy G3 are shown in Figs. 27, in which the amount of dissolved Ni is plotted as a function of Fe or Cu concentration in the test solution. Figure 27 is similar to Figs. 20–24, which showed electrochemical corrosion data. The amount of dissolved Ni is more or less constant for Fe and Cu concentrations below those of the baseline solution and increases sharply at the high concentration.

Crevice Tests—The effects of the concentration of several additives on localized corrosion have been studied further for the five alloys using a multiple crevice device [3]. The baseline solution and the concentrations of additives were the same as for the second series of tests. The tests were a carried out over a six-day period. After the crevice test, the weight loss was determined for each sample after removal of corrosion products, and the surface was photographed at various magnifications. Samples were 2.5 cm $\times$ 5.1 cm plates with a thickness between 0.05 and 0.1 cm.

In the baseline solution that contained all trace elements at a median level (Table 2), crevice corrosion was observed only for SS 317L (Table 9). Severe crevice corrosion was observed when the Fe level was raised from 45 ppm (baseline) to 2000 ppm, except for Ti, for which crevice

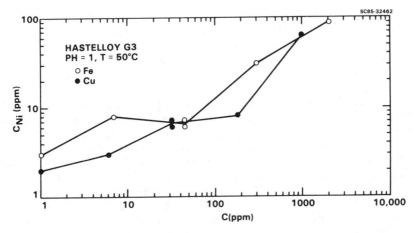

FIG. 27—*Concentration of dissolved Ni in test solution after exposure test for Hastelloy G3 as a function of Fe or Cu concentration, pH 1,* T = *50°C.*

TABLE 9—*Results of crevice corrosion tests (weight loss in mg).*[a]

Additive (Concentration)	Ferralium 255	Hastelloy G3	SS 317L	Monit	Ti Grade 2
Baseline	−0.2	−0.1	23	−0.3	0.3
Fe (2000 ppm)	58	93	490	126	0.2
Fe (300 ppm)	n.d	32	270	44	n.d.
Cu (1000 ppm)	0.1	0.3	410	−0.1	0.4
Cu (178 ppm)	0.1	−0.2	0.1	−0.6	0.3
Al (2000 ppm)	n.d.	0.4	n.d.	−0.5	n.d.
PO₄ (500 ppm)	n.d.	n.d.	7.7	n.d.	n.d.
I (100 ppm)	0.6	n.d.	n.d.	n.d.	n.d.
F (1000 ppm)	n.d.	n.d.	n.d.	n.d.	n.d.
Ca (1000 ppm)	n.d.	n.d.	60	n.d.	n.d.
Zn (1000 ppm)	n.d.	n.d.	66	n.d.	n.d.

[a]n.d. = not determined. Negative values correspond to a weight gain.

corrosion did not occur in any of these tests. At 300 ppm Fe, significant crevice corrosion was also observed for SS 317L, Monit, and Hastelloy G3. The SS 317L sample with a thickness of 0.07 cm was perforated after the six-day test. Addition of Cu at 1000 ppm produced severe crevice corrosion for SS 317L, but not for the other alloys. This result is somewhat surprising, since the previous tests had suggested that Cu and Fe induced pitting corrosion to about the same extent. Cu and Fe apparently affect the initiation and/or propagation stages of pitting and crevice corrosion to a different extent.

For SS 317L, additions of 500 ppm PO₄ reduced the extent of crevice attack (Table 9) as compared to the baseline solution, while 1000 ppm Zn and 1000 ppm Ca produced a modest increase. The negligible weight loss for additions of 178 ppm Cu needs clarification. It could be related to the change in corrosion mechanisms that has been discussed in relation to Fig. 24, in which a minimum of corrosion rates was found between 5 and 32 ppm Cu. Zinc sulfate, which is not one of the species studied in this project, was added to evaluate the possibility of cathodic protection. Additions of 2000 ppm Al had no effect for Hastelloy G3 and Monit; 100 ppm iodide did not change the low corrosion rate for Ferralium 255 observed in the baseline solution.

The tests in the baseline solution were run in triplicate because the first run did not show any localized corrosion and corrosion rates were very low. Solution analysis and other checks did not produce any evidence of why this result differed from that for the other two runs in the baseline solution.

This test series has confirmed most of the significant effects determined from the statistical analysis of the 32 runs at pH 1 of the previous series. For Ferralium 255, the inhibiting effect of iodide, which was predicted by the analysis of the first test series (Table 8), was not confirmed. The effect of Si was not tested. For Hastelloy G3, no definite effect of Al was observed. For SS 317L, additions of Ca did not produce the predicted accelerating effect. Phosphate, which was predicted to be an inhibitor for SS 317L and Ti, did not have a significant effect for these two materials. For Monit, Al additions did not accelerate corrosion rates, and Cr was not tested. For Cu and Fe additions, a threshold concentration has been determined above which these ions accelerate corrosion rates of all five alloys. Pitting occurs under these conditions for SS 317L, while crevice corrosion is the main corrosion mechanism for Ferralium 255, Monit, and Hastelloy G3. Ti is attacked only when high fluoride concentrations coincide with low Al levels.

Polarization Curves as a Function of Concentration and Exposure Time

The long-term R_p and E_{corr} measurements discussed above have shown that the nature and concentration of trace elements and the exposure time can have profound effects on the corrosion behavior of the five alloys studied. Except for Ti, the dominating corrosion process was localized corrosion in the form of pitting or crevice corrosion. To obtain additional information concerning the corrosion mechanisms and kinetics of the alloys corroding under these conditions, polarization curves have been obtained after exposure for one, three, or six days in the modified baseline solution (Table 2), and in the same solution with trace elements such as Cu or Fe at a high level for which significant effects had been determined in the statistical analysis (see Table 8).

The potentiodynamic polarization curves (0.6 V/h) have been recorded, displayed, and analyzed with the PAR software "softcorr" Model 332. The anodic sweep was reversed at 3 mA/ cm². The run was started and ended about 30 mV negative of the initial E_{corr}. After this curve had been taken, a cathodic curve was recorded starting about 30 mV anodic to the (new) E_{corr}. For each set of polarization curves, a new electrode was used. Two carbon rods served as counter electrodes, and a SCE as a reference electrode.

Only the experimental polarization curves will be discussed here to illustrate the effects of trace elements and exposure time. The analysis of these curves is given elsewhere [2,3].

As discussed above, SS 317L suffers from localized corrosion even in the baseline solution. This is reflected in Fig. 28a, which shows the anodic polarization curves as a function of exposure time. A broad hysteresis is observed on the return sweep in the cathodic direction. The protection potential E_{prot} lies below the initial E_{corr} for all three exposure times, which suggests that SS 317L will suffer from localized corrosion even in the absence of polarization as observed in the previous tests. An increase of the Cu concentration to 1000 ppm or the Fe concentration to 2000 ppm intensifies the localized corrosion even after only one day of exposure (Fig. 28b). The electrode is probably already pitted in this case, as can be seen by the high anodic currents and the much lower potentials at which the polarization is reversed. For longer exposure times, the surface is severely pitted and the polarization curves are those for actively corroding surfaces with anodic current densities of about 2 mA/cm² (Fig. 28c and d).

For Ferralium 255, effects of exposure time and additions of Cu and Fe similar to those in SS 317L have been observed (Fig. 29a–d). However, the shape of the forward and reverse anodic sweep is different insofar as the broad hysteresis is absent. After 3 and 6 days in the baseline

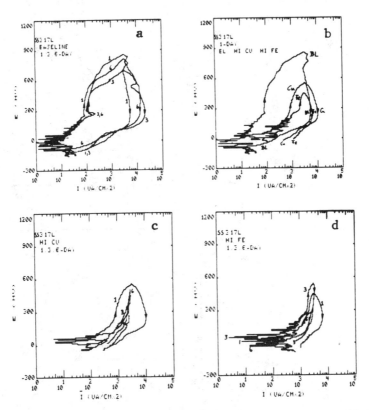

FIG. 28—*Potentiodynamic anodic polarization curves with reverse sweep for SS 317L:* (a) *effect of exposure time in baseline solution;* (b) *comparison of results in baseline solution, at 1000 ppm Cu, and at 2000 ppm Fe;* (c) *effect of exposure time at 1000 ppm Cu;* (d) *effect of exposure time at 2000 ppm Fe.*

solution, the surface dissolves in an active state at what appears to be a limiting current, which is the same for both directions of the sweep (Fig. 29a). The small peak at about $+250$ mV is probably due to oxidation of deposited Cu and/or Fe. A comparison of the effects of Cu and Fe after one day shows the much larger activating effect of Fe (Fig. 29b). For Cu additions, a strong increase of the anodic current is seen between 1 and 3 days (Fig. 29c). For 3 and 6 days, the surface dissolves in the active state at current densities between 0.5 and 1 mA/cm². For Fe additions, this active state occurs already after one day (Fig. 29d).

The polarization behavior of Monit (Fig. 30a–d) is more similar to that of Ferralium than that of SS 317L. The hysteresis is probably somewhat more pronounced in the baseline solution (Fig. 30a) and for Cu (Fig. 30c) and Fe (Fig. 30d) solutions. For the latter two cases, the anodic dissolution currents seem to be reduced for six-day exposures.

For Hastelloy G3, polarization curves have been recorded (Fig. 31a–d) that are similar to those of Ferralium 255 (Fig. 29a–d) as far as the effects of exposure time and Cu or Fe additions are concerned with the surface corroding in the active state.

The polarization curves for Ti grade 2 shown in Fig. 32a–d include the cathodic curves that exhibit a limiting current that is larger in the presence of Cu of Fe additions. Fe additions raise E_{corr} above that for the baseline solution or for Cu additions. The currents on the return sweep are smaller than on the anodic sweep even in the presence of Cu and Fe, suggesting some form

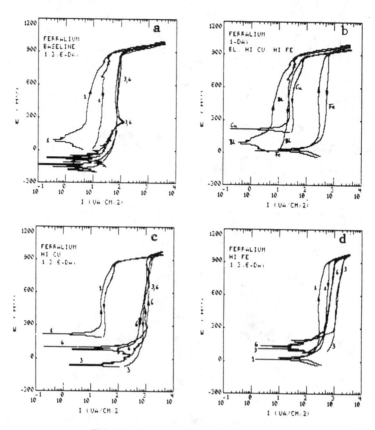

FIG. 29—As in Fig. 28, but for Ferralium.

of passivation. The anodic currents do not exceed 20 $\mu A/cm^2$. An increase of the fluoride concentration from 32 to 178 ppm does not have a significant effect on the anodic polarization curves (Fig. 32d).

Conclusions

1. Relatively minor variations of the chemistry of the test solutions used in this study can have drastic effects on the corrosion rates of the five alloys studied. For a given solution, high corrosion rates can occur for some alloys, while others remain virtually unattacked. This can be illustrated for the case of Ti, in which corrosion rates were 2.3 mm/y for a solution which contains high levels of F, Br, NO_3, Ca, Cr, and Si and low levels of I, PO_4, Cu, Mo, Al, Fe, and H_2S. For Ferralium 255, Hastelloy G3, and Monit, corrosion rates were less than 2.6 $\mu m/y$ in the same solution. When the high levels were changed to low levels and vice versa, Ti had the lowest corrosion rates.

2. That very large variations in corrosion rates were observed in the Plackett-Burman test series suggests that first- and higher-order interactions play a major role in determining the corrosion behavior. A foldover design was used at pH 1 to determine the main effects and species such as Cu and Fe, which are strong oxidizers have been identified as accelerators of corro-

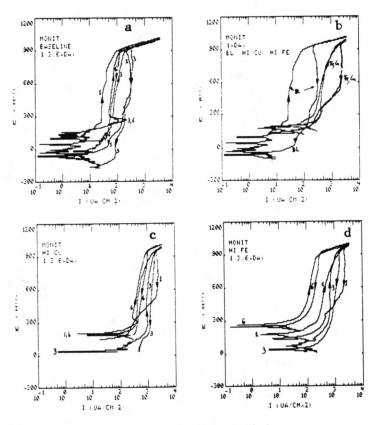

FIG. 30—*As in Fig. 28, but for Monit.*

sion. Interactions of F, Al, and pH strongly affect the corrosion behavior of Ti, but it is possible that these interactions also involve NO_3 and Br as accelerators and PO_4 and I as inhibitors.

3. SS 317L suffers from pitting and crevice corrosion in most solutions. Ferralium 255, Monit, and Hastelloy G3 are more resistant, but localized corrosion was observed at high levels of Cu or Fe. Crevice corrosion tests showed that at comparable concentrations, Fe is more effective in producing crevice corrosion than Cu.

4. Polarization curves obtained over six-day periods in a modified baseline solution containing one species at a high level have provided new information concerning the kinetics of corroded electrode surfaces. For SS 317L in the baseline solution, the typical anodic polarization behavior of a surface susceptible to pitting and crevice corrosion is observed. When high levels of Cu or Fe are present, the polarization curves reflect the high corrosion rates and the active behavior of pitted surfaces. For Ferralium 255, Hastelloy G3, and Monit, no active-passive transitions are observed, and the surfaces dissolve at high rates with a limiting current when the Cu or Fe concentrations are high. Ti remains in the passive state for the solutions for which polarization curves were obtained, since the fluoride concentrations were low.

5. Statistical correlations between the electrochemical data and results from solutions analysis, as well as determination of number and dimension of pits and crevices, have proven that the

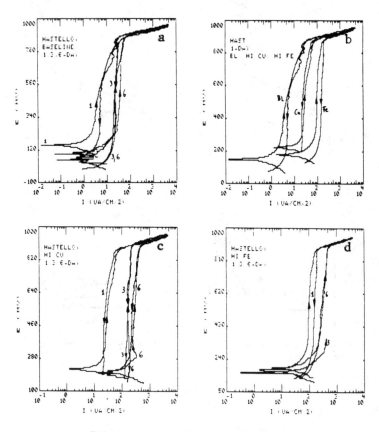

FIG. 31—*As in Fig. 28, but Hastelloy G3.*

electrochemical methods employed here give valid results while providing the important advantage that a continuous record of the corrosion behavior and its changes with time can be obtained when polarization resistance measurements are performed.

Acknowledgment

This project has been funded by EPRI under Contract No. RP1871-7. Many helpful and stimulating discussions with the EPRI Project Manager, B. C. Syrett, provided valuable inputs for the successful completion of the project. M. W. Kendig designed the computerized system for sequential measurements of the electrochemical data. K. Fertig provided input for the design of the experiments and prepared the statistical analyses.

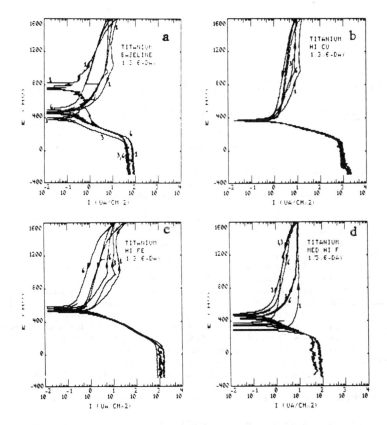

FIG. 32—*Potentiodynamic anodic polarization curves with reverse sweep for Ti, Grade 2:* (a) *effect of exposure time in baseline solution;* (b) *effect of exposure time at 1000 ppm Cu;* (c) *effect of exposure time at 2000 ppm Fe;* (d) *effect of exposure time at 178 ppm Fe.*

References

[1] Mansfeld, F., Jeanjaquet, S. L., Kendig, M., Raleigh, D. O., and Fertig, K., "The Effect of Trace Elements on Corrosion in Total Immersion and Wet/Dry Conditions," *Corrosion/85*, National Association of Corrosion Engineers, Boston, Paper No. 38, *Corrosion*, Vol. 42, 1986, p. 249.

[2] Mansfeld, F. and Jeanjaquet, S. L., "The Effects of Trace Elements on the Corrosion Behavior of FGD Materials," *Corrosion 1986*, National Association of Corrosion Engineers, Houston, Paper No. 362; Corrosion (in press).

[3] Mansfeld, F. and Jeanjaquet, S. L., "Effect of SO₂ Scrubber Chemistry on Corrosion," Final Report SC5834.67FR, Rockwell International, January 1986; EPRI CS-4847, October 1986.

[4] Koch, G. H., Thompson, N. G., and Means, J. L., "Review of the Effects of Trace Elements on the Corrosion of Alloys," EPRI Report on Research Project RP1871-6, EPRI CS-4697, Electric Power Research Institute, July 1986.

[5] Kendig, M., Jeanjaquet, S., and Mansfeld, F., in Proceedings of the Symposium, "Computer-Aided Acquisition and Analysis of Corrosion Data," The Electrochemical Society, *Proceedings*, Vol. 85-3, 1985, p. 60.

[6] Kendig, M., Jeanjaquet, S. L., and Mansfeld, F., 166th Meeting of the Electrochemical Society, Ext. Abstracts, Vol. 83-2, 1983, Abstr. No. 477.

[7] Mansfeld, F. and Kenkel, J. V., *Corrosion*, Vol. 33, p. 13.

[8] Dini, J. W. and Johnson, H. R., *Plating and Surface Finishing*, Feb. 1981, p. 52.

[9] Plackett, R. L. and Burman, J. P., *Biometika*, Vol. 33, 1946, p. 305.

Christian Fiaud[1]

Testing Methods for Indoor and Outdoor Atmospheric Corrosion

REFERENCE: Fiaud, C., **"Testing Methods for Indoor and Outdoor Atmospheric Corrosion,"** *The Use of Synthetic Environments for Corrosion Testing, ASTM STP 970,* P. E. Francis and T. S. Lee, Eds., American Society for Testing and Materials, Philadelphia, 1988, pp. 58–68.

ABSTRACT: Simulation tests for indoor and outdoor atmospheric corrosion appear essential to reliably predict the behavior of designs and materials in operating environments. Chamber tests are preferentially conducted for indoor corrosion. Some results are given concerning the sulfidation of copper and silver by H_2S (hydrogen sulfide) in the presence of additives such as NO_2 (nitrogen dioxide) or Cl_2 (chlorine molecule). It is shown that chemical reactions between gases are responsible for the acceleration of the rate of sulfidation of metals. Comparison with onsite measurements shows that several classes of test severities must be defined. In studying outdoor atmospheric corrosion, attention is presently focused on the use of electrochemical sensors, which allow the determination of times of wetness and corrosion rates. Chamber tests may be used in some special cases. An example is given for the behavior of metallic materials exposed to corrosion under storage conditions. The combination of chamber and electrochemical tests has led to some understanding of the mechanism of formation and spread of corrosion products for an Fe-Ni alloy. The alloy has a behavior intermediate between that of an active (iron-like) material and a self-passivated (stainless steel-like) one, depending on the aeration conditions of the surface.

KEY WORDS: corrosion, atmosphere, hydrogen sulfide, chlorine, nitrogen dioxide, storage

In spite of extensive work over more than 70 years on atmospheric corrosion, a number of questions remain. In particular, no satisfactory method for predicting the performance of materials in the atmosphere exists either as an on-site method or laboratory test. Long-term exposure methods require much time to determine the behavior of materials in atmospheric conditions. Shorter tests are better suited for assessing the best materials for a given environment. Two types of corrosive atmospheric environments have been distinguished, indoor and outdoor. In this paper, test methods for each type of environment are considered along with their main characteristics.

Indoor Atmospheres

Indoor atmospheres usually have low water vapor partial pressure with their relative humidity < 80%. They also have a low amplitude of temperature variation, i.e., no condensation can occur on the metallic surfaces (time of wetness = 0).

Only adsorbed water layers are found at the surface of metals indoors. In some cases, very hygroscopic corrosion products may develop and promote water condensation, but this is the exception, and because of the very thin layers of adsorbed water molecules, electrochemical processes are not likely to proceed to a great extent. As a matter of fact, electrochemical methods are seldom used to study indoor atmospheric corrosion.

[1]Professor, C. N. R. S. U. A. 216, Groupe Corrosion, Universite, Paris, 6, 11, rue P. et M. Curie, 75005 Paris.

In indoor concentrations polluting species such as gases or solids are at low concentrations as compared with outdoor concentrations. Only sulfur-containing contaminants (H_2S, COS, etc.) may be found in equivalent concentrations (Table 1). The most important parameters to consider are the nature and concentrations of polluting gases. Insofar as the metals investigated are mainly copper, silver, and generally metals used for electric or electronic applications, sulfiding gaseous species are of prime importance, but synergistic effects with other gaseous contaminants may exist, as will be shown later in the paper.

Outdoor Atmospheres

Outdoor atmospheres are characterized by periodic wetting of the metallic surfaces from water condensation, relative humidities $> 80\%$, fast temperature variations, and precipitation. Gas pollution levels are higher in outdoor atmospheres than in those indoors (Table 2). Accumulation rates of corrosive species and acidity levels (H^+) are also much higher [1]. Very little is known about accumulation rates for different locations and types of atmospheres, yet this parameter, together with the time of wetness, essentially determines the corrosiveness of a given site [2].

SO_2 (sulfure dioxide) plays an important role in outdoor atmospheric corrosion because of its high solubility in layers of water condensed at metallic surfaces. As a matter of fact, SO_2 is often the only polluting gas introduced in an atmospheric corrosion test. Recent work indicates, however, that other gases, such as NO_x (nitrogen oxides), Cl_2 (chloride molecule), etc. strongly affect the behavior of SO_2. For example, in the presence of NO_2 (nitrogen dioxide), a very important increase in the corrosion rate of several metals is observed compared to the effect of SO_2 alone [3,4]. This increase is attributed to a rise in the time of wetness of the metal, and/or to the oxidative effect of NO_2 increasing the concentration of the S_{VI} species.

TABLE 1—*Comparison of indoor and outdoor concentrations* [10].

Polluting Species	Indoor C_i, $\mu g/m^3$	Outdoor C_e, $\mu g/m^3$	$C_i / C_e \times 100$
SO_2	12.2	17.6	69
NO_2	26	43.5	60
H_2S	0.48	0.48	100
Cl_2	0.2	1.5	13

TABLE 2—*Typical indoor and outdoor concentrations of species associated with acid rain* [1].

Component	Indoor	Outdoor
Sulfate	6	20
Nitrate	1.2	4
Chloride	0.08	0.3
Ammonium	1.5	5
Calcium	0.3	6

NOTE: Urban concentrations, $\mu g/m^3$.

Indoor Atmospheric Corrosion Testing

As it is now recognized that mixtures of gases must be used to simulate the behavior of metals placed in atmospheric environments, investigations were undertaken to obtain results concerning the effect of gas mixtures such as H_2S (hydrogen sulfide)/NO_2 and H_2S/Cl_2 on the indoor corrosion (tarnishing) of silver and copper. The experimental device (Fig. 1) has been described elsewhere [5]. Gases are fed by means of permeation tubes, and the relative humidity rate is regulated through the adjustment of dry and wet air fluxes. Corrosion products are measured by a coulometric method [5].

H_2S-NO_2

Figures 2 and 3 show the reactivity of silver and copper to H_2S/NO_2 mixtures. The reaction of silver with H_2S in the absence of NO_2 leads to the formation of Ag_2S (silver sulfide), whereas several tarnishing products are found for copper. These include Cu_2S (copper sulfide), copper I and copper II oxides, and a mixed CuO (copper oxide)/Cu_2S compound.

When added to H_2S, NO_2 induces faster growth of silver sulfide, except at the lowest H_2S concentrations, for which NO_2 has no effect (Fig. 4). When the reaction rate is increased (H_2S) = 0.1 ppmv and NO_2 = 0.1 ppmv), the kinetics, which was parabolic in the absence of NO_2, becomes progressively linear as the NO_2 concentration increases, with an upper limit for the effect of the NO_2 concentration.

The same acceleration is obtained for copper sulfidation, but neither a critical lowest H_2S concentration is found, nor a highest NO_2 concentration.

NO_2 has an oxidative effect against H_2S, leading to the formation of elemental sulfur through a reaction such as:

$$H_2S + 2NO_2 + O_2 = S + 2HNO_3 \tag{1}$$

The very low reaction rate observed on silver at very low H_2S concentration (0.02 ppm) is due to the negligible amount of sulfur formed, whatever the concentration of NO_2 is.

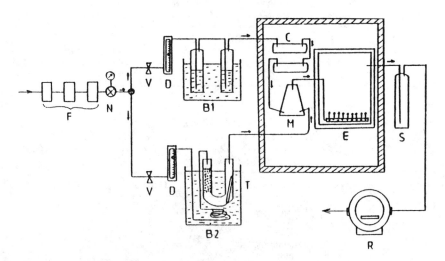

FIG. 1—*Environmental test device: F = filters, N = pressure regulator, V = needle valve, D = flowmeters, B₁ and B₂ = thermostated baths, C = condensors, M = mixing chamber, E = pyrex reaction chamber, T = permeation tube, and R = gas meter.*

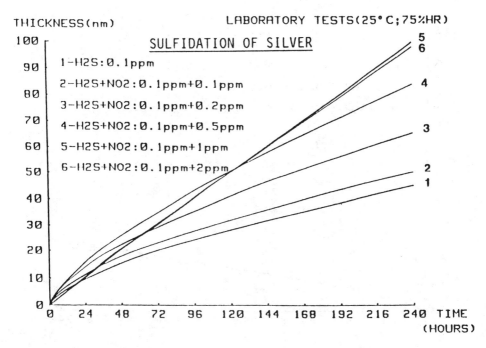

FIG. 2—*Kinetics growth of silver in H_2S and H_2S/NO_2 environments.*

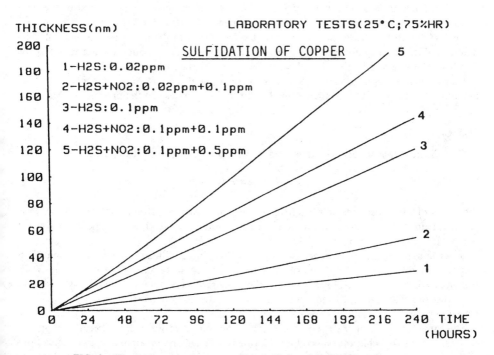

FIG. 3—*Kinetics growth of copper sulfide in H_2S and H_2S/NO_2 environments.*

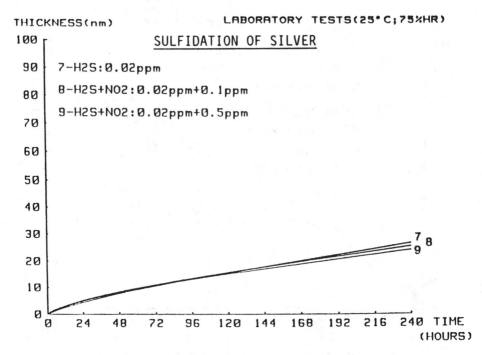

FIG. 4—*As Fig. 2, at low H_2S concentration.*

With a medium H_2S concentration (0.1 ppm), two limits must be considered in the NO_2 concentration: a lower limit below which no elemental sulfur is formed, and an upper limit beyond which all the hydrogen sulfide available for silver sulfidation is transformed into elemental sulfur. This explains the slow transition from a parabolic kinetics law controlled by the H_2S/Ag interaction to a linear kinetics law controlled by the S/Ag interaction.

An irreversible dissociation of H_2S occurs on copper in absence of NO_2. This means that the amount of elemental sulfur available for the sulfidation increases in the presence of NO_2.

H_2S-Cl_2

H_2S-Cl_2 mixtures are also characterized by a reaction between the two gases:

$$H_2S + Cl_2 = 2HCl + S \qquad (2)$$

the extent of which depends on the partial water vapor pressure. At low relative humidity, the reaction tends to be equilibrated. At high relative humidity ($> 60\%$), water vapor displaces the equilibrium towards the right.

Silver is not reactive to hydrogen chloride, so that, in conditions in which Reaction (2) is complete, and $(Cl_2)/(H_2S) < 1$, no silver chloride is found when silver is exposed in H_2S/Cl_2 mixtures, whereas some AgCl is detected in the same ratio conditions at a relative humidity $< 30\%$ (because Reaction (2) is equilibrated), or for ratio conditions such as $(Cl_2)/(H_2S) = 1$ (because some Cl_2 remains in excess).

The same influence of the $(Cl_2)/(H_2S)$ ratio is not observed for copper, because copper is reactive to hydrogen chloride and copper chloride is detected on the samples at all compositions of the gas mixture.

As expected from Reaction (2), the sulfidation of both metals is accelerated by the presence of chlorine (Figs. 5 and 6).

To take advantage of these results in adjusting a corrosion test, it is necessary to make correlations with on-site tests. Data in the field are now available from different sources [6, 7], and the authors undertook some investigations on copper and silver exposed in indoor conditions.

Silver and copper plate samples were exposed in four different indoor locations (French telephone centers) and analyzed at various exposure times within a 2-year period. For silver, Ag_2S is always detected as a tarnishing product. The kinetics of growth is parabolic. AgCl (silver chloride) is also found in all locations, except those that were far from coasts. The growth kinetics of AgCl is linear. The results indicate that the behavior of silver in indoor conditions may be simulated by a test involving a H_2S/Cl_2 mixture at low concentration levels (< 0.5 ppmv), with a $Cl_2)/(H_2S)$ ratio greater than 1 (AgCl excepted), and relative humidity $= 75\%$.

Such a test does not hold for copper. For the same exposure conditions, no copper sulfide forms on copper on the four sites, yet copper chloride was always detected. As suggested by laboratory experiments at very low H_2S concentrations, this behavior is probably due to the preexisting Cu_2O layer, which protects the metal against sulfidization. Several classes of severities must be envisaged for a metal such as copper, depending on the polluting levels of the gases on the site. This conclusion meets Abbott's recommendation [6]. He defines four test severities for copper, with a 0 class corresponding to no test at all in the case of less corrosive atmospheres.

Outdoor Atmospheric Corrosion Testing

Very few chamber tests for the simulation of outdoor corrosion have been proposed, despite the fact that many laboratory studies do exist in this field. Two main reasons probably explain this lack of realistic chamber tests:

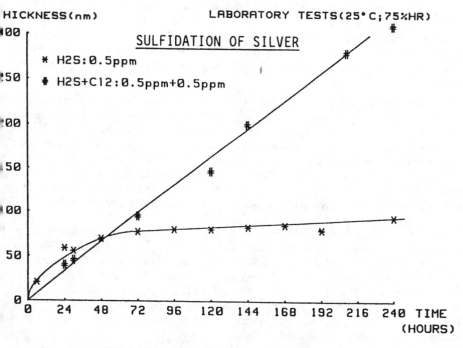

FIG. 5—*Corrosion products on stacked Fe-36Ni plates.*

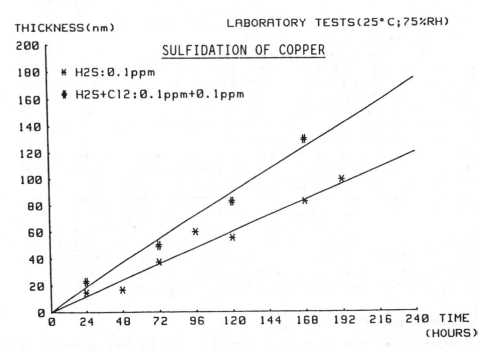

FIG. 6—*Influence of dissolved oxygen concentration on the passivation characteristics of Fe-36Ni alloy.*

1. less well-defined needs as compared to the needs of the electronic industry for protection against indoor corrosion, and

2. technical difficulties in the conception of a test. In particular, water condensation on samples is necessary in principle, and is not easy to achieve because the polluting gases are carried away by the condensates on chamber walls, and a subsequent drop in the concentration of these gases occurs.

Experiments in the field are rather focused on site tests, either for the characterization of a metal's resistance to corrosion on a given site, or for the estimation of the corrosiveness of a given site. This last approach seems more promising because it may lead to a rapid evaluation of the risks of degradation of a material if some checking for this material has been established as a function of the characteristics of the environment. In this connection, electrochemical sensors are very useful in determining times of wetness and corrosion rates. The miniaturization of the sensors is a promising recent improvement that allows direct determination of the time of wetness of either metallic or nonmetallic materials [8], or accurate determination of the corrosion rate of materials [9].

One of the most frequently encountered outdoor atmospheric corrosion problem is the development of corrosion products on stacked sheets during storage or transportation. This type of corrosion does not lead to deep deterioration of the metal, but rather, to surface deterioration. The phenomenon is clearly related to temperature variations, which allow the condensation of water layers between the metal sheets. Atmospheric pollutants may play a role in producing the detected corrosion products—often chlorides and sulfates—but these pollutants are probably not essential in initiating the corrosion process. The phenomenon is frequently observed on such materials as aluminum, galvanized steel, iron alloys, etc.

A study of the corrosion of Fe-36 Ni metal sheets in storage conditions was conducted by simulating the phenomenon in a corrosion chamber, and by performing electrochemical tests on the thin electrolyte layer.

The corrosion products developed under natural storage conditions occur as plates 0.2 to 2 mm long and 1 to 2 μm deep (Fig. 7). The laboratory simulation tests were conducted on 30 × 40 × 2 mm samples, either suspended alone, or placed into the chamber in a pile of at least 8 samples. The experimental atmosphere was a mixture of air, water vapor, and eventually, sulfur dioxide. The total duration of the tests was 1000 h; samples were periodically removed from the chamber and examined to detect the beginning of the attack. The morphology of the corrosion products was very similar to that observed on on-site exposed specimens when the following parameters were used in the chamber test: relative humidity = 90%, T = 323 K, no SO_2. Less than 50 h were necessary to detect corrosion products. With increasing exposure times, the corrosion sites did not spread, but new sites appeared so that the obtained pattern was between localized and uniform corrosion. Suspended samples showed the same behavior as piled ones, at least if their surface state was a rather rough one, e.g., a 80 or 320 grit paper polish. At 600 grit polishing, the suspended samples showed no trace of corrosion, whereas black spots appeared on stacked ones. As it was also observed that corrosion products preferentially developed along the polishing grooves, it was concluded that corrosion preferentially starts in the regions where the adsorbed or condensed water collects, which are also regions where oxygen does not have free access.

The influence of the addition of SO_2 (8 ppm) to the experimental atmosphere produced a drop in the induction time for the formation of corrosion products. At 60% relative humidity, the corrosion products were detected after a 200 h test, whereas they were not detected after 2000 h in the absence of SO_2.

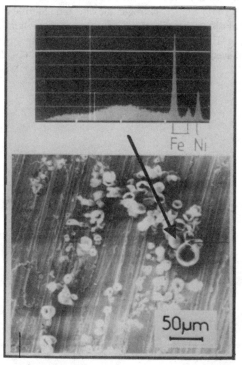

FIG. 7—*Localization of corrosion products on Fe-36Ni electrodes in conditions of differential aeration.*

To explain these results, some electrochemical tests were conducted on the Fe-Ni alloy. $I = f(E)$ potentiokinetics anodic curves in a bulk solution of 0.1 M sodium sulfate showed that the alloy is self-passivated only in fully oxygenated solutions, and not in naturally aerated ones (Fig. 8). This probably explains the preferential initiation of the attack between stacked sheets or at the bottom of the polishing grooves, because oxygen access is more difficult at the corresponding metal surfaces. To explain the localization of the attack on the metal sheets, a simple galvanic test was imagined in which two metallic samples were immersed in the bulk Na_2SO_4 solution and electrically connected. One sample had free access to dissolved oxygen at its surface, while the other was isolated under a Teflon disc (Fig. 9). Because of a differential aeration between the two samples, corrosion developed as expected on the unaerated sample, yet the corrosion products remained localized at some distance from the edge of the disc, with no products at the center of the disc (Fig. 9). The galvanic effect resulting from the differential aeration is obviously limited by the high resistance of the thin layer of electrolyte. A current and potential distribution is likely to exist within this layer. This effect may explain the localization of the attack frequently observed on the metal coils.

In summary, the corrosion of the Fe-36 Ni alloy sheets observed in storage conditions is mainly due to the presence of condensed water layers between the sheets, and insufficient access of oxygen to these layers, which leads to local depassivation of the alloy. In contrast to the classical form of localized corrosion (pitting), the attack does not progress deeply, but rather, from spot to spot on the whole portion of the alloy for which the above conditions are fulfilled.

Conclusion

Indoor atmospheric corrosion may be simulated by chamber tests in which mixtures of gases are introduced, but care must be taken in the definition of the experimental procedure. An example is given for silver and copper. Nitrogen dioxide and chlorine accelerate the sulfidation of both metals somewhat differently. On silver, the effect of the pollutants depends on the ratio of their partial pressures (H_2S/NO_2 and H_2S/Cl_2), but this not so on copper. This behavior is attributed to the role played by elemental sulfur in the sulfidation of the two metals.

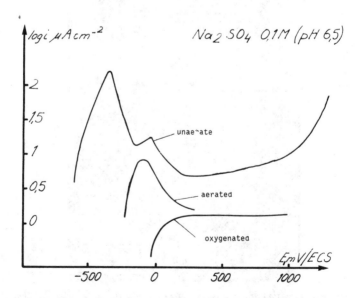

FIG. 8—*Kinetics growth of silver sulfide in H_2S and H_2S/Cl_2 environments.*

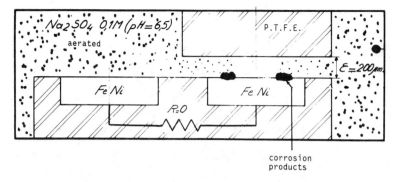

FIG. 9—*Kinetics growth of copper sulfide in* H_2S *and* H_2S/Cl_2 *environments.*

In practice, several test severities will probably have to be defined for copper. A unique test may be used for silver. Outdoor corrosion testing is best conducted using electrochemical sensors. However, in some cases, simple chamber tests may be used. This is the case for the simulation of the corrosion of Fe-36 Ni sheets stacked for storage or transportation. The corrosion of the alloy is due to the presence of unaerated condensed water layers between the foils. The process starts by depassivation of weak points of the surface, and extends laterally rather than medially.

References

[1] Sinclair, J. D. and Weschler, C. J., "Proceedings of the Symposia on Corrosion Effects of Acid Deposition and Corrosion of Electronic Materials," *The Electrochemical Society*, Vol. 86, No. 6, p. 258.

[2] Sinclair, J. D., Psota-Kelty, L. A., and Weschler, C. J., *The Electrochemical Society*, Vol. 85, No. 6, p. 328.

[3] Leygraf, C., Zakipour, S., and Portnoff, G., *The Electrochemical Society*, Vol. 85, No. 6, p. 307.

[4] Svedung, O. and Johansson, L. G., *The Electrochemical Society*, Vol. 85, No. 6, p. 318.

[5] Fiaud, C., Safavi, M., and Vedel, J., W.U.Korr. Vol. 35, 1984, p. 361.

[6] Abbott, N. H., *Materials Performance*, Vol. 24, No. 8, 1985, p. 46.

[7] Guinement, J. and Fiaud, C., *13th International Conference on Electric Contacts*, Lausanne, France, 1986.

[8] Sereda, P. J., Croll, S. G., and Slade, H. F., *Atmospheric Corrosion of Metals*, STP 767, American Society for Testing and Materials, Philadelphia, pp. 267–285.

[9] Mansfeld, F., Jeanjaquet, S. L., and Kendig, M. W., "Proceedings of the Symposia on Corrosion Effects of Acid Deposition and Corrosion of Electronic Materials," *The Electrochemical Society*, Vol. 86, No. 6, p. 239.

[10] Rice, D. W., Cappel, R. J., Kinsolving, W., and Laskowski, J. J., *Journal of the Electrochemical Society*, Vol. 127, No. 4, 1980, p. 891.

DISCUSSION

R. Christie[1] *(discusser's question)*—In using the indoor atmosphere test does the author notes that gas is absorbed by the chambers at best initially to a reduced concentration. Does this affect the reproducibility of the test? Has the author tested electroplated tin or tin alloys in the H_2S/Cl test chamber?

C. Fiaud (author's closure)—Polluting gases are in fact adsorbed on the walls of the chamber in the first moments of the test, and a drop in the concentration of the gases may be established at the outlet of the chamber for a time duration that depends on the level of concentration of the gases. In the given experimental conditions, an equilibrium between inlet and outlet concentrations is generally attained 4 h after the beginning of the test. At very low concentrations of gases, the metallic samples (and occasionally the nonmetallic parts of samples) may also contribute to a decrease in the mean concentrations of the active gases in the chamber. For some reproducibility of the test, a constant sample surface/chamber volume ratio has to be maintained.

No investigations were made on tin or tin alloys.

[1]GEC First Research Center, East Lane, Werbley, Middok, HA9 7PP, United Kingdom.

Thomas P. Murphy[1] and Graham A. Pape[1]

Review of Corrosion Studies of Metal Containers Using Synthetic Media

REFERENCE: Murphy, T. P. and Pape, G. A., **"Review of Corrosion Studies of Metal Containers Using Synthetic Media,"** *The Use of Synthetic Environments for Corrosion Testing, ASTM STP 970,* P. E. Francis and T. S. Lee, Eds., American Society for Testing and Materials, Philadelphia, 1988, pp. 69–78.

ABSTRACT: Corrosion phenomena in metal containers differ from those encountered in, for example, pipeline or bridge construction, in that they can be apparent at very low corrosion rates. Thus, contamination of the contents of the package can occur with relatively low metal uptakes, rendering them unsuitable for use. Dissolution of tin or iron from tinplate and of aluminium from body or end stock are examples of this contamination. Sulphide stains may be unsightly, but are not harmful. Perforation of the container may allow the product to escape, leading to secondary corrosion or contamination, and associated losses; or to the ingress of contaminants, leading to flavor changes or microbial spoilage. In practice, the incidence of problems is very low.

Corrosion testing frequently uses the actual products to be packed; this is the ultimate test. Synthetic systems have been used to simulate complex natural food in particular, but also to clarify the behavior of formulated products. Test methods involve electrochemical studies, coupon testing, and test packing. The behavior of coatings for cans may also be studied in this way. Performance is assessed by considering metal pick-up and incidence of swelling or perforation. Controlled-environment chambers are used in accelerated testing.

Characterization of materials also involves using synthetic media in laboratory testing, e.g., measurement of tin oxide levels in tinplate. The properties of containers and components may be checked using synthetic or, occasionally, natural materials.

While such procedures are helpful in clarifying corrosion processes and studying components or materials, they have limitations. Ideally, the test procedure should correspond to the products to be packed under anticipated conditions of use. Nevertheless, the use of synthetic systems contributes to the optimization of the final container/product combination.

KEY WORDS: corrosion, synthetic, metal containers, materials characterizations, tinplate, aluminum, food materials, lacquers

Corrosion in containers differs markedly from that in major structures such as bridges and pipelines. The fundamental process—the conversion of metal to an oxidized state is unchanged—but the practical consequences are different, as is the scale of change. What, then, are the requirements for containers; what corrosion phenomena arise; and how can synthetic media be used to study them?

Requirements

Although "containers" can be extended to include anything up to a supertanker, we are concerned with smaller items, up to a maximum of, say, 25 litres in capacity, as are used in the packaging industry. Packaging must:

---

[1]Principal scientist and senior scientist, Metal Box PLC, Wantage, Oxon, OX129 BP, United Kingdom.

(1) contain the product,
(2) protect the product,
(3) deliver the product, and
(4) not cost too much.

Corrosion phenomena must be considered with the first two requirements in mind. The final requirement is, of course, a constraint on the possible approaches. Corrosion science is an economic as well as a technical science. Designing for conditions and shelf-lives not anticipated wastes resources.

Corrosion Phenomena

The corrosion rates that pose problems in containers can be much lower than those for bridges. A few tenths of a millimeter of penetration would be enough to perforate the wall of a can within a year. Again, the uptake of a few parts per million (ppm) of metal may render the contents of a can unsuitable for use. Let us briefly examine the metals involved and the effects of their corrosion before considering product types and synthetics.

The metals employed in container manufacture are aluminum alloys, and steel coated with tin, as in tinplate, or with a chromium/chromium oxide layer, as in tin free steel (TFS or ECCS). Some uncoated steel (black plate) is used for drums. Solder metals, tin, lead, and alloys are used, although in diminishing proportions as welded and two piece seamless cans become more common. Lacquers (enamels) and varnishes may be applied to the metals for external decoration or protection from internal attack.

Attack on internal steel puts iron into solution and may also lead to perforation of the can, allowing the escape of the product or the ingress of chemical or microbial contamination. The attack is aggravated where small areas of anodic steel are coupled to large areas of cathodic tin. Fortunately, in many food products the complexing action of fruit acids makes tin a protective anode. Thus, plain tinplate is often acceptable. In some cases, steel is anodic and lacquered plate is required.

The dissolution of tin is not harmful, although color or flavor changes may arise. The accompanying generation of hydrogen in the cathodic process can cause swelling of the can. Though harmless, in the kitchen this is not distinguishable from swelling caused by CO_2 arising from microbial action.

Attack on aluminum containers, often in their easy-open ends, can lead to perforation and loss of product. In this context, the value of the aluminum oxide rupture potential (breakdown potential) relative to the rest potentials of the can metals is important (Figs. 1 and 2).

Solder attack is encountered less nowadays, as fewer soldered cans are made. Again, leakage is possible with corrosion of the sideseam material.

Metal pick-up can be important, as a few ppm of iron can alter the flavor of a cola drink or render an aerosol starch unsuitable. Aluminum pick-up at the level of a few parts per million in soft drinks can cause color changes.

Other corrosion phenomena can cause the formation of sulphides of iron and tin. Peas and meat products, for example, can produce this effect. While completely harmless, the sulphide stains on the metal surface may be unsightly. A similar unsightly effect can be produced by the detinning of the tinplate below the lacquer.

External attack may take the form of rusting, induced by condensation or by cannery water; or of etching of the tin if conditions, e.g., pH, are appropriate. When aluminum is attacked, it may develop unsightly white or black stains. Though undesirable, these phenomena are relatively minor, unless perforation of the container occurs.

Charbonneau [1] has provided a very useful review of the major types of corrosion encountered in containers.

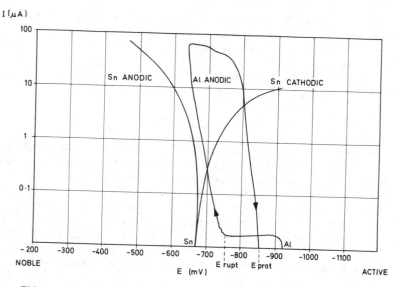

FIG.1—*Polarization curves for Al and Sn in milk product (pitting risk high).*

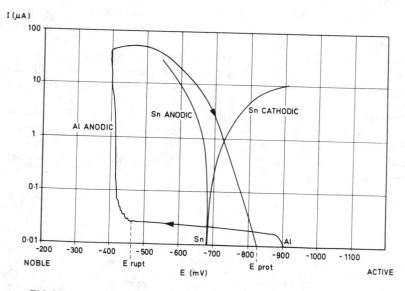

FIG.2—*Polarization curves for Al and Sn in milk product (pitting risk low).*

Corrodant Types and Synthetics

Studies of the corrosion effects encountered in containers, as exemplified above, often employ the actual products to be packed. The use of synthetic media can play an important part in the understanding of corrosion phenomena arising with real products, whether natural, e.g., food or beverages, or formulated.

The application of synthetics may take various forms. The media may be truly synthetic, simulating the behavior of real corrodants; model systems may be used, in which simplified versions of a product are built up; additions may be made to typical major ingredients; and natural systems may be modified.

Some synthetics are intended to evaluate the properties of materials or components rather than directly mimicking a real corrodant.

Formulated Products

With products formulated for the domestic/industrial market, good control of compositions is possible. Synthetic systems may be used to examine the effect of specific ingredients or of formulation changes. Many materials in this category—oven cleaners, bathroom sprays, industrial cleaners—basically are comprised of alkaline aqueous solutions with additives such as surfactants or sterilants. Heinzte and Braun [2] have described studies using alkaline solutions, principally sodium carbonate and bicarbonate, that deal with the corrosion of tin on tinplate, which is mild steel coated with tin. The importance of pH and the effects of ions such as halides, silicate, phosphate, and hypochlorite are discussed. Attack on steel is usually limited unless specifically aggressive species are present. Breakdown of aerosol propellants [3] can produce aggressive ions.

Sheppard and Su [4] have studied aerosol can corrosion mechanisms, considering the effect of propellant in pressurized containers. They too used alkaline media, ammonium hydroxide solution in tap, or de-ionized water with sodium phosphate added. At Metal Box, we have found that the effect of propellant is important, though we have not always had to pressurize the cell. Sodium hydroxide solutions have with addition of various surfactants, proved useful simulants, as Table 1 shows.

Costa and Culleré [5] employed sodium hydroxide solutions with nitrite and chloride added in their study of tin dissolution. Sodium hydroxide again figured in the studies of tin dissolution carried out by Dickinson and Lotfi [6]. The latter workers used rotating disk and impedance techniques, while Costa and Culleré employed couple current measurements.

Dissolution of tin in sodium hydroxide leads to sodium stannate. Abd El Rehim and his colleagues [7] recently published some results regarding the breakdown of tin passivity in stannate solution. Their work indicates the importance of halide ions, showing how these can interfere with passivity; repassivation can occur with the precipitation of a basic oxychloride, an insoluble corrosion product.

TABLE 1—*Lead dissolution from coupled and uncoupled samples in 3% caustic soda after 2 weeks.*

3% Caustic Soda Plus Surfactants	Lead Pick-up (ppm) from	
	Pb	Pb/Fe
NaOH alone	446	325
NaOH + ADE	561	290
NaOH + RE601	614	489
NaOH + BV	521	212
NaOH + ME	279	195
NaOH + C2M	567	255

Tin dissolution is the main corrosive effect encountered in this class of product. At the higher pH values, aluminum is not widely used; and the steel of tinplate or tin-free steel (TFS) is not greatly attacked in the absence of substances promoting pitting, although some dissolution of the chromium oxide layer is possible.

Hydrogen generated cathodically may swell the container as tin dissolves, although this tends to be a slow process. The risk of hydrogen swell is less for these products than in more acid media.

Natural Products

Acid media are encountered in many foods and beverages and occasion the employment of synthetic media as corrodants. With formulated products, as described earlier, the ease of control of composition permits great use of complete products. Model systems using some of the ingredients, described by Murphy for antifreeze systems [8], can also be usefully employed.

For natural products, however, the situation may be much more complex, with more factors involved. The corrosive behavior of, say, orange juice or tomatoes may depend upon the strain of fruit; where it is grown; soil condition; rainfall; sunshine; fertilizer; pesticide; time of harvest; postharvest treatment; and filling and processing conditions. This makes simplification welcome.

Among the main corrosive ingredients in foods are the natural acids, such as citric, malic, oxalic, and tartaric. These have formed the basis of many synthetic systems used to investigate can corrosion. Their corrosive action has been studied and they form the basis for investigating the effects of other materials, e.g., pesticide residues. Elder [9] has given a useful summary of some reactions of organic acids with various metals; but the reactions with tinplate have concerned most workers, as this has been the material of choice for most food packaging.

Corrosion Effects with Particular Container Materials

Detinning

The ability of the fruit acids to form complexes, especially with tin, is important, and several authors have dealt with this property. Willey [10] has described the effects of tin ion complexation on the relative potentials of tin, steel, and tin-iron alloy in food media. He has used synthetic systems with perchloric and hydrochloric acids, as well as the fruit acids, and found that complexing activity can markedly influence corrosion. Sherlock and Britton [11] have considered the effects of complex formation on tin dissolution rates, and found it an important factor. Gouda and her colleagues [12] have reported on the corrosion of tin in the complexing acids citric, oxalic, and tartaric acid, although the researchers' conclusion that control is exerted by the complexing of Sn^{2+} ions may underestimate the influence of cathodic action in the overall process.

Willey has pointed out the importance of cathodic depolarizers, and other authors, e.g., Montanari, Milanese, and Massini [13] have indicated the controlling influence of the cathodic reactions. This cathodic control accords with our experience and with actual practice. Corrosion of packed tinplate cans, whether containing synthetics or real food, depends greatly upon the cathodic reaction. Control of the availability of depolarizers, or of the exposed steel and tin-iron alloy (since tin is a poor cathode for hydrogen) can control the corrosion. Such depolarizers can be oxygen, controlled by limiting the air content of the filled can, or natural dyes, such as anthocyanins present in the food. The latter can act directly on the corrosion process. Their reduction can lead to undesirable color changes. For these reasons cans that might otherwise be used plain (uncoated) may be lacquered. The reactions of tinplate with food media are very important and many authors have studied them [14-16].

Residues

Other substances may also enter into the process. Thus, fertilizer residues may be important—in particular, nitrate. The effect of nitrate ions on detinning has been studied widely using real food media or synthetics or both [17-20]. Kolb has presented a comparison of the effects in synthetic and real media [21]. From the studies, the general view is that the influence of nitrate becomes important at low pH.

In addition to fertilizer residues, residues of pesticides may be present. These can influence the corrosion process, and various authors have described work with model solutions of acids [22-26]. The dithiocarbamates appear active in promoting attack on steel of tinplate. This is believed to take place via the formation of carbon disulfide.

Sulfide Staining

Other sulphur-bearing materials in certain meat and vegetable products can cause staining through sulfide formation. Black iron sulfide and purple tin sulfide can be formed. These are harmless but unsightly. Cysteine solutions have been used to study this process [27] and metabisulfite can also be used. Modification of the oxide film on tinplate can enhance its resistance to staining and alter its corrosion behavior, both plain or lacquered.

Aluminum Corrosion

Oxide film stability is a dominant feature of aluminum corrosion. Nagayama and co-workers have described the dissolution of aluminum oxide films in oxalic acid [28]. They distinguished a field-dependent pore deepening process and nonfield-dependent pore widening. We have observed such effects in formulated products.

Though contamination by aluminum ions is to be avoided, most concern has focused on the possibility of perforation, which may arise if localized rupture of the oxide film can take place. Pourbaix has indicated [29] that the film is stable from about pH 4 to pH 9 but other factors can play a part. Uhlig and Bohni have reported on how the stability of the film can be influenced by, for example, ions of more noble metals like iron or copper, and by the presence of aggressive species like chloride ion [30]. Chloride can arise from various sources, such as the water used in the formulation of soft drinks or by reaction in the can—e.g., in propellant solvolysis as described by Uchida and Wagatsuma for alcohol-related systems [31]. Whitman [32] has shown the influence of humidity and trace amounts of anions, particularly chloride and sulfate, on the externally initiated stress corrosion cracking of aluminum easy-open-end scores.

Gimbitskaya and her colleagues have applied mathematical methods to the study of aluminum corrosion in ternary mixtures of water, acetic acid, and sodium chloride [33]. The combination of chloride with organic acids or salts is a favorite synthetic, with citrate/chloride mixtures employed by Horst and English [34] to study the critical pitting potential of 3004 aluminum alloy for cans. Acetic acid/chloride combinations more closely resemble pickle systems [35] while citric acid/chloride can simulate some soft drinks, especially isotonic ones. We have found it useful to "doctor" soft drink formulations to take account of local water variations.

Temperature, too, can be important and Cantoni et al. have used sodium chloride, acetic acid, and citric acid to study the effects on aluminum corrosion of temperatures ranging from the ambient temperature to −18°C [36]. At low temperatures, the attack was reduced, a phenomenon we have observed with gravy/meat products containing chloride. Raising the temperature can be harmful, as we have found when studying the change of rupture potential in milk products (Figs. 1 and 2). In circumstances like this, although synthetics may be useful, it is prudent to confirm findings using real media.

Chrome-Coated Steel

Real products have generally been used to evaluate chrome coated sheet (TFS or ECCS). Landgraf [37] has described the performance of this material for various product types. Some Italian work [38] suggests that, while adequate for less acidic packs, TFS may be more vulnerable than tinplate for food products of lower pH. This seems plausible considering the complexing action of food acids on tin. Hottenroth [39] has studied the behavior of chromed steel in synthetic media, and Dembrovskii and colleagues have used lactic acid as well as beans to study iron and chromium dissolution from chromed sheet by γ-spectroscopy [40]. Despite the absence of tin, the use of TFS materials in containers, either alone or, for example, as ends on tinplate bodies, is growing. The material is almost invariably used lacquered.

Lacquered Materials

Lacquering of can materials alters their corrosion performance. By covering the metal, the lacquer isolates it from the product, reducing interaction. Treated as inert barriers, lacquers can alter the ratio of exposed areas of can metals [41,42]. This can change corrosion patterns.

However, lacquers are not totally inert; reactions can happen through them. The performance of lacquered materials has been studied. Over the years, we have found it useful to employ a synthetic medium of citric acid, sugar, and dye, with in-can electrodes as described by Uchida [31], to study lacquered can corrosion using polarization resistance (PR) measurements. Attempts to correlate to metal pick-up yielded a relationship in the form:

$$\frac{1}{PR} = A[Sn] + B[Fe] + C \tag{1}$$

where [Sn] and [Fe] represent concentrations of tin and iron dissolved, and "constants" A, B, and C are rather ill-defined. We found it more useful to monitor changes in PR with time as an indication of lacquer system degradation, rather as AC impedance changes can be used. Scantlebury and co-workers[2] have employed such techniques with coated sheet using sodium sulfate and citric acid solutions. This work indicates the importance of pore blocking and unblocking, depending on the solubility of corrosion products. While such methods are often applied to coated metals, the coatings used are usually much thicker than can lacquers. Among the various papers [43–45] on lacquered materials, Koehler's work [46] is very relevant.

Component/Material Evaluation

Just as protective systems can be studied using synthetic media, materials and components can be evaluated. Synthetic media can be used to carry out studies on systems ranging from a complete lacquered container to the base steel stock.

Tinplate, the most widely used material, is fairly complex, from the inside out comprised of mild steel, the tin iron alloy $FeSn_2$, metallic tin, tin oxide, perhaps containing chromium species, a film of mill oil and, if present, lacquer. These regions can influence the performance of the container in various ways. The integrity of the lacquer coating controls the access of the corrodant to the metal, and can be assessed using the Enamel Rater test [47]. This employs sodium chloride or sulfate solution with surfactants as an electrolyte between the can and a central electrode. Variants have been developed to test can ends, and to reveal the sites of metal exposure. Acidified copper sulfate is used to indicate metal exposure on deformed lacquered samples to demonstrate flexibility of the lacquer.

[2]Scantlebury, J. D., Hepburn, B. J., and Gowers, K. R., private communication.

The adhesion of the lacquer is important, and mixed phosphate buffers have been used to examine the resistance of the lacquer to cathodic undermining on tinplate. Loss of tin oxide can give rise to lacquer detachment and Koehler [48] has dealt with this phenomenon of cathodic disbondment. TFS is generally more resistant than tinplate in this context.

The tin oxide film is important and, with its associated chromium species, it has been the object of considerable study [49–52]. Britton's method for oxidizable chromium produced by the commercial "passivation" treatment employs phosphate buffers, and variants have been developed [53]. We find that for the cathodic coulometry of tin oxide, 0.001 M hydrogen bromide solution is effective after deaeration.

The tin itself is measured using anodic stripping according to the classic Kunze and Willey method [54]; both free tin and tin-iron alloy can be measured. The alloy is of great importance in controlling the cathodic reaction by limiting diffusion of atomic hydrogen into the base steel. Its continuity is measured in the ATC (Alloy-Tin Couple) test developed by Kamm and co-workers [55]. Here, grapefruit juice with additions of other substances serves as a synthetic for studying the couple current between tinplate detinned to the alloy and pure tin. The resulting values can help classify plate according to its likely resistance to detinning [56]. Walpole and Harden have presented other results on alloy exposure employing sodium carbonate and borate buffers as electrolytes [57]. The exposure of underlying steel has been studied electrochemically by various authors using synthetic systems [58–60]. Different media have been used with a common potentiostatic procedure. Our preference has been for the use of ammonium thiocyanate. The Pickle Lag test or Specification for General Requirements for Tin Mill Products ASTM A 623-83 is used to assess the quality of the steel base itself, employing hydrochloric acid [6M].

The resistance to external rusting is assessed by exposure in sulphur dioxide chambers and by salt spray testing. Method of Salt Spray Testing ASTM B 117 has been used as a means of screening internal lacquer systems for use in packing highly aggressive general-purpose products. Humidity cabinets are also employed to promote filiform corrosion on lacquered materials.

Future Trends

Lacquered materials will be used more widely in the future as increasing efforts are made to reduce metal uptake. More use of TFS materials requiring lacquer application is expected. Changes in tinplate, too, are expected, with greater use of lower tin coating weights and the use of plate having nickel as well as tin in its surface layers [61–64]. Welding has replaced soldering in many areas and two-piece cans will also be featured widely. The use of metal-plastics composite containers will provide a whole new range of questions to answer. All of these developments are likely to provide opportunities for the application of synthetic media.

Conclusions

Synthetic media are useful in studying container corrosion to demonstrate important basic principles like the complexing action of fruit acids, cathodic control of detinning, and importance of rupture potentials, and to evaluate materials. They can provide consistency in experiments in which the real corrodant may be variable. Though they can be used to simplify the situation with natural products in particular, there is always a need to support tests using synthetics with studies on the actual products to be packed.

Acknowledgment

The authors wish to thank the Directors of Metal Box plc for permission to publish this paper.

References

[1] Charbonneau, J. E., *Proceedings of the Symposium "A to Z of Container Corrosion,"* Chicago, IL, 31 Feb. 1982, pp. 3-13.

[2] Heintze, K. and Braun, F., *Werkstoffe u Korrosion,* Vol. 7, No. 12, 1956, pp. 716-723.

[3] Root, M. and Maurey, M. J., *Journal of the Society of Corrosion Chemistry,* Vol. 18, 1959, p. 402.

[4] Sheppard, E. and Su, L. S., *Corrosion (NACE),* Vol. 31, No. 6, 1975, pp. 192-197.

[5] Costa, J. M. and Culleré, *Corrosion Science,* Vol. 16, 1976, pp. 587-590.

[6] Dickinson, T. and Lotfi, S., *Electrochimica Acta,* Vol. 23, No. 6, 1978, pp. 513-519.

[7] Abd El Rehim, S. S., El Samah, A. A., and El Sayed, A., *British Corrosion Journal,* Vol. 20, No. 4, 1986, pp. 196-200.

[8] Murphy, T. P., *Proceedings of the 6 SEIC (Ferrara),* Sept. 1985, pp. 1227-1239.

[9] Elder, G. B., *Processing and Industrial Corrosion,* 1975, pp. 247-254.

[10] Willey, A. R., *British Corrosion Journal,* Vol. 7, No. 1, 1972, pp. 29-35.

[11] Sherlock, J. C. and Britton, S. C., *British Corrosion Journal,* Vol. 7, No. 3, 1972, pp. 180-183.

[12] Gouda, V. K., Rizkalla, E. N., El Wakah, A., and Ibrahim, E. M., *Corrosion Science,* Vol. 21, No. 1, 1981, pp. 1-15.

[13] Montanari, A., Milanese, G., Massini, R., and Cassara, A., *Pitture e Vernici,* Vol. 60, No. 7, 1984, pp. 47-51.

[14] Mergey, C. and Hanusse, H., *British Corrosion Journal,* Vol. 12, No. 2, 1977, pp. 103-107.

[15] Hancox, J. H., Sherlock, J. C., and Britton, S. C., *British Corrosion Journal,* Vol. 7, No. 3, 1972, pp. 222-231.

[16] Massini, R., *Industria Conserve,* Vol. 51, No. 4, 1976, pp. 268-272.

[17] Andrae, W., *Verpackung,* Vol. 23, No. 4, 1982, pp. 143-148.

[18] *Industria Conserve,* Vol. 47, 1972, pp. 100-104.

[19] Marsal, P., *Bulletin of the CRFB (Thionville),* 1977, p. 14.

[20] Catala, R. and Carbanes, J., *Revista Agroquimica y Technologia de Alimentos,* Vol. 21, No. 3, 1981, pp. 341-352.

[21] Kolb, H., *Verpackungs Rundschau Technologische Wissenschafts Beil,* Vol. 6, 1974, pp. 41-47.

[22] Marsal, P., *Influence des produits phytosantaires sur la corrosion des boites en fer blanc,* CIPC Kecskemet, Paris, 1970.

[23] Board, P. W., Holland, R. V., and Elbourne, R., *Journal of the Science of Food and Agriculture,* Vol. 18, 1967, p. 323.

[24] Board, P. W., Holland, R. V., and Britz, D., *British Corrosion Journal,* 238, 1968.

[25] Hallaway, M., *Biochimica Biophysica Acta,* Vol. 36, 1959, p. 538.

[26] Cheftel, H., *Corrosion and Anticorrosion,* Vol. 7, No. 4, 1959, p. 125.

[27] Marsal, P., Darre, J. M., and Birck, J. C., *Bulletin of the CRFB (Thionville),* 1979, pp. 8-15.

[28] Nagayama, M., Tamura, K., and Takahashi, H., *Corrosion Science,* Vol. 10, 1970, pp. 617-627.

[29] Pourbaix, M., *Atlas des Equililbres Electrochimiques CEBELCOR,* Pergamon Press, Oxford, England, 1966.

[30] Uhlig, H. H. and Bohni, M., *Journal of the Electrochemical Society,* Vol. 116, No. 7, 1969, pp. 906-910.

[31] Uchida, T. and Wagatsuma, K., *Bosei Kanri,* Vol. 24, No. 9, 1980, pp. 23-28.

[32] Whitman, J., *Proceedings of the 32 Annual Meeting of the Society of Soft Drink Technology,* May 1985, pp. 95-108.

[33] Gimbitskaya, I. V., Volkova, M. E., and Beloglazov, S. M., *Izv Vyssh Uchebn Zaved Khim Tekhnol,* Vol. 17, No. 8, 1974, pp. 1262-1264.

[34] Horst, R. L. and English, G. C., *Materials Performance,* July 1980, pp. 13-17.

[35] Maercks, O. and Maercks, I., *Verpackungs Rundschau Technologische Wissenschafts Beil,* Vol. 22, 1971, p. 59.

[36] Cantoni, C., Merlino, P., and Maccapani, M., *Imballagio,* Vol. 25, No. 221, 1974, pp. 3-6.

[37] Landgraf, R. G., *Proceedings of the Symposium "A to Z" of Container Corrosion,* 3 Feb. 1982, pp. 35-39.

[38] Azzerri, N. and Baudo, G., *British Corrosion Journal,* Vol. 10, No. 1, 1975, pp. 28-32.

[39] Hottenroth, B., *Verpackungs Rundschau Technologische Wissenschafts Beil,* Vol. 24, No. 3, 1973, pp. 19-23.

[40] Dembrovskii, M. A., Florianovich, G. M., Sherstyannikov and Levyanto, S. I., *Konservnaya i Ovoshchesushil'naya Promyshlennost',* Vol. 5, 1974, pp. 26-29.

[41] Massini, R., *Industria Conserve,* Vol. 49, No. 2, 1974, pp. 80-94.

[42] Murphy, T. P. and Walpole, J. F., *Aerosol Report,* Vol. 11, 1972, pp. 525-541.

[43] Scantlebury, J. D. and Sussex, G. M., *Corrosion Control by Org Coatings (NACE),* Lehigh, 1980, p. 235.

[44] Kleniewski, A., *British Corrosion Journal*, Vol. 10, No. 2, 1975, pp. 91-98.
[45] Sherlock, J. C. *Verpackungs Rundschau*, Vol. 28, No. 6, 1972, pp. 810-817.
[46] Koehler, E., *Corrosion Control by Organic Coatings NACE*, H. Leidheiser, Ed., 1981.
[47] Morgan, E., *Tinplate and Modern Can Making Technology*, Pergamon Press, Oxford, England, 1985, Chapter 6.
[48] Kochler, E. L., *Corrosion (NACE)*, Vol. 40, No. 1, 1984, pp. 5-8.
[49] Britton, S. C., *British Corrosion Journal*, Vol. 1, 1965, p. 91.
[50] Britton, S. C., *British Corrosion Journal*, Vol. 10, No. 2, 1975, pp. 85-90.
[51] Rauch, S. E. Jr. and Steinbicker, R. N., *Journal of the Electrochemical Society*, Vol. 120, No. 6, 1923, p. 735.
[52] Carter, P. R., *Journal of the Electrochemical Society*, Vol. 108, No. 782, 1961.
[53] Aubrun, P., Rocquet, P., and Penneva, G., *Bulletin of the CRFB (Thionville)*, 1975, pp. 1-26.
[54] Kunze, C. T. and Willey, A. R., *Journal of the Electrochemical Society*, Vol. 99, No. 9, 1952, p. 354.
[55] Kamm, G. G. and Willey, A. R., *Beese and Krickl Corrosion (NACE)*, Vol. 17, No. 2, 1961, p. 106.
[56] Mergey, C., *British Corrosion Journal*, Vol. 19, No. 3, 1984, pp. 132-138.
[57] Walpole, J. F. and Harden, G. D., *Proceedings of the 2nd International Tinplate Conference*, London, Oct. 1980.
[58] Tsurumaru, M., Nunokawa, A., and Suzuki, Y., *Proceedings of the 2nd International Tinplate Conference*, London, Oct. 1980, p. 348.
[59] Rozenfeld, L., Katser, I. M., Galkin. D. L., and Frolova, L. V., *Zashchita Metallov*, Vol. 11, No. 1, 1975, p. 109.
[60] Murphy, T. P. and Smith, H., *Proceedings of the 3rd International Tinplate Conference*, London, Oct. 1984, p. 299.
[61] Lempereur, J. and Renard, L., *Proceedings of the 3rd International Tinplate Conference*, London, Oct. 1984, p. 185.
[62] Allouf, R., *Proceedings of the 3rd International Tinplate Conference*, London, Oct. 1984, p. 247.
[63] Moriyama, H., Shimizu, N., Fujimoto, T., Nomura, V., and Onoda, I., *Proceedings of the 3rd International Tinplate Conference*, London, Oct. 1984, p. 214.
[64] Mochizuki, H., Nakahouji, H., Ogata, H., Ichida, T., and Irie, T., *Proceedings of the 3rd International Tinplate Conference*, London, Oct. 1984, p. 227.

DISCUSSION

D. McIntyre[1] (discussion questions)—Regarding the stress-corrosion cracking failure of the aluminum can be shown in your slides.

(1) What was the alloy?
(2) What was the cracking agent?
(3) Is there any cyclic effect (corrosion fatigue)?
(4) How can this problem be prevented?

T. P. Murphy and G. A. Pape (authors' closure)—

(1) 5182.
(2) cannery water containing both chloride and nitrate.
(3) unknown but doubtful.
(4) adequate drying of ends before shrink wrapping.

[1]Cortest Laboratories, 11115 Mills Rd., Suite 102, Cypress, TX 77429.

Anselm T. Kuhn,[1] Peter Neufeld,[2] and Trevor Rae[3]

Synthetic Environments for the Testing of Metallic Biomaterials

REFERENCE: Kuhn, A. T., Neufeld, P., and Rae, T., **"Synthetic Environments for the Testing of Metallic Biomaterials,"** *The Use of Synthetic Environments for Corrosion Testing, ASTM STP 970*, P. E. Francis and T. S. Lee, Eds., American Society for Testing and Materials, Philadelphia, 1988, pp. 79–97.

ABSTRACT: Synthetic environments used for testing biomaterials are considered. In terms of the environment they seek to simulate, these can be subdivided into saliva substitutes and body fluid substitutes. In respect of the former, it is seen that most formulations used in the past cannot even be taken into solution. In both cases, evidence is presented to suggest that use of electrochemical corrosion test methods or in-vivo use of the same methods gives results that are in error because of simultaneous anodic oxidation of organic species, such as simple redox-type amino acids. The correlation between in-vivo and in-vitro corrosion rates is examined. Agreement is far from satisfactory in many cases.

KEY WORDS: synthetic environments, biomaterials, saliva substitutes, body fluid substitutes, corrosion

The developing skills of orthopaedic, thoracic, and dental surgeons have led to a steady increase in the variety of man-made components inserted in the body. These components may be aids to promote healing and union of fractures. They may be replacements for worn or nonfunctioning articulating joints. They may be inserts in the cardiac system. Most commonly, they will be associated with a deficient dentition.

Surprising though it may be service in the body in any of the functions described above is a most demanding duty, and the great majority of materials commonly used in engineering fail either mechanically, or as a result of corrosion, when used in this environment. Sometimes, when fretting occurs, the two failure modes are associated. Indeed, the number of materials presently accepted as being suitable for use in the body, is remarkably small.

It is vital that such biomaterials do not fail prematurely. At the very lowest level, an implant may cost less than $100. The associated surgery, however, could cost a 100 times this amount. Use of cheap materials is almost invariably a false economy.

But quite apart from such crude financial arguments, the failure of an implant or other biomaterial can be a most serious matter. Leaving aside observations of Scales [1] or French et al. [2] that corrosion of implants is frequently associated with discomfort, even acute pain, there remain the consequences of releasing into the body amounts of metals, such as cobalt, chromium, and nickel, in concentrations capable of provoking allergic reactions and possibly

[1]Associate dean, Faculty of Science and Technology, Harrow College of Higher Education, Northwick Park, Harrow HA1 3TP, United Kingdom.

[2]Principal lecturer, Department of Chemical Engineering, Polytechnic of the South Bank, Borough Rd., London SE1, United Kingdom.

[3]Senior research investigator, Orthopaedic Research Unit, University of Cambridge, Addenbrookes Hospital, Hills Rd., Cambridge, United Kingdom.

even more serious effects, based on the proven mutagenic and carcinogenic properties of these species [3–5]. Nor should the biomechanical consequences of failure be overlooked, whether these stem from the loosening of an implant or its fracture, both of these being possible consequences of corrosion.

For all these reasons, it is clear that candidate metallic biomaterials should be subjected to rigorous corrosion (and other) tests, before acceptance by the dental or medical professions.

Corrosion tests have most certainly been conducted, and there are well over a 100 publications dealing with the corrosion of actual or potential dental or surgical biomaterials. ASTM has itself published several volumes [6–8] concerned with this subject. Surveying all of these publications, one notes that there are at least some papers (see, for example, Ref 9), mainly but not exclusively in the dental or medical journals, where the techniques in the corrosion experiments have been poor. Some of the more common errors will be discussed here, but a hallmark of such publications is the dearth or absence of references to work in the mainstream of corrosion science. The two cardinal principles of test procedures bear restating:

1. They must be internally valid, that is to say they constitute a physically and chemically, as well as a biologically valid measurement.

2. The results given by any test procedure must have relevance to "in-vivo" conditions, either in terms of a close matching of the physico-chemical environments or on the basis of some prior validation, derived from a similarity of corrosion rates determined in either case.

It is the contention here that so complex is the problem of corrosion in the biological environment that very close collaboration between corrosion scientists and biologists or biochemists is essential. The starting point in devising an appropriate test must be a sound knowledge of the physico-chemical environment to which the implant will be subjected. This will vary according to the location of the implant, while dental prostheses are themselves exposed to a wide variation of conditions. The various chemical environments, as well as the mechanical forces (stress, strain, and friction) involved, are discussed below. A further point that must be emphasized is that a consideration of the test environment cannot be divorced from the choice of the experimental technique, for reasons that will be made apparent.

The Physico-Chemical Environment of Biomaterials In Situ

Introduction

The physico-chemical environment to which biomaterials are exposed in the body, varies greatly, depending on the location of the particular component. Sometimes, two extremities of the same component will encounter totally different environments, as for example a hip-joint. Indeed, apart from the fact that all components are exposed to a mean temperature of 37°C (with many dental materials experiencing considerable transitory excursions from this temperature in the range 5 to 50°C), it is difficult to identify any other common factors. From this, it should be clear that no single test regime is likely to be capable of serving in a universal role. Some of the more important environments within the body will therefore be singled out for examination.

The Oral Environment

In terms of the volume of published work and probably the financial implications, the oral environment is arguably the most important. Most metallic components in the mouth are bathed in saliva, the composition of which should therefore be considered. Many workers have cited Jenkins' book, *Physiology of the Mouth,* where the composition of saliva is considered in detail [10]. It should be noted that the salivary composition as reported by Diem [11] differs by a factor two, for example, in the value of chloride ion concentration. The most thorough analysis of the situation is probably by Darvell [12] who states, quoting Mandel [13]:

The fluids secreted by the parotid submaxillary sublingual and minor salivary glands have been shown to vary considerably from one to another, to be complex in composition, and to be affected by:
1. Type, intensity and duration of stimulus
2. Time of day
3. Diet
4. Age and sex
5. A variety of diseases & pharmacological agents

From this, it is clear that no single test solution can simulate the variety of conditions that occur in the oral cavity, and the truth of this will be recognized by any dental practitioner who sees a particular alloy being corrosion-resistant and tarnish free in the mouth of one patient, while failing from these causes in another.

Nonoral Environment

Many implants are used in the body outside the oral environment as orthopaedic devices. Virtually all of these are load-bearing and are likely to be subjected to varying degrees of fretting, fatigue, and crevice corrosion. Even an ostensibly static implant, such as a bone plate, can be subjected to all of these. Almost inevitably, slight motion occurs between the fixing screws and the plate, as indicated by wear marks around the countersink when the screws are removed, indicative of fretting or fretting corrosion. Such implants are also subject to fatigue, while crevice corrosion around the countersink is well known.

Total joint replacements are subjected to similar actions, and are highly stressed in a cyclical manner. They articulate against another foreign material and are usually anchored into position with a polymethylmethacrylate cement, which may again become a site for fretting, fatigue, and crevice corrosion. The precise composition of the fluids in which such implants are immersed, is not known, although the composition of the more important body fluids has been well-studied; a summary of three relevant fluids is shown in Table 1. At this stage, it can only be assumed that fluid bathing a static bone plate, for example, approximates to interstitial fluid, while that forming around a total joint resembles synovial fluid, though this will depend, to some extent, on whether the synovium remains after surgery.

TABLE 1—*Composition of selected components of three body fluids. (Based on data from Documenta Geigy Scientific Tables, Ciba-Geigy, K. Diem, and C. Lentner, Eds., 7th ed.).*

Component	Interstitial Fluid, mg L^{-1}	Synovial Fluid, mg L^{-1}	Serum, mg L^{-1}
Sodium	3280	3127	3265
Potassium	156	156	156
Calcium	100	60	100
Magnesium	24	...	24
Chloride	4042	3811	3581
Bicarbonate	1892	1880	1648
Phosphate	96	96	96
Sulfate	48	48	48
Organic Acids	245	...	210
Protein	4144	15 000	66 300

As can be seen from Table 1, interstitial and synovial fluids as well as serum are similar in composition with the exception of their protein contents. As this appears to play a significant role in corrosion, as discussed below, it would seem to be essential that this factor is taken into account in laboratory studies of corrosion.

A comparison of the compositions of the so-called "physiological salt solutions" (Table 2), which have been overwhelmingly used in corrosion studies of metallic biomaterials, with those listed in Table 1, shows the former to be only a first approximation to the real thing. An improved approach might be to use a tissue culture medium of which there are many varieties [14], some of which have been designed for the growth of specific cell types. They differ mainly in their content of biologically important components such as amino-acids and cofactors. Their inorganic salt content resembles that of the solutions in Table 1. In addition, a culture medium would typically contain an inorganic or organic buffer to maintain a pH of 7.4, up to 20 amino acids (to a total concentration of 750 mg/L) and vitamins and cofactors (total concentration of 1 mg/L). Finally, a 10% supplement of animal serum is often added, giving a total protein concentration in the region of 6 g/L, a value between that of interstitial and synovial fluids. Such a standard tissue culture medium thus provides a more realistic model of the physiological environment in which to measure corrosion of metallic biomaterials.

One minor difficulty arises in respect of maintaining sterility. Bacteria grow extremely well in such media. Microbial spoilage can be easily avoided, however, either by use of appropriate sterile techniques or antibiotics. The latter is the simplest in practice, though is open to criticism in that it introduces yet another potentially electro-active species.

Corrosion Testing Procedure for Assessment of Metallic Biomaterials

Introduction

It will be suggested here that the apparent corrosion rate data obtained for a given sample may depend not just on that sample and the corrosion environment to which it is exposed, but also on the technique employed. This might seem surprising, but serves only to underline the fact that not all corrosion techniques are appropriate to a given problem.

The corrosion of metallic biomaterials has been studied by both electrochemical and nonelectrochemical methods. Extremely sensitive though the former techniques are, they impose a perturbation on the system, if only a small one, and to this extent, can be used only to estimate corrosion rates, and not to measure them. Some of the most widely used methods in the study of corrosion of metallic biomaterials are listed on the next page:

TABLE 2—*Composition of some "physiological" solutions.*[a]

Composition	Hanks	Ringers	Tyrodes	Cigada
NaCl	8.0	9.0	8.00	8.74
KCl	0.4	0.42	0.2	. . .
CaCl$_2$	0.14	0.24	0.20	. . .
NaHCO$_3$	0.35	0.20	1.00	0.35
MgCl$_2$6H$_2$O	0.10	. . .	0.10	. . .
Na$_2$HPO$_4$ · 2H$_2$O	0.06	. . .	. . .	0.06
NaH$_2$PO$_4$	0.10	. . .	. . .	0.06
MgSO$_4$ · 7H$_2$O	0.06	. . .	0.05	. . .
Glucose	1.00	. . .	1.00	. . .
pH	7.4	7.4	7.4	7.4

[a] In grams per litre.

1. Electrochemical methods:

   steady-state potentiostatic
   cyclic voltammetry (potentiodynamic scanning)
   current-time at constant potential
   polarization (resistance/impedance)

2. Nonelectrochemical methods:

   weight loss (usually at open circuit)
   reflectance change
   chemical analysis of corrosion products
   optical observation of corroding surface

Each of these techniques has its own strengths and weaknesses and some are more appropriate than others to the study of biomaterials corrosion. It is axiomatic that biomaterials corrode very slowly, and the measurement of such low corrosion rates is inevitably more difficult than work involving more labile materials. These very low corrosion rates of biomaterials must be seen as the first major experimental difficulty. The second and widely neglected effect relates to the length of time taken for passivation of many biomaterials to approach completion. Thus a number of authors have shown that even after 1000 h, alloys, such as the Co-Cr series, are incompletely passivated [15], as are the stainless steels [16]. In spite of this, the majority of published corrosion studies of such materials are based on potentiodynamic scans or linear polarization methods, which have been completed in an hour or so. That is in a hundredth of the time required for near-complete passivation. Hoar pointed out long ago [17] that the equations widely used in linear polarization methods were an approximation, and that in DC polarization measurements, time effects are frequently observed in passivating systems. Today, AC methods are far more widely used. Time effects are not usually seen with these techniques, not because they are absent, but simply because the method obscures them. A recent publication [18] considers this problem at some length. Apart from a failure to complete measurements without regard to time for passivation, the shape of the current-voltage relationship, where shown, can be revealing. Some authors [9] have shown "Tafel Plots" essentially linear over more than 1 V and with a slope of 1200 mV/decade. Such results, which have no basis whatsoever in theory, are often the hallmark of incorrect implementation of experimental procedure and (in contrast to data from validly conducted work) allow no mechanistic conclusions to be derived.

The final point in discussion of the use of electrochemical methods is that the validity of corrosion rate data deduced from these methods assumes that no Faradaic processes other than those involved in corrosion are taking place to any significant extent. Were this not to be the case, and we shall show grounds for supporting this fear, the results would be in error. Coexistence of other anodic reactions is the most likely, and this would result in an overestimate of the corrosion rate. Williams and Clark [9] have expressed concern on this point though only with respect to species Fe(II)/(III) (without suggesting whence these might emanate) and apparently overlooking the much more real danger from anodic oxidation of organic species.

The correct implementation of nonelectrochemical corrosion methods is likewise not foolproof. All comments relating to the need for establishing properly passivated (or at the least defined) surfaces apply equally here, and there are publications based, for example, on solution analysis, where the time scale of the experiment is not explicitly stated and by inference appears to fall well short of the above criteria. Several workers have studied metals in finely divided powder form, in most cases to take advantage of their very high specific surface area. Wood et al. [19] showed how the metal surface area : solution volume parameter could critically affect results, and there are some workers [9] who do not appear to have given thought to this, though using very small solution volumes. In the limit, it is clear that this would lead to meaningless results. Two further comments are called for in respect to work based on metal powders. First, those authors [9] who do not calculate the surface areas involved, deny other workers the opportunity to relate such results to existing work. Second, there is a suggestion, that data derived from powdered metals may not necessarily be extrapolated to the same metal in massive form. Such fears are based on restrictions placed, in the former case, on the size and topography of the anodic and cathodic regions, normally adjacent to one another in free corrosion, as well as the fact that the high degree of curvature of small particles imparts to them, a surface energy greater than that of their massive counterparts.

Physical Parameters in Artificial Test Procedures

Given the test environment (chemical composition and temperature) as well as the technique employed, the immediate physical environment of the test specimen is a factor of very great importance that is often overlooked. Thus it is extremely difficult to conduct a corrosion test in which adventitious crevice effects are absent. Suspending a sample by means of a piece of thread tied around it, laying a sample on the floor of a beaker, the deposition of any form of debris on the surface of a sample—all of these can lead to increased corrosion, rates resulting from crevice formation. Greene and Francis [20] have demonstrated the magnitude of the spurious crevice currents that are the result and have discussed in detail the proper means of holding a specimen. Experimental difficulties of this kind have been reviewed by Hayes [21]. Some workers have deliberately introduced a crevice configuration into their tests of biomaterials [22] (see also a review [23]), a perfectly legitimate procedure. An area of uncertainty relates to the hydrodynamic regime to which corroding samples are exposed. There is circumstantial evidence, for example in relation to the Co-Cr alloys that their rate of passivation is related to the vigor with which the solution is agitated, while for Ni-Cr alloys in seawater the fact has been established [24]. This effect, may well stem from changes of near-surface pH, arising from reduction of oxygen [25]. It should be kept in mind that, in-vivo, there is virtually no convective movement of solution.

Fretting and abrasion effects (including fretting corrosion) such as those which might be found in the oral environment or in prosthetic joints) have been studied by only a few workers, either using classical "friction and wear" equipment [22,26–31] or [33] using the abrasive action of particles such as silicon carbide (SiC) in suspension. Under such regimes, the attrition rate of all materials is greatly increased. The extent to which such severe conditions correspond to those actually found in-service, has not been considered in depth. A few workers [33,34] have also examined the effect of scratching the surface of their test specimens during the corrosion tests. This can lead to momentary increases of current of a factor, $\times 10$, presumably reflecting the breakdown of passivity and a result, the implications of which, both in terms of corrosion rate and the ingestion of metallic species, appears not to have been properly digested, taking the occlusal surface of Ni-Cr alloy jacket crowns as a case in point. There has been a limited amount of work on corrosion-fatigue [36–38] and also stress-corrosion in vivo and in vitro [39].

Formulation of a Test Solution

It will now be accepted that no single test solution can be expected to satisfactorily simulate the wide range of conditions encountered by a metallic component inserted in the body, and the near-ubiquitous use of the standard physiological solutions (Table 2) as corrosion electrolytes, regardless of the ultimate location in the body of the test specimen, is therefore to be deprecated. We shall consider here the two situations of greatest interest, namely, the oral environment and the soft-tissue environment.

Testing of Components in Simulated Oral Environments

There have been three main approaches to the problem of testing metallic components for service in the oral environments. These are

(1) use of genuine human saliva,
(2) use of an artificial saliva, and
(3) use of simple test solution.

Use of Human Saliva

Several workers have reported corrosion results based on human saliva. Normally, these are based on "pooled" samples to which a number of individuals contribute. (The "pooling" is presumably to minimize effects arising from individuals whose salivary composition falls far outside the mean). Newman et al. [40], for example, describe how such pooled saliva was autoclaved and then frozen until use. Finkelstein [41] has reported the corrosion of amalgams in human saliva, but in greater depth than Newman. He describes how the saliva was collected and stored in ice-baths for not more than 2 h to prevent spoilage. A portion of this was centrifuged for 20 min at 1200 g and 3.5°C, the supernatant being decanted from the sediment. Both fresh and decanted specimens were then gradually frozen to −5°C for several days and then defrosted. Finkelstein's subsequent studies could thus call on saliva specimens with four distinct histories, namely

(1) fresh whole saliva,
(2) fresh supernatant saliva,
(3) defrosted whole saliva, and
(4) defrosted supernatant saliva.

Additional studies based on Ringer's solution were also reported. The results of this work revealed substantial differences in results from these five different solutions. Based on potentiodynamic scans, whole fresh saliva and its defrosted counterpart showed little difference while the two supernatant samples (fresh and defrosted) likewise appeared to behave like one another over most of the potential range, though giving rise to higher currents in the corrosion region. The passivation peak current was the same in all cases.

Use of the potential-step method revealed more pronounced differences between the various solutions, with whole saliva giving very irregular current-time transients, in which the downward trend was marked by sudden current excursions, similar to those caused when pitting occurs. In contrast, the supernatant solutions gave smooth current-time curves. Finkelstein [41] ascribed these differences in current-time behavior to the presence of particulate matter in the whole saliva, suggesting the effects were due to movement of sediment particles through solution. It was stated that motion of these particles could be observed both in the reaction vessel and the salt-bridge under the influence of applied potential. It is not easy to see how such motion could have affected the *i-t* data, except that Pini [42] has shown that suspended particles enhance mass-transport effects, while the formation of crevices where such particles rest on or adhere to a surface was mentioned earlier. In a second series of experiments, the authors compared the corrosion of amalgams in supernatant solution with that in Ringer's solution. In the latter, corrosion was up to ×1000 times faster in the active region, with corresponding changes in the open-circuit corrosion potential.

Commenting on the electrochemical data, Finkelstein suggests that chloride ion concentration may be lower in genuine saliva than in Ringer's solution. He also raises the possible action of protinaceous materials in saliva acting as corrosion inhibitors. The lower chlorine concentration, it is suggested, would tend to move the rest potential in an anodic direction, while formation of a protinaceous layer, acting as a corrosion inhibitor, would have the opposite effect, shifting the potential cathodically. The authors concluded this study with an X-ray and elec-tron-optical examination of the corrosion products. The former revealed no differences between products formed in Ringer's solution on the one hand and human saliva on the other. Scanning electron microscope (SEM) studies did show morphological differences, and these authors con-clude, citing a rather inaccessible study on the use of human saliva, with a plea for further study of substitute salivas. A separate point, made by Darvell [12] relates to bacterial involvement and resulting chemical differences. Suppression of bacterial action is possible either by addition of a bactericide (so introducing still further complications) or by filtration together with working under sterile conditions. Filtration would probably also remove high-viscosity organic material, the effects of which are mentioned elsewhere in this paper.

Use of Artificial Saliva

Many workers, starting from sources, such as Jenkins [10] or Diem [11], have formulated an electrolyte with the aim of simulating human saliva. Some of these formulations are listed in Table 3. Finkelstein [40] shows a similar table and quotes the work of Marek and Hoffman [43] who explored the corrosion of dental amalgam as a function of artificial saliva composition. Darvell [12] quotes still other formulations. This author alone, appears to have highlighted some of the major problems in making up solutions of simulated saliva. He points out that many of the compositions shown by him, or listed in Table 3, have been advocated without regard to the solubilities of the various components, and that in many cases, even after many weeks, some of the constituents in such formulations have failed to dissolve. He further singles out three separate problems in formulation of an artificial saliva. The first, is the fact that most salivary secretions are supersaturated with respect to hydroxyapatite and often other calcium phos-phates. Passage of saliva from the secreting gland into the oral cavity, where the pH is somewhat higher, causes precipitation. The second problem relates to the role of dissolved carbon dioxide (CO_2) in maintaining saliva pH. Loss of CO_2 into the atmosphere will cause an increase in solu-tion pH. (On this point, Darvell does not discuss the possibility of continuous bubbling of CO_2-N_2 or other mixtures for pH maintenance). The third issue raised by him relates to the viscosity of saliva. To the best of our knowledge, there have been no attempts to reproduce this particular property, whose effect would probably be a retardation of corrosion with increasing viscosity. Darvell [12] reports on attempts to develop a satisfactory artificial saliva. A series of

TABLE 3—*The composition od some artificial salivas, g L^{-1}.*

Components	De Micheli [35]	Angelini [58]	Brune [15][a]	Tani [59]
NaCl	0.700	...	0.40	...
KCl	1.200	1.47	0.40	1.47
NaHCO$_3$	1.500	1.25	...	1.25
K$_2$HPO$_4$	0.200	0.19[b]	...	...
NA$_2$HPO$_4$	0.260	...	0.78[c]	...
KSCN	0.330	0.52	...	0.52
Urea	0.130	...	1.0	...
CaCl$_2$ · 2H$_2$O	...	...	0.80	...
Na$_2$S · 9H$_2$O	...	...	...	0.19
Lactic acid	...	...	...	0.90
pH	...	to 6.7 using lactic acid 5–6		7

[a]Composition of Fusayama [60] with omission of 4-g mucin. Meyer et al. use same as Brune but with NaH$_2$PO$_4$ · H$_2$O = 0.69 g [59].
[b]As monohydrate.
[c]As dihydrate.

17 solutions was prepared, each containing one or more of the following: phosphate, chloride, thiocyanate, bicarbonate, urea, citrate, urate, other halogens, lactate, ammonium ions, and sulfate ions. These solutions were evaluated as corrodents for amalgams, using a reflectance method, concerning which, certain reservations have been expressed [44]. The tests enabled the actions of individual components of the solutions to be identified. On the basis of this work, Darvell arrived at a formula for an artificial saliva, which is prepared immediately before use from three stock solutions whose composition is shown in Table 4.

Darvell points out that thiocyanate could not be included in Solution A, owing to acid-induced decomposition, while its inclusion in Solution B would have resulted in precipitation of sodium hydrogen urate. The working solutions were prepared by pipetting appropriate quantities of stock solution in the ratios shown in parentheses and in alphabetical order into distilled water. This minimized premature loss of CO_2. Solution pH was then adjusted by addition of phosphoric acid to a value of 6, which though more acidic than most other prescriptions was that deemed to be the most realistic representation of unstimulated saliva. Darvell also comments that on standing, or in use, the above solution developed a cloudiness, which was found to be due to the presence of bacterial rods, both gram negative and positive, although this could be prevented by filtration through a 0.45-μm filter. Apart from the question of viscosity, the other unsatisfactory aspect reported by Darvell is a slow drift of pH to higher values, as a result of CO_2 loss. As suggested above, it is felt that bubbling of CO_2-N_2 mixtures could be used to prevent this.

Use of Simple Test Solutions

The philosophy here is to use a simple test solution, which while it makes no attempt to mimic saliva, contains at least one of the more important species in saliva and as such can be used for ranking or screening purposes without any of the problems associated with the complex makeup of more realistic solutions. The use of sodium chloride (NaCl) alone, though reported, is undesirable since it is unbuffered and its pH will change quite rapidly in use. Acidified hydrochloric acid (HCl) (perhaps too severe) and buffered NaCl are preferable. Aqueous solutions of sodium sulfide have also been used, but the poor stability of this [45] together with absence of chloride ions, which must be the most important single component, weigh against its use.

TABLE 4—*Composition of stock solutions.*

SOLUTION A ($\times$100)

NaH$_2$PO$_4$, 56 g
NaCl, 150 g
NH$_4$Cl, 22 g
Trisodium acetate dihydrate, 2.2 g
Lactic acid, 7 g
Distilled water to 1 L

SOLUTION B ($\times$50)

Urea, 20 g
Uric acid, 1.5 g
NaOH, 0.4 g
Distilled water to 2 L

SOLUTION C ($\times$100)

NaHCO$_3$, 60 g
NaSCN, 20 g
Distilled water to 1 L

Gas Purges

Many workers purged their solutions with gas, either during measurements or immediately beforehand. Every conceivable option appears to have been used including air, oxygen, nitrogen, and argon, absence of purging, use of a layer of oil to exclude air and the use of gas mixtures. The most common error is to use oxygen (or air) purges for those electrochemical techniques where an external source of power is applied, for example, potentiostatic or potentiodynamic scans. There is no reason to suppose this will alter the actual rate of anodic dissolution at any given potential. In terms of the observed current, an error will be introduced because of oxygen reduction, and this can be very significant if the test electrode is a metal or alloy (for example, silver or gold) on which this reaction is fast and the corrosion rate slow. It is of course good practice to pass oxygen or air during currentless potentiometric studies.

Gas mixtures (Table 5) usually include CO_2 (whose partial pressure will control pH) and oxygen with a partial pressure simulating that found in vivo. CO_2 is not electrochemically active to any significant extent and thus does not give rise to spurious effects.

With techniques where the specimen potential is not externally controlled, variation of oxygen partial pressure can affect the rest potential and thus the corrosion rate. The value of CO_2 can thus be of some importance. Cahoon and Hill [46] have reported perhaps the most detailed matrix of gas mixtures, based on "low," "medium," and "high" concentrations of oxygen in mixtures with carbon dioxide. Assuming the pH was held constant, small differences would seem to show, but their significance was not clear, nor did any firm trends emerge.

The Testing of Metallic Biomaterials for Surgical Use

Until quite recently, metals and alloys of interest as biomaterials were tested in one of the "Physiological Solutions" listed in Table 2, the composition of which resembles body fluids except for the absence of organic species other than sugars. Only more recently have test solutions included species such as amino acids, proteins, or actual serum. Table 6 is a summary of the complex effects observed when such species are added. Observed corrosion rates are sometimes larger, sometimes smaller, when these additional components are included in the test solutions. Most of these effects, as listed in Table 5, are not commented on in the original papers.

TABLE 5—*Gas mixtures used in corrosion testing of biometals.*

Gas Mixture	Conditions	References
CO_2-O_2-Ar	CO_2-O_2 at 20-60-mm Hg pH hold at 7.6	Cahoon [37]
CO_2-O_2-Ar	$O_2-5-200$-mm Hg,CO_2 8 to 100-mm Hg, pH 7.33 effect on i-V data	Cahoon [62]
$O_2 + CO_2$	O_2 10 to 300 mm, CO_2 8 to 100 mm, pH 7.2 to 7.4, effect on pollution data	Cahoon [46]
$O_2-N_2-CO_2$	PO_2 0.7 to 19 mm complex effects	Sutow [63]
N_2-CO_2	Affects corrosion, stainless steel but simulates pH change	Ogundele [64]

TABLE 6—*Effect of proteins/amino acids/serum on metallic corrosion.*

System	Observations	Author
Co—Cr and variation stainless steels in PSS, sweat, blood	sweat the most aggregate medium	Samitz [65]
316 LVM, 0.9% NaCl 1% and 10% serum	serum accelerates corrosion rate	Brown [26]
316 LVM fretting	serum corrosion less than NaCl	Brown [29] [66]
Cu & Ni in ringers with and w/o additions	cystine enhances Cu passivan depresses Ni passivan, alanine bov Pl albumen no effect	Svare [51]
Ti and alloys ringers with and w/o amino acids	amino acids have no significant effect	Solar [67]
316, Ti mixed metals ringers with and w/o proteins in fretting mode	all organics inhibit corrosion	Brown [28]
Ag—Sn—Hg artificial saliv with/without hog mucin	no significant difference	Ross [49]
Ti—6Al—4V lactated ringers and bovine plasm	slight additional passivan with plasma	Aragon [68]
Ti—6Al—4V, Ti—Ni hanks with/without protein	cysteine causes much faster corrosion of Ti—Ni slight effect on Ti—Al—V tryptophan no effect	Speck [69]
Co, saline with serum albumen	serum accelerates corrosion	Williams [18]
Al, saline, human blood, human plasma	NaCl most aggress, then plasma then blood	Salvarezza [70]
Al, Co, Cu, Cr, Mo, Ni, Ti, Co—Cr, plus or minus serum, albumin, fibrinogen	Al, Ti, unaffected Cr, Ni, slight increase Co, Cu, large increase Mo, large decrease	Williams [71]

Discussion of Test Solutions for Corrosion Studies

The test solutions described above can be examined in a number of ways. That proposed formulations should be readily soluble and also of reasonable stability, has been emphasized by Darvell [12] who quoted the poor stability of thiocyanate in slightly acid media. Beyond this, we may examine the individual components of proposed solutions both in terms of their effect on corrosion and their electrochemical behavior, that is, whether they behave as redox species under the prevailing conditions, and if so, to what extent such redox behavior will contribute a current to the overall reading, thereby falsifying electrochemical measurements.

Commentary on Presently Used Test Solutions

The components of solutions described in Tables 1 through 3 can be classified, in electrochemical terms, as follows:

(1) active corrosion agents,
(2) corrosion inhibition agents,
(3) neutral species (neither one nor two), and
(4) electroactive species.

Active Corrosion Agents

Foremost among these is the chloride ion, whether seen as a complexing agent or a depassivating species, which by competing with chemisorbed oxygen or hydroxide (OH) radicals, increases the corrosion rate. There is, in most corroding systems, a direct correlation between chloride ion concentration and corrosion rate, over a wide range of concentration.

The second active species is the hydrogen ion. It will be seen that pH values ranging from "5 to 6" [15] to 7.4 have been reported, with occasional use of even more acid solutions. The ×100 fold range in acidity which these figures imply cannot be without significance in terms of corrosion rate.

A third possibly active agent is any species which can act as a ligand or otherwise chelate the metal. By a lowering of the free energy for the dissolution process such compounds can promote corrosion. Sulfate ions, though less aggressive than chloride in this context, must also be included under this heading.

Corrosion Inhibition Agents

Phosphates are widely used as corrosion inhibitors and might be expected to act in this way when included in the composition of test solutions. Calcium and magnesium can probably be seen as neutral in their action. There is always the possibility that they might precipitate as the hydroxide or carbonate, so forming an insoluble and mildly protective layer on the metal surface. Such precipitation results from the near-surface pH change [25] as oxygen is reduced, so consuming hydrogen ions. The magnitude of this effect is dependent on the corrosion rate and because this is low for biomaterials, the effect (which is important, for example, when steel corrodes in seawater) is probably not significant here.

Chelating agents can often function as corrosion inhibitors and thus appear in this category as well as the previous one.

Neutral Species

Sodium and potassium may be considered as neutral species except insofar as they contribute to the overall electrical conductance of the solution, so facilitating the corrosion process.

Electrochemically Active Species

These are defined as species (whatever their classification under above), which can take part in a Faradaic process, that is, be anodically oxidized or cathodically reduced in the potential range covered under typical experimental conditions. If they do so react, the associated current will be added to (anodic) or subtracted from (cathodic) the corrosion current (anodic dissolution) where this is recorded.

To what extent do organic species of the types listed in the various tables fall into this category? The answer to this takes one into the realm of electro-organic chemistry, which is itself divided into synthetic applications, and those dealing with fuel-cell research. The latter is of greatest relevance here, because it is concerned with the anodic oxidation of organic species at relatively low anodic potentials. On the other hand, fuel cell electrodes (in order to optimize this function) tend to be of high surface area, and often include noble metals. On both counts, they would be expected to be more active catalytically than the passive metal surfaces encountered in biomaterials. In the absence of any strictly relevant data, readers will have to reach their own conclusions, using the data in Table 7 to guide them. Inspection of this table shows clearly that a number of species also found in synthetic or natural salivas or in body fluids can be anodically oxidized at or close to the potentials taken up by metals at open circuit in the body, and in some cases, at very similar concentrations as occur in vivo. The only discrepancy relates, as emphasized above, to the nature of the metal, which is mostly gold or platinum whereas biomaterials (dental alloys being an important exception) are stainless steel or similar transition-metal alloys or titanium, all of which passivate. A priori, one would expect reaction rates on the latter surfaces to be slower, and Kuhn et al. [47] have shown how the rate constant for the oxidation of one particular species (ethene) depends on the nature of the anodic metal. This study would allow at least an estimate of the correction factor involved (Table 7) and also shows results for

TABLE 7—*Some examples of anodic oxidation of organic compounds.*

Compound	Anode	Current Density	Comments	References
Glucose	Ru/Pt	250 μA/cm^2 at 0.0 V SHE	biofuel cell	Sandstede [72]
Ammonia	Pt/Pt	25 mA/cm^2 at 550 mV	key to amino-acid activity?	Vielstich [73]
(CH$_2$OH)$_2$	Pt/Pt	10 mA/cm^2 at 650 mV SHE		Vielstich [73]
Lactic acid	Pt/Pt	10 mA/cm^2 at 1.0 V SHE	lower pot'ls not reported	Horanyi [74]
Urea	Pt/Pt	500 μA/cm^2 −200 to +200 mV SCE	Krebs–Ringer electrolyte (glucose data in same paper)	Giner [75]
Redox proteins	RuO$_2$	about 10 μA/cm^2 at +0.2 V SCE	about 2 mM protein concentration	Harmer [76]
Sulfide anion[a]	Ag	2 μA/cm^2	diffusion controlled current	Kaiser [48]
Cu, w/w/o citrate	Cu	. . .	accelerates corrosion but thiocyanate inhibits	Mostafa [77]
Cysteine, cystine graphite + metal phthalocyanine		both species oxidized and reduced	. . .	Zagal [78]

[a] Inorganic species.

one inorganic species, found in saliva-sulfide ion. Credit is due to Kaiser et al. [48] for recognizing the electrochemical activity of this species in relation to corrosion testing.

The findings of these authors also raise the wider issue of mass transport. No study of any corrosion process can be said to be complete unless these effects are investigated, yet in spite of this self-evident truth, we have located only a single reference where this parameter was considered. Ross et al. [49] reported on the potentiostatic corrosion of dental amalgam under stirred and unstirred conditions. In the former case, thanks to improved mass transport of oxygen, the characteristic shape of the current-voltage plots was completely altered in the active-passive region.

There is one further interesting observation to support the concern that corrosion rate measurements may be corrupted by the presence of Faradaic reactions associated with electrochemically active species in solution. Gettleman et al. [50] report on linear polarization measurements in vitro (Ringers solution) and in vivo (in the mouth of a baboon). They studied a range of metals and alloys, but their most remarkable result related to pure gold where a corrosion rate of 0.3 μA was found in vitro, a ten times larger value being found in vivo. At the relatively low potentials found under their conditions, it seems most unlikely that the higher current observed in vivo was entirely caused by corrosion, and the authors themselves were clearly concerned by the discrepancy and suggested the existence of a noncorrosion related Faradaic reaction. In terms of error, the difference is enormous, and the well-known freedom from corrosion of high-purity gold dental alloys suggests the result must indeed have been an artefact.

The Role and Action of Serum, Amino Acids, and Proteins in Test Solution

General

In recent years, workers in biomaterials science have increasingly recognized that absence of proteins, amino acids, or serum from test solutions, represented a major discrepancy between test conditions and those encountered in-service. As a result, there is a growing body of corrosion literature where such species were added to the more traditionally used test media. Table 6 summarizes some results found with and without such additions and shows the somewhat discrepant nature of the findings, addition of these compounds (at least apparently) sometimes enhancing, sometimes retarding corrosion. As before, we can examine the activity of these compounds under the following headings:

(1) as complex formers (ligands),
(2) as chelating agents,
(3) as electroactive species,
(4) as redox reagents, and
(5) as adsorbents, with film formation.

It will be recognized that certain species may fall into more than one of the above categories.

Complexes of Metals with Amino Acids

Many of the metals used as biomaterials are members of the first transition series and share common chemical properties. One of the most important of these, in the present context, is their ability to form coordination complexes with organic species. Thus many amino acids and proteins readily form complexes with cobalt and nickel. The interaction of metals with biologically important molecules is treated in depth in a series edited by Sigel [52], Volumes 1, 2, and 9 being relevant to the present situation. The work of Pettit and Hefford [53] is specially relevant, these authors classifying simple amino acids in terms of their side-chain, with glycine and alanine being essentially noncoordinating, serine and threonine weakly so and glutamic acid or lysine strongly reactive.

The miscellany of effects in Table 6 calls for comment. The effect of serum in reducing corrosion rate of stainless steel under fretting conditions is confirmed by several workers. It seems unlikely that this stems from a lubricating action (serum is not known as such) but could result from the known fact that any film formation that reduces metal-to-metal contact is known to have beneficial effects. Serum may well work in this way, with the local heat generated by friction leading to breakdown and denaturation of the serum protein with film formation. A possible clue lies in an observation reported by Brown and Merritt [26]. In the presence of serum, not only was the wear-rate much less, but apparently there was an almost total absence of ferric hydroxide. Remembering how effective "Jewellers Rouge" (iron oxide) is as an abrasive, the observation may be significant. Whether it stems from the ability of the sera to complex the otherwise insoluble iron hydroxide, or whether it is an "electrochemical effect" where the potential of the sample was made less anodic, is open to speculation. Perhaps the most striking results reported in Table 6 are those reported by Svare et al. [51] where addition of cystine dramatically increases the anodic current at a nickel surface. Analysis of the data is hampered by the fact that units of current are apparently unspecified. The current enhancement is seen in both active and passive regions, and on a simplistic argument, where the metal is already reacting anodically in the active condition, added species would not make it still more active. This leads to the strong suspicion that what is being seen is simply anodic oxidation of a redox protein, which cystine is, supported by the data of Harmer and Hill (Table 7).

Space does not permit more detailed comment, but close inspection of many of the papers listed in Table 6 reveals a multitude of unanswered questions. The pre-treatment of sera, or the fraction used, is frequently unspecified while absence of any mention of filtration procedures, or reference to the avoidance of the many forms of degradation that these can undergo, is worrying.

A far more serious issue is raised by the work of Woodman et al. [19] who used nonelectrochemical methods to follow corrosion of simple transition metals in powder form. They attempted to accelerate the overall corrosion process by running, in parallel, experiments that were identical save for the fact that some had ×10 and others ×100 the amount of metal (expressed in terms of surface area) in the control series. While such increases in surface area did result in greater concentrations of dissolved metal, this increase in metal ion concentration was but a fraction of the increase in surface area. Whether this was the result of all available proteins in solution having been scavenged, or whether some form of equilibrium was set up, is not clear. The very definite message stemming from their work is that the ratio of metal surface area to volume or concentration of organic species, must be considered.

Correlations Between In-Vivo and In-Vitro Tests

In the last resort, a correlation between in-vivo and in-vitro data is the only basis for reassurance as to the validity of a test procedure. The picture is confused (as Table 8 shows) and agreement is poor not only quantitatively but even as to the absolute direction (in vivo faster than in vitro or v.v.). At least a partial explanation for this stems from the methodological shortcomings (of which examples have been given above) of much of the in-vitro data. That a hundred-fold span is quoted for in-vitro rates of stainless steel corrosion probably results from those who ignored the findings of Greene et al. [15] who demonstrated how corrosion rates for such alloys could decrease by this order of magnitude, as a function of time, as the alloy passivated. The apparent very large discrepancy in data for corrosion of cobalt can again be explained in similar terms, and consultation of the original work [55] demonstrates the rate at which onset of passivation in-vivo reduces the rate of the corrosion process, a phenomenon that the in-vitro study [18] notably failed to reproduce. It should be explained here that the authors of the in-vitro co-study (who did not seek to relate their results to the in-vivo work) did not cite an actual corrosion rate, and the figure quoted has been obtained by us, making certain necessary assumptions in order to do so. Turning to in-vivo data, it is widely accepted that animal experiments of all kinds produce a far greater scatter of results than experiments conducted in a laboratory. The $\times 3$ discrepancy reported by Revie et al. [54], as between different dogs, is typical. There is at least one report [56] that suggests that the corrosion rate of implanted species is a function of their shape. A final point is due to Koegel and Black [57] who point out that, in vivo, the corrosion

TABLE 8—*Comparisons between in-vivo and in-vitro corrosion rates.*

System	Observations	Author
In-vivo (oral cavity of Baboon) in-vitro artificial saliva (Tani–Zucchi) linear	in-vivo corrosion current about $\times 5$ higher for Au, Au alloys S/S, Ti, Co—Cr, Ni—Cr at 1.72, and 1000 h (alternative Faradaic Rxn Polarization suspected in vivo)	Gettleman [49]
304 S/S Ti implanted in dogs and rabbits or in isotonic NaCl linear polarization	unsterilized S/S–good agreement; sterilized S/S agreement fair; Ti sterilized agreement poor; Dog 1 rate $\times 3$ Dog 2 rate	Revie and Greene [54]
S/S, Co—Cr and Ti, potentiometric. Hanks and goat implant, 0 to 71 days	S/S 71-day Hanks = 0.31-V goat = 0.17-V Co—Cr 71-day Hanks 0.39-V goat = 0.31-V Ti 35 days Hanks = 0.33-V goat = 0.45-V	Hoar and Mears [34]
316L S/S in vivo by analysis	in-vivo rates from 1.5 to 1000 in-vitro 90 to 7000 ($ng/cm^2/day$) no agreement	Smith [79]
Co (in dog)	0.09 $mg/cm^2/day$ at 35 days (still decreasing with time)	Colangelo [55]
Co (in vitro)	approximately 7 $mg/cm^2/day$ estimated by us from data in [18]	Williams [19]

rate of biomaterials decreases, almost without exception, as a function of time. Any in-vitro test that does not reproduce this behavior (and there are many) must be open to question.

Conclusion and Recommendation

On the basis of this review, the following conclusions and recommendations are made:

1. There is evidence that certain organic species, notably serum, accelerate the corrosion rate of at least some metals under static (that is, nonfretting) conditions, and that such species should be included in artificial test environments for evaluation of biomaterials.

2. The use of electrochemical methods for corrosion testing, especially in the presence of complex organic species, simple sulfides, and other possibly electrochemically active compounds should be avoided if such compounds, by their anodic oxidation, were shown to interfere with the measurement. The importance of this is probably greatest with noble metals, less with passivating metals.

3. Recognizing the slowness of the passivation process for many alloys used as biomaterials, the value of short-term experiments (less than 1000 h) is open to question, in respect of these materials.

4. The use of electrochemical techniques conducted in simple electrolytes, such as buffered NaCl, afford a valid means of ranking or screening biomaterials, subject to 1 to 3 above.

5. The importance of correct specimen mounting to avoid undesired crevice formation and the role of solution agitation effects should be borne in mind.

6. The formation, in vivo, of a soft or hard tissue capsule around an implant, must retard corrosion. It should be possible to simulate this in laboratory corrosion tests, by application of some sort of sheath around the test specimen, and this approach should be investigated.

References

[1] Scales, J. S., *Corrosion of Biomaterials,* Tape-Slide Lecture, A. T. Kuhn.
[2] French, J. G., Cook, S. D., and Haddad, R. J., *Journal of Biomedical Materials Research,* Vol. 18, 1984, pp. 817–828.
[3] Heath, J. C. and Freeman, M. A. R., *The Lancet,* 20 March 1971, pp. 564–566.
[4] Deutman, R. and Mulder, T. J., *Journal of Bone Joint Surgery,* Vol. 59A, 1977, pp. 862–865.
[5] Rae, T., *Journal of Bone and Joint Surgery,* Vol. 63B, 1981, pp. 435–440; and Vol. 57B, 1975, pp. 444–450.
[6] Syrett, B. C. and Davis, E. E., *Corrosion and Degradation of Implant Materials, STP 684,* American Society for Testing and Materials, Philadelphia, 1979, pp. 229–244.
[7] Fraker, A. C. and Griffin, C. D., Eds., *Corrosion and Degradation of Implant Materials: 2nd Symposium, STP 857,* American Society for Testing and Materials, Philadelphia, 1985.
[8] Luckey, H. A. and Kubli, F., *Titanium Alloys in Surgical Implants, STP 796,* American Society for Testing and Materials, Philadelphia, 1982.
[9] Williams, D. F. and Clark, G. C. F., *Journal of Materials Science,* Vol. 17, 1982, pp. 1675–1682.
[10] Jenkins, G. N., *Physiology of the Mouth,* Blackwells Scientific Publications, Oxford, England, 1970.
[11] Diem, K., *Documenta Geigy Scientific Tables,* 6th Edition Geigy, Macclesfield, United Kingdom, 1970.
[12] Darvell, B. W., *Journal of Oral Rehabilitation,* Vol. 5, 1978, p. 41.
[13] Mandel, I. D., *Journal of Dental Research,* Vol. 53, 1974, p. 246.
[14] Paul, J., *Cell Tissue Culture,* 5th ed., Churchill Livingston, Edinburgh, United Kingdom, 1976.
[15] Brune, D., Hultquist, G., and Leygraf, C., *Scandanavian Journal of Dental Research,* Vol. 92, 1984, pp. 262–267.
[16] Greene, N. D. and Jones, D. A., *Corrosion Science,* Vol. 1, 1966, pp. 345–353.
[17] Hoar, T. P., *Corrosion Science,* Vol. 7, 1967, p. 455.
[18] Gonzalez, J. A., *Corrosion Science,* Vol. 25, 1985, pp. 519–530.
[19] Woodman, J. L. and Black, J., *Biomaterials 1980,* G. D. Winter and D. F. Gibbons, Eds., Wiley, New York, 1982, pp. 245–250.
[20] Greene, N. D. and France, W. D., *Corrosion,* Vol. 21, 1965, pp. 275–280.

[21] Hayes, M., *Techniques in Electrochemistry, Corrosion and Metal Finishing,* A. T. Kuhn, Ed., Wiley, New York, 1987.

[22] Brown, S. A. and Simpson, J. P., *Journal of Biomedical Materials Research,* Vol. 15, 1981, pp. 867–878.

[23] Kuhn, A. T., *Biomaterials,* Vol. 2, 1981, pp. 68–77.

[24] Tuthill, A. H. and Schillmoller, C. M., paper presented at Ocean Science and Engineering Conference, Washington, DC, 1965, cited as Shreir, L. L., *Corrosion,* Vol. 1, Newnes Butterworth 1979, London, pp. 4–128.

[25] Kuhn, A. T. and Chan, C. Y., *Journal of Applied Electrochemistry,* Vol. 13, 1983, pp. 189–207.

[26] Brown, S. A. and Merritt, K., *Biomaterials, Medical Devices and Artificial Organs,* Vol. 9, 1981, pp. 57–63.

[27] Cook, S. D., et al., *Biomat. Med. Dev. Art. Org.,* Vol. 11, 1983/4, pp. 281–292.

[28] Brown, S. A. and Merritt, K., *Journal of Clinical Applications of Biomaterials,* A. J. C. Lee and T. J. Albrecktsson, Eds., Wiley, New York, 1982, pp. 195–201.

[29] Brown, S. A. and Merritt, K., *Journal of Biomedical Materials Research,* Vol. 15, 1981, pp. 479–488.

[30] Harrison, A., *Journal of Biomedical Materials Research,* Vol. 9, 1975, pp. 341–353.

[31] Gettleman, L., *Biomaterials, Medical Devices and Artifical Organs,* Vol. 7, 1979, pp. 191–198.

[32] Khidirov, S. S. and Musaev, G. G., *Elektrokhimya,* Vol. 21, 1985, pp. 698–701.

[33] Wright, S. R., Cocks, F. H., and Gettleman, L., *Corrosion,* Vol. 36, 1980, pp. 101–103.

[34] Hoar, T. P. and Mears, D. C., *Proceedings of the Royal Society,* Vol. 294A, 1966, pp. 486–502.

[35] de De Micheli, S. M. and Riesgo, O., *Biomaterials,* Vol. 3, 1982, pp. 209–212.

[36] Bapna, M. S. and Lautenschlager, E. P., *Journal of Biomedical Materials Research,* Vol. 9, 1975, pp. 611–621.

[37] Cahoon, J. R. and Holte, R. N., *Biomedical Materials Research,* Vol. 15, 1981, pp. 137–145.

[38] Wheeler, K. R. and James, L. A., *Journal of Biomedical Materials Research,* Vol. 5, 1971, pp. 267–281.

[39] Bundy, K. J., Marek, M., and Hochman, R. F., *Journal of Biomedical Materials Research,* Vol. 17, 1983, pp. 467–487.

[40] Newman, S., *Journal of Biomedical Materials Research,* Vol. 15, 1981, pp. 615–616.

[41] Finkelstein, G. F. and Greener, E. H., *Journal of Oral Rehabilitation,* Vol. 4, 1977, pp. 347–354.

[42] Pini, G., *Electrochimica Acta,* Vol. 22, 1977, pp. 1423–1425.

[43] Marek, M. and Hochman, R. F., *50th Session of the International Association of Dental Research,* Las Vegas, NV, 1972.

[44] Kuhn, A. T., *Techniques in Electrochemistry, Corrosion and Metal Finishing,* A. T. Kuhn, Ed., Wiley, New York, 1987.

[45] Kuhn, A. T. and Kelsall, G. H., *Journal of Applied Chemical Biotechnology,* Vol. 33A, 1983, pp. 406–410.

[46] Cahoon, J. R. and Hill, L. D., *Journal of Biomedical Materials Research,* Vol. 12, 1978, pp. 805–821.

[47] Kuhn, A. T., Wroblowa, H., and Bockris, J. O. 'M, *Transactions of the Faraday Society,* 1967, pp. 1458–1465.

[48] Kaiser, H. and Popp, W., *Proceedings of the 8th International Congress of Metallic Corrosion,* 1981, pp. 76–81.

[49] Ross, T. K., Carter, D. A., and Smith, D. C., *Corrosion Science,* Vol. 7, 1967, pp. 373–376.

[50] Gettleman, L., Cocks, F. H., and Darmiento, L. A., *Journal of Dental Research,* Vol. 59, 1980, pp. 689–707.

[51] Svare, C. W. and Belton, G., *Journal of Biomedical Materials Research,* Vol. 4, 1970, pp. 457–467.

[52] Sigel, H., *Metal Ions in Biological Systems,* Marcel Dekker, New York, 1979.

[53] Pettit, L. D. and Hefford, J. W., *Metal Ions in Biological Systems,* Vol. 9, Chapter 6, pp. 173–212.

[54] Revie, R. W. and Greene, N. D., *Journal of Biomedical Materials Research,* Vol. 3, 1969, pp. 465–470.

[55] Colangelo, V. J. and Greene, N. D., *Journal of Biomedical Materials Research,* Vol. 1, 1967, pp. 405–414.

[56] Wood, N. K. and Kaminski, E., *Journal of Biomedical Materials Research,* Vol. 4, 1970, pp. 1–12.

[57] Koegel, A. and Black, J., *Journal of Biomedical Materials Research,* Vol. 18, 1984, pp. 513–522.

[58] Angelini, E. and Zucchi, F., *Surface Technology,* Vol. 21, 1984, pp. 179–181.

[59] Tani, C. and Zucchi, F., *Minerva Stomatologica,* Vol. 16, 1967, pp. 710–717.

[60] Meyer, J.-M., *Corrosion Science,* Vol. 17, 1977, pp. 971–982.

[61] Fusayama, K. and Nomoto, S., *Journal of Dental Research,* Vol. 42, 1963, pp. 1183–1189.

[62] Cahoon, J. R. and Chaturvedi, M. C., *Medical Instruments,* Vol. 7, 1973, pp. 131–135.

[63] Sutow, E. J. and Pollack, S. R., *Journal of Biomedical Materials Research,* Vol. 10, 1976, pp. 671–693.

[64] Ogundele, G. I. and White, W. E., *Corrosion and Degradation of Implant Materials: 2nd Symposium, STP 859,* American Society for Testing and Materials, Philadelphia, pp. 117–135.

[65] Samitz, M. H. and Katz, M. A., *British Journal of Dermatology*, Vol. 92, 1975, pp. 287–290.
[66] Brown, S. A. and Merritt, K., *Journal of Bone and Joint Surgery*, Vol. 63B, 1981, p. 105.
[67] Solar, R. J. and Pollack, S. R., *Journal of Biomedical Materials Research*, Vol. 13, 1979, pp. 217–250.
[68] Aragon, P. J. and Hulbert, S. F., *Journal of Biomedical Materials Research*, Vol. 6, 1972, pp. 155–164.
[69] Speck, K. M. and Fraker, A. C., *Journal of Dental Research*, Vol. 59, 1980, pp. 1590–1595.
[70] Salvarezza, R. C. and de Mele, H. H., *Journal of Biomedical Materials Research*, Vol. 19, 1985, pp. 1073–1084.
[71] Clark, G. C. F. and Williams, D. F., *Journal of Biomedical Materials Research*, Vol. 16, 1982, pp. 125–134.
[72] Sandstede, G., *From Electrocatalysis to Fuel Cells*, University of Washington Press, Seattle, WA, 1972.
[73] Vielstich, W., *Fuel Cells*, Wiley Interscience, New York, 1970.
[74] Horanyi, G., *Journal of the Electroanalytical Chemistry*, Vol. 117, 1981, pp. 131–138.
[75] Giner, J. and Colton, C. K., *Journal of the Electrochemistry Society*, Vol. 126, 1979, pp. 1687–1693; de Mele, M. F. L., Videla, H. A., and Avia, A. J., *Journal of the Electrochemistry Society*, Vol. 129, 1982, pp. 2207–2211.
[76] Harmer, M. A. and Hill, H. A. O., *Journal of Electroanalytical Chemistry*, Vol. 189, 1985, pp. 229–246.
[77] Mostafa, M. Y. and Mourad, M. Y., *Journal of Electroanalytical Chemistry*, Vol. 130, 1981, pp. 221–228.
[78] Zagal, J. H. and Herrera, P., *Electrochemica Acta*, Vol. 30, 1985, pp. 449–454.
[79] Smith, G. K. and Black, J. P., *Corrosion and Degradation of Implant Materials: Second Symposium*, STP 859, American Society of Testing and Materials, Philadelphia, 1985, pp. 223–250. ·

DISCUSSION

J. M. Sykes[1] (written discussion)—You have emphasized the problems involved in carrying out meaningful electrochemical studies in vitro using complex environments. Aren't there going to be even greater difficulties in making electrochemical measurements in vivo? Have worthwhile in-vivo measurements been made?

A. T. Kuhn, P. Neufeld, and T. Rae (authors' closure)—A number of workers have reported "in-vivo" electrochemical measurements. We may conveniently divide these into potentiometric (open circuit) and externally driven experiments. We see those in the former class (which also include potential versus time measurements) as being simple, foolproof, and informative. Examples of these, some of which have been cited, include

• Hoar and Mears, *Proceedings of the Royal Society*, Vol. 294A, 1966, pp. 486–510 (potential versus time measurements on goat tibia).

• Brown and Simpson, *Journal of Biomedical Materials Research*, Vol. 15, 1981, pp. 867–878 (potential versus time measurements on fractured goat tibia).

• Potential measurement of various metallic components in the oral cavity (reference not known).

Turning to the second category of driven electrochemical measurements, we would guess that there have been some 30 or so studies, some using linear polarization in vivo to measure corro-

[1]University of Oxford, Department of Metallurgy and Science of Materials, Parks Rd., Oxford (OX1 3PH).

sion, others, using cyclic voltammetry as an analytical tool to follow in-vivo changes in chemical composition of body fluids. Among those related to corrosion of biomaterials, the work of Gettleman et al. on baboons was cited by us. We might also mention:

- Colangelo et al., *Journal of Biomedical Material Research*, Vol. 1, 1967, pp. 405–414 (overview of linear polarization and implants in dogs).
- Revie and Greene, *Journal of Biomedical Material Research*, Vol. 3, 1969, pp. 465–470 (comparison of in-vivo and in-vitro) and references cited therein.
- Steinemann and Perren, *Transactions of the 3rd Annilitial Metallurgical Society for Biomaterials*, Vol. 1, 1977, p. 122 (linear polar in vivo for rabbits with aluminum, titanium, vanadium, iron, cobalt, nickel, copper, zirconium, niobium, molybdenum, silver, tantalum, platinum, and gold).

The majority of more recent in-vivo electrochemical studies have been pharmacologically oriented, for example, using platinum electrodes implanted in the brain, where the technique is primarily an electroanalytical one. In the corrosion studies, the actual extent of agreement between in-vivo and in-vitro results can be best gauged by inspection of the Gettleman paper and that of Revie and Greene and is very much "hit-and-miss" being good in some cases, very poor in others, as shown in Table 8, for cobalt, for example, most probably because of methodological shortcomings. We are of the opinion that the problems of interference caused by presence of sundry electroactive species in-vivo, present a continuing difficulty to would-be users of this technique if their intention is to obtain accurate corrosion rates, though the problem of interference will be more severe with the electrocatalytically active metals (silver, gold, and platinum) and less with others (Ti, stainless steel). In a qualitative sense, inspection of the data cited above does afford a real and useful insight into the continuing passivation process in vivo, and though potentiometric measurements might have provided a similar insight; we do not decry the method.

James F. D. Stott,[1] *B. S. Skerry,*[2] *and R. A. King*[1]

Laboratory Evaluation of Materials for Resistance to Anaerobic Corrosion by Sulfate-Reducing Bacteria: Philosophy and Practical Design

REFERENCE: Stott, J. F. D., Skerry, B. S., and King, R. A., **"Laboratory Evaluation of Materials for Resistance to Anaerobic Corrosion by Sulfate-Reducing Bacteria: Philosophy and Practical Design,"** *The Use of Synthetic Environments for Corrosion Testing, ASTM STP 970,* P. E. Francis and T. S. Lee, Eds., American Society for Testing and Materials, Philadelphia, 1988, pp. 98-111.

ABSTRACT: Reliable laboratory test methods for the evaluation of the performance of materials against anaerobic corrosion by sulfate-reducing bacteria are not readily apparent in the literature. Traditionally, such testing has been carried out in small scale "batch cultures" using filled and stoppered vessels or small cells contained in anaerobic jars. The problem with such tests is that they give much lower rates of corrosion than those frequently experienced in the field. This is almost inevitable, as the total quantity of hydrogen sulfide produced is small compared with the surface area of the test coupons, and quickly becomes denuded. From our knowledge of the microbiology and electrochemistry of anaerobic corrosion by sulfate-reducing bacteria, any realistic test must meet certain conditions. The duration must be long enough to allow the stable crystallographic forms of metal sulfide to form. These tend to be nonprotective and cathodic to the metal. The vessel should be sufficiently large to present a realistic volume of test medium to exposed specimens. It should contain a medium that supports the growth of sulfate-reducing bacteria, but excludes organic constituents, which act as inhibitors. It should operate in a semicontinuous culture mode to maintain the organisms in an active state of growth and sulfide production. It should be maintained at a temperature which is compatible with the active growth of the sulfate-reducing bacteria. Air should be effectively excluded. Large anaerobic culture vessels (20-L capacity) have been constructed containing both aqueous and soil environments. These have been used in a number of tests involving various environments. Electrochemical monitoring by linear polarization, electrical resistance, and advanced electrochemical techniques has been incorporated. Results of a 250-day test involving cast iron pipe sections are outlined. The test regime described provides a reliable method for the evaluation of the likely long term anaerobic corrosion behavior of any selected materials and associated corrosion protection methods.

KEY WORDS: sulfate-reducing bacteria, *desulfovibrio,* anaerobic corrosion, evaluation, sulfide, hydrogen sulfide, corrosion testing, cast iron, performance (materials)

The disposal of electrons at the cathode of a corrosion cell in the absence of oxygen at neutral pH normally results in the rapid stifling of the corrosion process because of the hydrogen overpotential. However, natural anaerobic conditions, such as wet clay soils or stagnant, polluted waters, have long been known to be "aggressive" towards buried iron and steel [1,2]. In such

---

[1]Group manager and manager of technology development, Corrosion and Protection Centre Industrial Services, University of Manchester Institute of Science and Technology, P.O. Box 88, Sackville St., Manchester M60 1QD, United Kingdom.
[2]The Sherwin-Williams Co. Research Center, 10909 S. Cottage Grove Ave., Chicago, IL 60628.

cases, iron sulfides occur as a result of corrosion, both as an adherent corrosion product, for example, on pipes, and also in the surrounding environment. From the occurrence of iron sulfides, the presence of the sulfate-reducing bacteria may be deduced.

The black corrosion products thus formed liberate hydrogen sulfide (H_2S) on acid treatment, which distinguishes them from black iron oxide (magnetite). The corrosion products often are loose and, when dislodged, pits lined with bright metal, which correspond to areas of anodic dissolution, are visible. With cast iron, graphitization has sometimes been seen, in which the iron becomes dislodged, leaving only the graphite structure and corrosion product.

The mechanism of anaerobic corrosion caused by the sulfate-reducing bacteria has always been controversial. Booth and co-workers [3–6] produced much laboratory evidence in support of a cathodic depolarization theory. The situation has been reviewed by Miller and his co-workers [7–9] who are inclined to the view that some direct depolarization of steel is carried out by the bacteria, although some process associated with the presence of iron sulfide is probably more important quantitatively.

There are several iron sulfides that are known to be cathodic to steel. King and Miller [7] consider the main roles of the bacteria to be:

(1) generation of sulfides by their growth,

(2) "regeneration" of fresh iron sulfide (FeS), enabling it to remain cathodic to iron, and

(3) depolarization of the FeS cathode, thus bringing fresh surfaces constantly into contact with steel by their movement.

Theoretically, it is easier to depolarize FeS than it is to depolarize steel, since atomic hydrogen is usually quite strongly adsorbed onto steel surfaces.

Costello [10] produced a modified theory in which hydrogen sulfide is seen as the major cathodic reactant:

$$H_2S + e \rightarrow HS^- + \frac{1}{2} H_2 \qquad (1)$$

with the high current density inside pits sustained by the very large surface areas of FeS available as the cathode.

Theories such as the above, in which anaerobic corrosion is explained purely in terms of inorganic chemical reactions, now appear widely accepted by corrosion scientists and engineers.

Direct stimulation of the anodic reaction by bacterially produced sulfide has been suggested [11], but is considered important only at the start of the corrosion process [7]. There seems little evidence to corroborate other suggested corrosion mechanisms of the sulfate-reducing bacteria [12,13].

In general, the testing of materials other than iron and steel for anaerobic corrosion or biodeterioration by the sulfate-reducing bacteria has been neglected. Gilbert, [14] however, reported severe corrosion of copper pipes by sulfate-reducing bacteria. The widespread belief that heavy metals are toxic to sulfate-reducers and are, therefore, immune to attack is not unheld [15]. Tiller [16] has also reported two case histories of failure of stainless steel caused by sulfate-reducing bacterial attack and Puckorius [17] has reported extensive failures of AISI 304 stainless steel condensers caused by sulfate-reducing bacteria.

To interpret field observations better, formulation of a reliable laboratory method based upon near-natural conditions, which may be used to test materials for anaerobic corrosion caused by sulfate-reducing bacteria, is required.

Pipelines and other metal objects buried in the ground and structures erected in estuaries frequently show sulfide corrosion. Booth estimated in 1964 [25] that at least 50% of corrosion failures of underground pipes in the United Kingdom are caused by bacterial corrosion. Pipeline corrosion is most severe in wet clay or clay loam of about neutral pH value. Progressive pitting corrosion is common; cast iron pipes with wall thickness of 6.3 mm have occasionally become perforated within a year of installation under these conditions, and perforation within 4 years is quite common. Such corroded objects, if examined immediately after removal from the soil, have a black corrosion product that frequently smells of hydrogen sulfide (or that liberates hydrogen sulfide on treatment with acid). Sulfate-reducing bacteria can usually be detected in the corrosion product in such instances, and in the soil in the vicinity of the pipe, in very much larger numbers than the "background" count for that district.

Local anaerobic conditions favoring the growth of sulfate-reducers can readily arise under heavy mixed microbial growth in industrial situations, such as open recirculating cooling water systems [26], the paper making industry [27], or heat exchangers cooled by river water [28]. Catastrophic corrosion of condensers by marine sulfate-reducers in a once-through system on the Arabian Gulf has been described by Temperley [29]. Ships berthed in estuaries in which they rest on bottom mud at low tide have been reported to suffer from this type of corrosion [30].

By far the greatest manifestation of corrosion problems by sulfate-reducers has been in the oil and gas industry. The problems are reviewed in a book produced by National Association of Corrosion Engineers [31].

Design Philosophy

Traditionally, the testing of materials against anaerobic corrosion caused by sulfate-reducing bacteria has been carried out in small scale "batch cultures" using filled and stoppered vessels or small cells contained in "anaerobic jars" [4]. The problem with such tests is that they give much lower rates of corrosion than those frequently experienced in the field. This is almost inevitable, as the total quantity of hydrogen sulfide produced is small compared with the surface area of the test coupons. It is also consumed rapidly by the corrosion reaction. After approximately 1 or 2 weeks, the organisms may die in batch culture once the nutrient supply is exhausted. Under these conditions, the system becomes inert. Moreover, short-term tests do not allow sufficient time for the stable crystallographic forms of FeS to form. This may take several months, according to King and Miller [7], at which time corrosion rates may accelerate rapidly as the early, protective, mackinawite (FeS_{1-x}) film changes to nonprotective, iron-rich, smythite (Fe_3S_4) and pyrrhotite ($Fe_{1-x}S$) [18]. Smythite and pyrrhotite tend to spall and not to reform. This latter stage is never reached in batch cultures.

Aware of the major inadequacies of batch culture testing, several workers have used semicontinuous cultures, in which a quantity of test medium is replaced with fresh medium, which maintains the organisms in an active state of growth and sulfide protection [19-21]. Invariably, however, such testing has been carried out in small vessels where the ratio of surface area of specimens to volume of medium has been very much greater than is the case with buried or immersed objects in natural environments. Such tests have also often been relatively short.

A further difficulty with laboratory testing of materials against anaerobic corrosion by the sulfate-reducing bacteria lies with the composition of the test medium. Traditional liquid media used for the growth of these organisms almost invariably contain yeast extract at a level of 1 gL^{-1} (1000 ppm) [22]. Adams and Farrar [23] have shown this additive to be a relatively effective corrosion inhibitor, a finding supported by our own preliminary tests. However, the use of yeast extract in corrosion test media for cultivation and growth of sulfate-reducing bacteria has continued, even though a solution of inorganic trace elements may substitute for it at very low concentrations [24].

To achieve a satisfactory laboratory simulation of corrosion caused by the sulfate-reducing bacteria, the authors contend that any vessel or test apparatus for the cultivation of sulfate-reducing bacteria should incorporate certain features:

• It should be sealed completely from the atmosphere and be absolutely airtight as the ingress of oxygen will severely inhibit the growth of these obligately anaerobic organisms. Oxygen ingress, after growth and sulfide production has been initiated, will result in the deposition of elemental sulfur within the test vessel. Since elemental sulfur is quite soluble in hydrogen sulfide containing-waters, its presence would tend to change the nature of the corrosion reactions.

There should, preferably, be provision for inert gas sparging to remove any initial oxygen from the test medium when the test is initiated. Inert gas sparging after initiation of microbial growth is to be avoided because it will displace the hydrogen sulfide produced by the organisms.

• It must be constructed of materials that are resistant to attack by hydrogen sulfide.

• It must be sufficiently large to present a realistic volume of test medium to the exposed specimens.

• It must be maintained at a temperature which is compatible with active growth of the sulfate-reducing bacteria (298 K to 303 K in most instances, although up to 333 K may be necessary for the freshwater and soil thermophilic species *Desulfotomaculum nigrificans*). It is, therefore, highly desirable to incorporate some form of thermostatically controlled heater.

• It must contain a medium which supports growth of the sulfate-reducing bacteria. The most commonly used media are those of Postgate [22] or the American Petroleum Institute RP38. These media are comprised of mixtures of mineral salts in water with a carbon source (usually sodium lactate). For corrosion test applications, the first batch of medium used to initiate a test should contain a reducing agent, which acts as an oxygen scavenger, such as sodium thioglycollate with ascorbic acid. Corrosion test media should also contain a substantial amount of ferrous iron. This serves as a good indicator of the growth of sulfate-reducing bacteria through the intense blackening it imparts to the medium when hydrogen sulfide is produced. Sodium chloride may be added up to a level reflecting the environment the test is required to simulate, provided that a suitable salt-tolerant strain of sulfate-reducing bacteria is chosen.

• Inlet and outlet lines for medium replacement should be provided with suitably airtight valves.

• For general corrosion testing, autoclaving or other forms of sterilization of equipment are unnecessary provided that a sufficiently large inoculum of sulfate-reducing bacteria is used to initiate the test, and that the medium is used immediately after making it up. The regime will provide a mixed culture containing organisms other than sulfate-reducing bacteria, but the sulfate-reducers will predominate under the test conditions.

The objective of the present work is to construct a laboratory test rig that satisfies all of the above requirements, to provide a laboratory test to evaluate materials against anaerobic corrosion by sulfate-reducing bacteria. The test procedure was designed to be a reasonably realistic simulation of an aggressive natural environment, and one that allowed electrochemical corrosion monitoring techniques to be applied to metallic specimens. The vessel was designed to run relatively long-term tests (e.g., 6 months), in which meaningful corrosion rates may be obtained.

Experimental Procedure

Equipment

Several semicontinuous culture vessels based upon the above philosophy have been built. Two such rigs are illustrated photographically in Fig. 1, and one rig is shown diagramatically in Fig. 2. Each test apparatus is comprised of a toughened glass cylindrical column section (Corning QVF), 0.3 m in diameter by 0.3 m in height, having an approximate volume of 20 L.

A top plate (A) and a base plate (B) were fabricated from 1-cm thick unplasticized PVC. Prior to testing, each vessel was assembled using 10-mm diameter steel bolts (C), QVF collars (D), and QVF neoprene gaskets (E). A PVC drainage tap (12.7-mm diameter) (F) was fixed to the underside of the base plate for periodic draining of the test medium to allow fresh nutrient batches to be added. The top plate was fitted with three 316 grade stainless steel lockhead units (6.4-mm diameter) (G), sealed using "Viton" rubber "O" rings to which two 'Nypro' 316 grade

FIG. 1—*Twenty-liter vessel for testing materials against sulfate-reducing bacteria.*

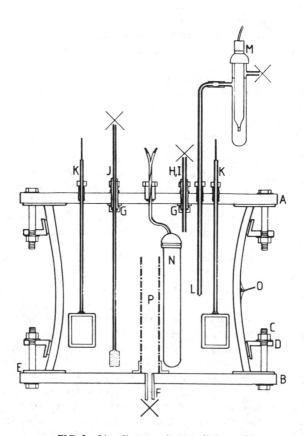

FIG. 2—*Line diagram of twenty liter vessel.*

stainless steel ball-valves were attached (H,I), the first to facilitate periodic nutrient refilling, and the second to allow periodic release of gas pressure generated in the vessel and act as an inert gas outlet during initial degassing. A third bulkhead (J), was fitted with a needle-valve used to regulate the deaerating inert gas supply for the initial 24-h period of each long-term exposure test. Drilled nylon units sealed with "Viton" rubber compression "O" rings were used to suspend the test specimens (K) in the anaerobic chamber, thus permitting electrochemical monitoring. A similar compression seal arrangement was used to connect a saturated calomel reference electrode (M). A thermostatically controlled heater of a type commonly used in aquarium tanks (N) was sealed in the center of the top plate. The vessel could thus be controlled to ±2 K of the desired temperature as monitored by a thermocouple arrangement attached to the external wall of the glass vessel (O). Mounting the heater element internally in the rig allowed a radial heating pattern that ensured similar test conditions in each part of the anaerobic environment around the annulus of specimens.

In one version of the rig, which was designed to hold mud or waterlogged soil to a height of about 0.2 m, a 3 cm diameter tube 0.2 m in length (P) was attached to the bottom plate around the drainage tap (F). This allowed nutrient medium in the upper 0.1 m or so of the vessel to be periodically drained and replaced without loss of soil or mud from around the test specimens. It also ensured that the heater was always surrounded by liquid rather than solid soil.

Specimen Preparation

Materials

Although any engineering material and corrosion protection system may be tested, results are reported for electrodes fabricated from ductile cast iron piping used for underground potable water transmission in the United Kingdom. In practice, external corrosion caused by sulfate-reducing bacteria is an important contributory factor in the long-term deterioration of such materials. The centrifugally cast, ductile iron pipe material selected for study was supplied by the Water Research Council, Swindon, England. It was selected from a batch of pipes on the basis of typical chemical composition and metallurgical properties. The material had the following characteristics:

1. Dimensions: 117 mm OD, 6 mm wall thickness.
2. Chemical analysis (%): C(3.54), Si(2.55), Mn(0.39), S(0.01), P(0.062), Mg(0.030).

The material was characterized metallographically by optical microscopy. Externally, the cast iron structure assumed a normal ferrite matrix of equiaxed shape, containing spheroidal graphite nodules. A double-layer surface oxide was apparent (~0.02 mm thick), comprised of a mill-scale oxide from the initial casting, with a second oxide probably arising from a subsequent heat treatment annealing process.

The cast iron material was studied in two surface conditions. First, samples were exposed either in the factory finished state (i.e., with an intact bitumen seal over the millscale surface oxide) and, second, samples were exposed with the bitumen and the mill-scale abraded to expose the bright cast iron matrix. The latter treatment was used to simulate mechanical damage that, in practice, often occurs during delivery, laying, or backfilling of the line. In all cases, test coupons (60 mm × 50 mm) were dry cut from the pipeline section. Mechanical abrasion was conducted using a hand-held "Makita" power grinder. Two coats of a twin pack coal tar epoxy known to be resistant to sulfate-reducing bacteria (Leigh's Epi-grip L611) were subsequently applied to mask all except a 40 mm × 30 mm area of the external pipeline surface on each coupon.

Procedure

Anaerobic Environment

Two anaerobic sulfate-reducing bacterial environments were selected for study. The first was a laboratory-prepared simulation using an aqueous medium nutrient solution based upon Postgate's medium B [22]. The composition of the modified medium is given in Table 1. Although not used throughout the entire test period, the cultures were started by adding the following constituents: L-ascorbic acid (0.1 gL^{-1}), sodium thioglycolate (0.1 gL^{-1}). The inoculum used comprised a sulfate-reducing bacteria culture isolate taken from soil and grown for 6 days on an enrichment medium (Postgate's B). This inoculum contained cells typical of *Desulfovibrio vulgaris/desulfuricans* organisms. The number of cells in the inoculum was of the order of 10^9 per ml. This nutrient solution was replaced every 10 to 30 days during each test.

The second environment was a naturally water-saturated anaerobic soil taken from an estuary within the dockland region of Manchester. Approximately 20 L of soil were dredged mechanically from a depth of 7 m of water in an area of low-flow. Natural biodegraded humus was retained with the soil sample. A natural sulfate-reducing bacteria count of 1×10^5 ml^{-1} was recorded using ASTM Test Method for Sulfate-Reducing Bacteria in Water and Water Formed Deposits D 4412-84. This was deemed representative of a very aggressive natural environment, to which metallic materials could be exposed in practice. In one of the laboratory rigs, 12 L of the soil were used with the test electrodes completely submerged. The rig was topped up to its capacity of 20 L with 8 L of the aqueous nutrient food, inoculated as described above. Following completion of the test program, the resistivity, pH, and redox potential of the soil were measured. A summary of these parameters is given in Table 2.

TABLE 1—*Composition of aqueous medium, pH adjusted to 7.5 ± 0.3 units.*

Constituent	Concentration
KH_2PO_4	0.5 g/L
NH_4Cl	1.0 g/L
Na_2SO_4	2.7 g/L
$CaCl_2$	0.18 g/L
$MgSO_47H_2O$	2.0 g/L
$FeSO_47H_2O$	0.15 g/L
Sodium lactate (60 to 70% solution)	4 mL

TABLE 2—*Summary of anaerobic test environment conditions.*

Environment	Temperature, K	pH	Sulfate-reducing Bacterial Counts, Cells, mL^{-1}	H_2S Concentration, $mg\ L^{-1}$
Aqueous medium	298 ± 2	7.1	10^8	380
Estuarine Soil[a] plus aqueous medium	298 ± 2	8.4	10^7	430

[a]Measured soil characteristics—Resistivity 52 Ohm · cm, redox potential −0.014 V (versus normal hydrogen electrode), natural SRB count 10^5 mL.

Specimen Arrangements

Variously sized and shaped electrodes were tested, instrumented through the drilled top-plate arrangement. Results are reported for electrodes as described above that were located radially in matched pairs (either 6 or 12 pairs) within the environment chamber. Such electrodes were drilled, tapped, and fitted with 3-mm "Viton" rubber "O" rings and a 3-mm OD, 0.3 m long 316 grade stainless steel electrode connector encased in a sealed nylon tube to ensure that no electrolyte could reach the connector rod.

Corrosion Assessment

Each electrode was photographed before and after testing. Weight-loss data were collected for each specimen following a standard inhibited acid cleaning procedure (ASTM 72 5.7.2 Standard Recommended Practice for Preparing, Cleaning, and Evaluating Corrosion Test Specimens G 1-72 5.7.2). Corrosion products from selected samples were analysed by X-ray diffraction (Phillips X-ray diffractometer), using a goniometer method with a cobalt target set at 40 kV, 20 mA.

Linear polarization monitoring was conducted using a commercial instrument calibrated for two electrode use (Magna-Corrator 1120). Semicontinuous readings were taken for each electrode pair after steady state conditions had been attained (approximately 100 s) every week. Corrosion rates were recorded in "mils per year" after area correction factors had been applied.

Results and Discussion

Weight-loss Data

Weight-loss data obtained for the test electrodes are presented in Table 3. Whereas the abraded metal surfaces suffered considerable corrosion loss over a 230 to 250 day test period irrespective of test environment, the bitumen coating combined with the factory mill-scale surface resisted corrosion attack relatively successfully.

General corrosion rates have been calculated from these weight-loss data assuming a density of 7.86 gcm^{-3} for cast iron, according to the following equation:

$$\text{corr rate (mpy)} = \frac{\text{Wt loss (mg)} \times K}{\text{time (days)} \times \text{density (gcm}^3) \times \text{area (dm}^2)} \tag{1}$$

where

K = conversion factor to change material loss to mpy, (i.e. 1.44).

Although the calculated corrosion rates from weight-loss data provide useful assessments of the corrosion condition of individual electrodes, they in effect represent an integrated corrosion rate for the total test period. For example, abraded cast iron coupons exposed either to aqueous medium or to estuarine soil corroded at a rate of between 5.4 to 6.2 mpy over the 230 to 250 day test period. However, no information can be derived directly from weight loss measurements that can be related to instantaneous assessments of the corrosion condition of the cast iron material during the test period. Electrochemical corrosion monitoring is therefore necessary to derive such information. Nevertheless, it is apparent that the test methods developed produced relatively high weight-loss values over the test periods used.

After exposure periods in excess of 200 days, the weight loss data generated by abraded electrodes were consistent irrespective of the test environment (Table 3). However, after testing for up to approximately 125 days, the siderite and mackinawite both tended to transform by sulfi-

TABLE 3—*Weight-loss data and calculated general corrosion rates for ductile cast iron electrodes in test environments containing sulfate-reducing bacteria.*

Environment	Electrode Surface Condition	Test duration, days	Weight loss, mg	Calculated Corrosion Rate, mpy
Aqueous medium	abraded metal	119	68.7 86.5	1.0
Aqueous medium	abraded metal	232	854.6 792.1	5.4
Aqueous medium	bitumen sealed mill-scale	119	7.3 8.4	0.05
Aqueous medium	bitumen sealed mill-scale	232	35.9 67.5	0.7
Estuary soil	abraded metal	104	335.8 352.7	5.0
Estuary soil	abraded metal	146	487.0 641.2	5.9
Estuary soil	abraded metal	250	922.2 1123.1	6.2
Estuary soil	bitumen sealed mill-scale	104	32.3 37.1	0.5
Estuary soil	bitumen sealed mill-scale	146	25.0 25.0	0.3
Estuary soil	bitumen sealed mill-scale	250	53.2 169.9	0.7

dation under anaerobic conditions to greigite (Fe_3S_4), and smythite (Fe_3S_4) with eventual transformation to pyrrhotite ($Fe_{1-x}S$), which is particularly nonprotective. The interrelationships of these iron compounds have been established in some detail by Rickard [34]. The nature of the sulfide films formed or the iron surface play an important role in the severity of corrosion. Indeed, King et al. [35] have shown that the corrosivity of chemically prepared iron sulfide is a function of the sulfur stoichiometry.

Linear Polarization Resistance Monitoring

Information pertaining to corrosion rates during the test period was obtained by semicontinuous linear polarization measurements. Electrolyte conductivity was not a problem, since even the estuarine soil had a measured resistivity of only 52.2 ohm cm. Representative results for electrodes tested in the aqueous medium are given in Table 4. These data provide useful insights as to relative corrosion rates throughout the test period. The corrosion rate of the bare cast iron in estuarine soil increased over time to a value of about 20 mpy.

Some caution is necessary when considering the results in terms of absolute corrosion rate values, since the linear polarization technique is generally susceptible to certain sources of error [32]. Predictions of absolute corrosion rate values, even in ideal testing conditions, would not be expected to have an accuracy better than a factor of 33. During the present work, the factor of uncertainty for the cast iron electrode pairs falls approximately within this range if the data are compared with equivalent weight-loss data (Table 3).

TABLE 4—*Linear polarization and corrosion potential data for ductile cast iron in estuarine soil containing sulfate-reducing bacteria (abraded metal).*

Time, days	Indicated Corrosion Rate, mpy, ± 0.5	Corrosion Potential (mV Versus Saturated Calomel Electrode) ± 10
0	<1	−720
25	1	−720
50	2.5	−690
75	3	−650
100	6	−610
150	18	−610
200	21	−580
250	23	−560

The abraded cast iron material in the aqueous medium displayed considerably lower weight loss compared with equivalently exposed electrodes in the estuarine soil (90 mg in aqueous medium, 500 mg in estuarine soil, Table 3). During the present work, it was apparent that relatively long periods (>125 days) were necessary before these effects were fully realized.

Since the test regime effectively optimizes anaerobic corrosivity rather than actively accelerating corrosion processes, it is anticipated that the weight loss data obtained (~6 mpy averaged over the test period of 230 to 250 days) would represent worst case conditions in practice for this cast iron material at damaged regions and in the absence of any other corrosion prevention measures.

The additional presence of the bitumen sealant over the mill-scale oxide considerably improved, the corrosion resistance of the cast material to anaerobic sulfate-reducing bacteria. However, after testing irrespective of environment, the bitumen coating exhibited signs of deterioration. Moreover, some small areas of black corrosion product were visible on these electrode surfaces. These effects were observed in all exposure tests as illustrated by the electrode exposed for 250 days in the aqueous medium (Figs. 3 and 4).

X-Ray Diffraction Studies

X-ray diffraction data obtained from the cast iron corrosion products removed from the estuarine soil after 250 days exposure are presented in Table 5. These data indicate that the corrosion products formed in the laboratory simulation were comprised mainly of iron carbonate ($FeCO_3$), and iron sulfide (FeS), these structures being assigned on the basis of standard powder diffraction file data.

These results are generally in accordance with literature suggested corrosion product formation in natural environments, in which the primary compounds formed appear to be siderite ($FeCO_3$), which is protective, and mackinawite (FeS_{1-x}). The bitumen, though not designed to be a protective coating, reduced corrosion to very low rates. In general, the corrosion rates measured were less than 1 mpy and only approached about 5 mpy towards the end of each test.

Visual Examination

The appearance of an abraded metal and a bitumen coated specimen after 250 days exposure is given in Figs. 3 and 4, respectively.

Corrosion product morphology was similar and comprised of black, partially adherent material. When this material was dislodged using a scalpel blade, bright, shiny metal became visible. Localized corrosion attack was apparent, suggesting the onset of pitting corrosion.

FIG. 3—*Condition of abraded metal test coupon after 250 days exposure to the anaerobic soil test environment containing sulfate-reducing bacteria.*

TABLE 5—*X-ray diffraction data for corrosion producers formed on abraded iron surfaces exposed for 250 days to anaerobic soil containing sulfate-reducing bacteria.*

X-ray Peak Priority	2θ	$2\sin\theta$	$d = \dfrac{n\lambda^a}{2\sin\theta}$	Assigned Structure
1	36.9	0.633	2.83	$FeCO_3$ siderite/Fe
2	61.7	1.026	1.74	$FeCO_3$
3	49.0	0.829	2.16	$FeCO_3$
4	28.3	0.490	3.65	$FeCO_3$
5	54.6	0.917	1.95	$FeCO_3/FeS$
6	24.6	0.426	4.20	$FeO (OH)^b$
7	44.2	0.752	2.38	$FeCO_3/FeS$
8	72.0	1.176	1.52	$FeCO_3$
9	81.9	1.31	1.37	...
10	77.0	1.245	1.44	...

[a]Where $n = 1$, $\lambda = 1.7902 \times 10^8$ μm ($CoK\alpha$).
[b]Probably from slight oxidation of FeS during sampling.

FIG. 4—*Condition of bitumen-coated metal test coupon after 250 days' exposure to the anaerobic soil test environment containing sulfate-reducing bacteria.*

Conclusions

1. The experimental rig design produced anaerobic environments containing sulfate-reducing bacteria, which generated high metallic corrosion rates without the artificial acceleration of naturally occurring corrosion processes.

2. Flexibility of design allows any metallic material to be tested for its resistance to anaerobic corrosion. Suggested corrosion protective measures may also be assessed. Suitable test environments may be selected to represent anticipated operating conditions.

3. The results reported for the anaerobic corrosion behavior of ductile cast iron pipeline material are in accordance within the literature. Corrosion products were composed primarily of $FeCO_3$ and FeS, overlying bright metal, which were black in color and tended towards shallow pitting corrosion.

4. The experimental test regime provided information not only concerning corrosion rate, but also in relation to corrosion mechanisms. Results obtained by traditional weight-loss testing were corroborated by electrochemical monitoring.

5. The corrosivity of the environment containing sulfate-reducing bacteria may be variable depending on the test conditions. In general, the corrosivity increased after fresh nutrient addi-

tions and with increasing time, as the bacterial populations became better established and stable crystallographic forms of iron sulfide were laid down. A test period of less than about 125 days could lead to erroneous long-term conclusions regarding metallic corrosion susceptibility in environments containing sulfate-reducing bacteria.

Acknowledgment

The authors express their gratitude to WRC Swirdon for supporting this study and to their colleagues, particularly, Dr. R. D. Eden, for assistance during the work.

References

[1] Bengough, G. D. and May, R., *Journal of the Institute of Metals,* Vol. 32, 1924, p. 81.
[2] Wolzogen Kuhr, C. A. H. Von and Van der Vlugt, L. S., *Water (den Haag),* Vol. 18, 1934, p. 147.
[3] Booth, G. H. and Wormwell, F., *Proceedings of the 1st International Congress on Metallic Corrosion,* Butterworth, London, 1962, p. 341.
[4] Booth, G. H. and Tiller, A. K., *Transactions of the Faraday Society,* Vol. 56, 1960, p. 1689.
[5] Booth, G. H. and Tiller, A. K., *Transactions of the Faraday Society,* Vol. 58, 1962, p. 110.
[6] Booth, G. H. and Tiller, A. K., *Transactions of the Faraday Society,* Vol. 58, 1962, p. 2510.
[7] King, R. A. and Miller, J. D. A., *Nature,* Vol. 233, 1971, p. 491.
[8] Miller, J. D. A. and Tiller, A. K. in *Microbial Aspects of Metallurgy,* 1st Ed., J. D. A. Miller, Ed., Medical and Technical Publishing Company, Aylesbury, England, 1971, p. 61.
[9] Miller, J. D. A., "Economic Microbiology," Vol. 6, in *Microbial Biodeterioration,* 1st Ed., A. H. Rose, Ed., Academic Press, London, 1981, p. 149.
[10] Costello, J. A., *The Mechanisms of the Corrosive Effect of SRB,* Ph.D. thesis, University of Cape Town, South Africa, 1975.
[11] Wanklyn, J. N. and Spruit, C. J. P., *Nature,* Vol. 169, 1952, p. 928.
[12] Schaschl, E., *Materials Performance,* Vol. 19, 1980, p. 9.
[13] Iverson, W. P. and Olsen, G. J. in *Microbial Corrosion, Proceedings of NPL Conference,* March 1983, The Metals Society, London, 1983, p. 46.
[14] Gilbert, P. T., *Journal of the Institute of Metals,* Vol. 73, 1947, p. 139.
[15] King, R. A. and Stott, J. F. D., in *Microbial Problems and Corrosion in Oil and Oil Product Storage,* E. C. Hill, ed., Institute of Petroleum, London, 1984, p. 93.
[16] Tiller, A. K. in *Microbial Corrosion, Proceedings of NPL Conference,* March 1983, The Metals Society, London, 1983, p. 104.
[17] Puckorius, P. R., *Materials Performance,* Vol. 22, 1983, p. 19.
[18] Smith, J. S. and Miller, J. D. A., *British Corrosion Journal,* Vol. 10, 1975, p. 136.
[19] Booth, G. H., Shinn, P. M., and Wakerley, D. S. in *Comptes Rednus du Congress International de la Corrosion Marine et des Salissures, Cannes, 1964,* C.R.E.O., Paris, 1965, p. 363.
[20] Dittmer, C. K., "Microbiological Aspects of Pipeline Corrosion and Protection, Ph.D. thesis, University of Manchester, England, 1975.
[21] Bell, R. G. and Lim, C. K., *Canadian Journal of Microbiology,* Vol. 27, 1981, p. 242.
[22] Postgate, J. R., *The Sulfate-Reducing Bacteria,* Cambridge University Press, 1979.
[23] Adams, M. E. and Farrer, T. W., *Journal of Applied Chemistry,* Vol. 3, 1953, p. 117.
[24] MacPherson, R. and Miller, J. D. A., *Journal of General Microbiology,* Vol. 31, 1963, p. 365.
[25] Booth, G. H., *Journal of Applied Bacteriology,* Vol. 27, 1964, p. 174.
[26] Purkiss, B. E. in *Microbial Aspect of Metallurgy,* 1st Ed., J. D. A. Miller, ed., Medical and Technical Publishing Company, Aylesbury, England, 1971, p. 107.
[27] Soimajarvi, J., Pursiainen, M., and Korhonen, J., *European Journal of Applied Microbiology and Biotechnology,* Vol. 5, 1978, p. 87.
[28] Kobrin, G., *Materials Performance,* Vol. 15, 1976, p. 38.
[29] Temperley, T. G., *Corrosion Science,* Vol. 5, 1965, p. 581.
[30] Patterson, W. S., *Transactions of the North East Coast Institution of Engineers and Shipbuilders,* Vol. 68, 1951, p. 93.
[31] NACE *Manual of the Role of Bacteria in the Corrosion of Oil Field Equipment,* Book Item Number 52106, National Association of Corrosion Engineers, Houston, TX.
[32] Hladky, K., Callow, L. M., and Dawson, J. L., *British Corrosion Journal,* Vol. 15, 1980, p. 20.
[33] Callow, L. M., Richardson, J. A., and Dawson, J. L., *British Corrosion Journal,* Vol. 11, 1976, p. 132.
[34] Rickard, D. T., *Geology,* Vol. 20, 1969, p. 67.
[35] King, R. A., Miller, J. D. A., and Wakerley, D. S., *British Corrosion Journal,* Vol. 8, 1973, p. 89.

DISCUSSION

C. K. Tiller (discussion)—1. This is obviously a useful call for determining the performance of materials under truly bad conditions. However in the field cyclic effects occur. Can the author comment on whether he intends to adopt anaerobic/aerobic cycles, and if so what guidelines will be adopted to ensure the environment represents field conditions for soils, marine systems and possibly cooling water systems.

2. Rules of corrosion induced by bacteria will to some extent be governed by temperature. It is interesting to vary the temperature and carry out experiments at between 20° and 60°C, respectively.

3. Does the author consider whether the cell could be used to study our ecosystem rather than a specific biological utility?

J. F. D. Stott, B. S. Skerry, and R. A. King (authors' closure)—1. The method is potentially very versatile and could easily be adapted to form an anaerobic/aerobic cycling test in order to simulate the type of conditions that may prevail in polluted estuaries for example. The difficulty is not in adapting the test to suit such conditions but in obtaining accurate information about such cycles. I agree completely that cycling conditions where hydrogen sulfide is initially formed and is later oxidized to elemental sulfur during an aerobic phase provide a potentially extremely corrosive environment for many materials, particularly if that sulfur is later oxidized to sulfuric acid by sulfur-oxidizing bacteria of the genus *Thiobacillus*.

2. Indeed, we try to adopt the test to within ±2°C of the environment under consideration, hence the inclusion of the thermostatically controlled heater (see *N* in Fig. 2). Currently we are running tests on pipeline materials for use in the North Sea Oil Industry. These have been immersed in marine mud at a temperature of 4°C to simulate field conditions, and we are using a cold-room for this exposure test.

3. Yes it could, but it is our experience that anaerobic soils and waters containing biogenic hydrogen sulfide are remarkably stable ecosystems once an equilibrium has been reached.

For aerobic/anaerobic cycling systems this is not the case, and it is possible to get, for instance, oil degradation by *Pseudomonads* in an aerobic phase followed by sulfate reduction utilizing carbon sources from oil degradation and finally aerobic oxidation of sulfide to sulfur and consequent biooxidation to sulfuric acid. This type of cycle occurs in offshore oil storage tanks, concrete legs of offshore platforms, and even in sewer pipes. It is worth mentioning that we have built other models to simulate such ecosystems, which we call "sulfureta" in which these events occur at different depths in a vessel rather than in the same place at different times.

[1]National Physical Laboratory, Teddington, Middlesex TW11 OLW, England.

Edward C. Hill,[1] Graham C. Hill,[1] and David A. Robbins[1]

Comparison and Control of Microbial Spoilage of Metal-Working Fluids and Associated Corrosion

REFERENCE: Hill, E. C., Hill, G. C., and Robbins, D. A., "**Comparison and Control of Microbial Spoilage of Metal-Working Fluids and Associated Corrosion,**" *The Use of Synthetic Environments for Corrosion Testing, ASTM STP 970,* P. E. Francis and T. S. Lee, Eds., American Society for Testing and Materials, Philadelphia, 1988, pp. 112–119.

ABSTRACT: The microbial environment in the metal-working fluid in a machine tool is complex, as the dissolved oxygen content, Eh, and temperature differ at different locations in the system. Thus different species of microorganisms flourish or decline at different points in the fluid depending on their individual needs and preferences. Experiments using simple shake flasks or stagnant flasks in the laboratory can produce misleading results. A recirculation apparatus is described that allows both aerobic and anaerobic microbial growth to occur. It can be monitored and manipulated (for example by reinoculation or the addition of biocides) over a long period. Metal swarf is added to the apparatus and its corrosion can be rated visually. The corrosiveness of the fluid can also be determined at intervals using the standard metal chip test.

KEY WORDS: sulfide, metal-working fluids, preservatives, biocides

Microbial Ecology of Metal-Working Fluids Systems

Microbial spoilage of metal-working fluids is well documented and has recently been reviewed [1,2]. The actual microbiology of the fluid is usually determined by snatch samples of the flowing fluid at the work-piece. However, Hill [3] has pointed out that such samples are poor indicators of the types, numbers, and relevance of the organisms present. Machine tools exhibit at least three different ecological niches:

(1) The warm, aerated fluid at the work-piece,
(2) The fluid in the pipes, which passes continuously over active biofilm on the pipe walls, and
(3) The fluid in the sump, which exhibits marked gradients of temperature, oxygen concentration, Eh, and possibly, nutrients and pH.

Sediments and surface scums are additional, different environments. Microbes in biofilms and sediments appear only in small numbers in the circulating fluid samples and their significance may be missed.

Laboratory Simulators

Attempts to simulate metal-working fluid spoilage in the laboratory using simple shake flasks or permanently static flasks are certain to yield misleading results. A model must include an

[1]Director, laboratory manager, and scientific officer, respectively, Unit M22, Cardiff Workshops, Lewis Road, East Mooris, Cardiff CF1 5EG, United Kingdom.

aeration phase, a piped phase, an adequately sized sump, and an appropriate metal swarf. Such a model is described by Hill [4] and is illustrated in Fig. 1. Recirculation (and some aeration) is achieved by the air-lift device, but in later models, individual positive displacement pumps are used for each unit. Replication of the units allows various parameters or treatments to be compared. The organisms that develop and the environment established relate closely to those in machine tool coolant systems. Both aerobic and anaerobic bacteria can flourish.

Microbial Corrosion

Metal corrosion is influenced by microorganisms and involves a number of mechanisms, which may occur successively or simultaneously in almost any combination. These are:

(1) microbial degradation of anti-corrosive additives,

(2) corrosion from microbial products such as acids, ammonia, carbon dioxide, (solubilizes insoluble oxides and hydroxides) and sulfide,

(3) corrosion pitting from the establishment of differential aeration and concentration cells,

(4) interference (by competing for oxygen) with protective oxide films on aluminum and stainless steel,

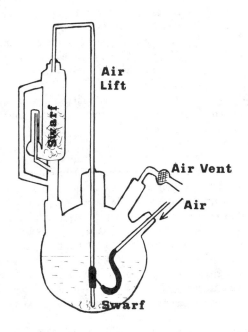

Recirculation Rig for

Coolant Testing

FIG. 1—*Metal-working fluid recirculation rig. Fluid is raised from the "sump" (and aerated) by air bubbles and passes into a side vessel containing swarf (representing the machine tool). This side vessel fills up and dumps back into the sump by a siphon action.*

(5) depolarization of the metal surface by removal of cathodic hydrogen, and
(6) hydrogen embrittlement.

The mechanisms have been reviewed by Miller [5], Hamilton [6], and others.

Sulfide Generation and Associated Corrosion

Hamilton [6] emphasized the devastating role of sulfide produced by anaerobic sulfate reducing bacteria. Hill et al. [7] pointed out that the term "sulfate reducing bacteria" (SRB) is misleading, as sulfide produced in metal-working fluids is more likely to be derived from the degradation of sulfonates, sulfur, or sulfurized oils and fats and hence the term "sulfide generating bacteria" (SGB) is preferred. These organisms are never present alone but are part of consortia established in biofilms and sediments. Any model system must allow these consortia to develop. The aerobic bacteria in the consortia produce essential nutrients for SGB and deplete oxygen, which is inhibitory to SGB.

Provided that the correct microbial ecologies are simulated, valid conclusions may be derived from corrosion studies. Corrosion may be determined merely by observing the metal swarf, by incorporating appropriate metal test coupons, or by determining the corrosiveness of the fluid.

Control of Sulfide Generation

Sulfide generation is not only important because of the aggressive corrosion it promotes, but also because of the potential (and sometimes deadly) health hazard it creates [7].

A number of possibilities exist to control the organisms that generate sulfide. They have recently been described by Hill et al. [7]. In brief, they include aeration (of transient benefit only), swarf removal, pasteurization, the addition of broad spectrum biocides, and the addition of chemicals with a specific activity against SGB.

Broad spectrum biocides are active against the microbial consortia as a whole and have been available for many years. The factors that control their useful life have frequently been stated [8]. As they all are somewhat toxic to man, the dosage must be at that critical level that affects microorganisms but not the work force. Unfortunately, the effective concentration in metalworking fluids falls progressively; in practice it is replenished either by an arbitrary top-up or by adding biocide when reinfection has been detected with Dip-slides. The former results in poor microbial control if the addition is too small, and the possibility of human or environmental toxicity if the addition is too great. Adding biocide in response to a Dip-slide result is always one step too late, as it demonstrates that infection has returned rather than that preservation has become inadequate. A simple device, the Echa Biocide Monitor (Echa Products, Cardiff, UK) now allows preservative/biocide concentrations to be monitored so that adjustments can be made to maintain a target concentration (Fig. 2) known to control the microbial consortium. Some species of SGB form resistant spores that are not susceptible to commonly used antimicrobial chemicals, but continuously suppressing less resistant members of the consortium stops nutrient synthesis and oxygen depletion. Thus, surviving SGB cannot flourish and promote sulfide corrosion.

An alternative is to ignore the consortia as a whole and instead use an anti-microbial chemical that acts only against anaerobic microorganisms such as SGB. The toxicity to man is then likely to be insignificant. Such a chemical is now available (Biocide A). Its effectiveness compared with a broad spectrum biocide (Biocide B) can be evaluated in model systems.

In the experiments with the models described here, Biocide B was used in one model with dosage regulated by the use of the Echa Biocide Monitor, and in another model, Biocide A was used. An alternative strategy was previously described in which the control of sulfide generation was attempted using simple gel models to investigate the optimum conditions for its generation,

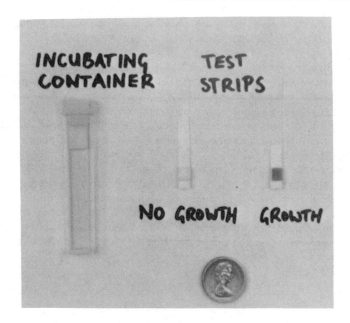

FIG. 2—*Biocide monitor strips. A 22 mm-diameter coin is shown for size comparison. Before incubation, the left-hand strip was dipped into a fluid containing sufficient biocide to inhibit the sensitive* Bacillus *on the strip. Before incubation, the strip on the right was dipped into fluid containing no biocide; the* Bacillus *grew and the growth indicating dye turned red.*

which proved to be low Eh, the absence of oxygen, the presence of heterotrophic bacteria, and a high swarf content, and these parameters were then manipulated [7, 9].

Methods

Three metal-working fluid recirculation rigs (described previously) were set up, each containing 3 L of an oil emulsion susceptible to sulfide spoilage. They were each inoculated with 10 ml of previously spoiled fluid plus other known sources of oil spoilage bacteria, yeasts, and fungi. The flasks and their side-arms contained metal swarf. The pumps were started and adjusted to flow rates of just over a liter per hour (varied between 1000 and 1200 ml/h). Each flask was reinoculated with 0.1 ml twice a week. The experiment ran for 74 days. One flask remained as a control, another was dosed with Biocide B at the start and then as indicated by frequent assays with Echa Biocide Monitor Strips, and the third flask was dosed with Biocide A initially and then as indicated by the Monitor Strips.

The Monitor Strips (Fig. 2) carry a small pad impregnated with spores of a sensitive *Bacillus*, nutrients, and a growth indicating dye. They are dipped in the fluid of interest for a few seconds to activate the nutrients and the microbial spores and are then incubated overnight at about 35°C. If the *Bacillus* germinates and grows, the pad turns red. If it is inhibited by the preservative/biocide, it remains white. The minimum inhibitory concentration (MIC) for any biocide can be determined very simply by testing known concentrations with the strips. For example, it is 50 ppm for Biocide B and 80 ppm for Biocide A. Hence, when testing fluids containing biocides, a range of dilutions are made and the results of strip tests noted. That dilution that just inhibits the *Bacillus* must contain the MIC. A simple calculation then gives the concentration in the undiluted fluid. Adjustments to this concentration can then be made to bring it back to the original target level. In this particular case, as far as possible, biocide concentrations were assayed twice weekly.

Microbial contamination was determined by conventional dilution plating techniques for heterotrophic bacteria and by streaking loopfuls of fluid onto Malt Extract Agar (Difco 0112-17.6) to recover yeasts and fungi semiquantitatively. A determination of the presence of SGB and nitrite reducing bacteria was made by using some simple new tube tests, the Sig Tests (Echa Products, Cardiff, UK). Their use has recently been described [7]. In essence, about 2 ml of the sample is transferred to the surface of a porous gel in a small stoppered tube and the tube is then incubated at an appropriate temperature. One gel turns black (Sig Sulfide Test) in the presence of SGB and the rate at which this occurs is proportional to the amount of SGB present. Heavy contamination can be detected overnight. Another gel (Sig Nitrite Test) semiquantitatively detects the presence of nitrite reducing bacteria, as this has significance for fluids which contain nitrite as a corrosion inhibitor. The stoppered tube containing the gel-plus-sample is incubated. If cavities appear in the gel, this indicates that nitrite has been reduced to nitrogen gas (Fig. 3).

Microbial assays were conducted twice weekly. Depletions of fluid were noted and adjusted as necessary.

FIG. 3—*Sig Nitrite test. Unused tube on left. The tube on the right had a few milliliters of infected fluid added before incubation. The bacteria present permeated the gel and converted the nitrite present into visible bubbles of nitrogen gas.*

Results

Large numbers of results were generated and, as this experiment was principally designed to demonstrate the use of model rigs, only a few examples are quoted (Tables 1 and 2). The activity of SGB and nitrite reducing bacteria was completely suppressed by the controlled use of Biocide B. However, the microbial ecology then shifted and fungal slimes developed. After cleaning this rig, it was restarted with a higher concentration of Biocide B (1000 ppm), which adequately controlled fungi and yeasts. The use of Biocide A almost completely suppressed SGB and largely suppressed nitrite reducing bacteria, but very large numbers of heterotrophic bacteria flourished.

Discussion

The continuous assaying of biocide concentration proved possible with the Biocide Monitor Strips, enabling precise top-up additions. However, the initial target level of Biocide B chosen was not adequate to control fungal growth in this particular fluid and the system fouled so badly with fungal slime that it could no longer be operated. It is interesting to note that before the system failed only small numbers of colony forming fungal units were detected, thus emphasizing the importance of visual and/or microscopic examination whenever any level of fungal contamination is detected. After cleaning, the biocide level was raised to 1000 ppm and fungi were thereafter no longer a problem.

SGB never flourished in this system. This could have been due to the direct action of Biocide B against SGB or the substantial inhibition of the heterotrophic microbes, which stimulate SGB

TABLE 1—*Microbial results from metal-working fluids' recirculation rigs; sample taken from flowing fluid.*

| | Number of Colony Forming, units/mL | | | |
Day	Heterotrophs	Yeasts and Fungi	SGB	Nitrite Reducing Bacteria
		CONTROL		
3	2.85×10^6	200	+ +	+ + + +
11	0.84×10^6	150	+	+ + +
14	6.4×10^6	700	+ +	not done
34	3.9×10^6	250	+ + +	+ + +
59	3.19×10^6	25 500	+ + + +	not done
		BIOCIDE B		
3	170	810	−	−
11	140	100	−	−
14	30	80	not done	not done
34	146 000	250	−	−
41		system cleaned to remove fungal slime		
59	38 800	−	−	−
		BIOCIDE A		
3	4.32×10^6	−	+	+ +
11	2.54×10^6	−	−	+
14	6.9×10^6	50	−	not done
34	30.0×10^6	−	+	+
59	9.3×10^6	not done	−	not done

NOTE: − is no reaction and + + + + is very severe reaction.

TABLE 2—*Biocide concentrations in metal-working fluid flowing in recirculation rigs.*

Day	BIOCIDE B, ppm	BIOCIDE A, ppm
3	>450	<70
11	>450	>100
34	<300	180
41	system cleaned	system not cleaned
59	>500	>190

NOTE: Biocide B concentration 500 ppm at start; adjusted on days 10, 20, 25, and 38; raised to 1000 ppm after day 41; adjusted on days 43 and 45. Biocide A concentration adjusted on days 6 and 46.

by removing oxygen, reducing Eh, and producing organic acids. Nitrite reducing bacteria were never detected in significant numbers in the Biocide B treated system. The system treated with Biocide A was virtually free of SGB, but heterotrophic bacteria flourished in greater numbers than in the control system.

Reduction of nitrite is an anaerobic process and as Biocide A has activity only against anaerobes, it was anticipated that nitrite reducing bacteria would be inhibited. This was verified, but as these bacteria are also able to grow aerobically without using their nitrite reducing mechanism, complete suppression of nitrite reducers with Biocide A is unlikely. However, from a practical viewpoint it is useful to be able to suppress both nitrite reduction and sulfide generation with a virtually nontoxic antimicrobial chemical. The Biocide Monitor Strips were somewhat difficult to interpret in the Biocide A-treated system, as the large numbers of heterotrophic bacteria masked the growth/inhibition of the sensitive *Bacillus* on the strips.

In the control system, large numbers of heterotrophic bacteria, SGB, and nitrite reducing bacteria flourished. Although yeasts and fungi were present, they did not foul the system to the same extent as in the Biocide B-treated system.

Conclusion

The models described simulated spoilage situations in metal-working fluids and were useful in comparing two biocides with widely different activities and in evaluating simple test kits. The Biocide Monitor Strips were useful tools for monitoring and adjusting biocide concentrations. The Sig Sulfide and Sig Nitrite tests detected two important groups of bacteria significant in corrosion. The controlled treatment of metal-working fluid with the broad-spectrum Biocide B suppressed all microorganisms at 1000 ppm but not at 500 ppm. The use of Biocide A at concentrations above 100 ppm controlled sulfide generation and nitrite reduction.

References

[1] Hill, E. C., in *Petroleum Microbiology,* R. M. Atlas, Ed., Macmillan Publishing Co., NY, 1984, Chapter 15, pp. 579–617.
[2] Rossmore, H. W., in *Comprehensive Biotechnology: The Principles, Applications and Regulations of Biotechnology in Industry, Agriculture, and Medicine,* M. Moo-Young, C. L. Cooney, and A. E. Humphrey, Eds., Oxford Pergammon Press, 1986, Chapter 14, pp. 249–269.
[3] Hill, E. C., in *Monitoring and Maintenance of Aqueous Metal-working Fluids,* K. W. A. Chater and E. C. Hill, Eds., John Wiley & Sons, Chichester, UK, 1984, pp. 97–112.
[4] Hill, E. C., *Tribology International,* Vol. 10, No. 1, 1977, pp. 49–54.
[5] Miller, J. D. A., in *Microbial Biodeterioration,* A. H. Rose, Ed., Academic Press, London, 1981, pp. 150–202.

[6] Hamilton, W. A., *Annual Review of Microbiology,* Vol. 39, 1985, pp. 195–217.
[7] Hill, E. C., Hill, G. C., and Robbins, D. A., in *Additives for Lubricants and Operational Fluids,* W. J. Bartz, Ed., Technische Akademie Esslingen, F.D.R., 1986, Vol. II, 11.10-1 to 11.10-7.
[8] *Biocides in the Oil Industry,* Institute of Petroleum, London, 1982, p. 87.
[9] Abu Shaqra, Q. M. and Hill, E. C., *Tribology International,* Vol. 17, No. 1, 1984, pp. 31–34.

William L. Silence,[1] *Stephen M. Corey,*[1] *and Juri Kolts*[2]

Alloy Ranking for Corrosion Resistance Laboratory Tests Versus Field Exposures

REFERENCE: Silence, W. L., Corey, S. M., and Kolts, J., **"Alloy Ranking for Corrosion Resistance Laboratory Tests Versus Field Exposures,"** *The Use of Synthetic Environments for Corrosion Testing, ASTM STP 970,* P. E. Francis and T. S. Lee, Eds., American Society for Testing and Materials, Philadelphia, 1988, pp. 120–131.

ABSTRACT: This paper reviews the relative ranking of the corrosion resistance of stainless steels and nickel-base alloys. Laboratory rankings for pitting resistance were determined by measuring critical pitting temperatures in chloride-containing solutions with ferric ion additions. The rankings were compared to field test data obtained from locations in coal-fired power generating flue gas desulfurization systems and in various applications in the pulp and paper industry. Excellent correlation was found. Laboratory test results for measuring uniform corrosion resistance do not correlate well with uniform corrosion rates from field racks retrieved from a wide variety of industrial applications.

KEY WORDS: pitting, field test, laboratory tests, pulp, paper, flue gas desulfurization (FGD)

Laboratory tests are used routinely to rank corrosion resistant alloys. These rankings are used both to predict field performance and also to evaluate newly developed alloys. Generally, the corrosion resistance of new alloys is designed for optimized performance in one or a number of laboratory environments. The time required for field exposures is usually long; therefore laboratory tests are generally used instead of field tests in the development of corrosion resistant alloys.

In materials selection, best-guess choices are often made on the basis of a number of laboratory tests thought most pertinent to the particular application. This paper will try to determine whether the use of these laboratory tests is in fact a sound basis for selecting corrosion resistant alloys.

Two types of corrosion phenomenon will be examined in this paper—pitting corrosion and general corrosion. The most common method of assessing pitting corrosion resistance used in this laboratory involves determining the critical temperatures for pitting in various environments. Common environments are ferric chloride, modified yellow death, yellow death, and green death solutions. The oxidizing agent in all of these environments is the ferric ion. These solutions are defined later.

---

[1]Research associate and corrosion engineer, respectively, Haynes International Inc., P.O. Box 9013, Kokomo, IN 46902-9013.

[2]Research associate, Conoco Inc., Ponca City, OK.

®FERRALIUM is a registered trademark of Langley Alloys Ltd.

®HASTELLOY is a registered trademark of Haynes International, Inc.

®TEFLON is a registered trademark of E.I. duPont de Nemours Co.

®20 Cb-3 (see Table 2).

A number of tests are used to assess the general corrosion characteristics of alloys. These tests include corrosion resistance in boiling solutions of 10% sulfuric acid, 1% hydrochloric acid, 65% nitric acid, 55% phosphoric acid, 85% phosphoric acid, and the oxidizing acid environment described in ASTM G 28-85, Practice A. Depending upon the oxidizing nature of the environment, one or more of these laboratory tests are used to estimate the corrosion resistance of alloys. For specific applications, data in those environments most closely simulating the actual field condition is obtained. This investigation shows that there is an excellent correlation between field tests and the prediction of localized corrosion on the basis of laboratory tests; however, the selection of alloys on the basis of uniform corrosion laboratory tests has to be done with discretion.

Materials and Procedure

All of the corrosion tests performed in this laboratory are done according to written instructions prepared for these tests. In this manner, tests performed many years apart have sufficient reproducibility for valid comparisons. While all of the tests performed are standardized within the laboratory, they may not be standard among different laboratories. Laboratory corrosion tests at or below boiling temperatures were conducted in separate, 1-L Erlenmeyer flasks. HASTELLOY® alloy C-276 autoclaves lined with TEFLON® were used to conduct tests above atmospheric boiling temperatures.

Tests for measuring resistance to uniform corrosion are the simplest. Coupons are weighed and placed into the environment after it has reached the desired temperature. After the test, the coupons are removed, cleaned, and weighed according to ASTM G 1. The weight difference is converted into a uniform thickness loss (e.g., mils/year or mm/year). ASTM G 31 was followed as a guide for laboratory immersion tests. Table 1 lists the standard test methods used in the laboratory evaluations.

ASTM A 262 and ASTM G 28-85, Practice A were used to determine corrosion resistance in an oxidizing environment.

To determine critical pitting temperatures, 24-h tests were performed. Environment temperatures were raised in 5°C (9°F) increments until the material pitted. ASTM G 46 was used as a guide for selecting procedures to identify and examine pits, and for determining the extent of pitting.

Media used for immersion pitting tests were as follows:

(1) green death solution (11.5% sulfuric acid (H_2SO_4) + 1.2% hydrochloric acid (HCl) + 1% cupric chloride ($CuCl_2$) + 1% ferric chloride ($FeCl_3$),

TABLE 1—*Standard test methods used in this evaluation.*

ASTM Designation	Full Name of Test
A 262	Detecting Susceptibility to Intergranular Attack in Stainless Steels
G 1	Recommend Practice for Preparing, Cleaning, and Evaluating Corrosion Test Specimens
G 28	Detecting Susceptibility to Intergranular Attack in Wrought Nickel-Rich, Chromium-Bearing Alloys
G 31	Laboratory Immersion Corrosion Testing of Metals
G 46	Examination and Evaluation of Pitting Corrosion
G 48	Pitting and Crevice Corrosion Resistance of Stainless Steels and Related Alloys by the Use of Ferric Chloride Solutions

(2) yellow death solution (4% sodium chloride (NaCl) + 0.01 M HCl + 0.1% $Fe_2(SO_4)_3$,
(3) 6% ferric chloride, and
(4) modified yellow death solution (4% NaCl + 0.01 M H_2SO_4 + 0.1% ferric sulfate $(Fe_2(SO_4)_3)$.

Tests were performed using the same procedure as for measuring uniform corrosion rates.

Alloys

Commercially available alloys, some of which were welded, were used in this investigation. Nominal chemical compositions for the several alloys are listed in Table 2.

Corrosion Test Racks

This investigation examined over 100 corrosion test racks which were exposed in a number of field test sites. Many of these test sites had produced problems in existing alloys; therefore, corrosion test racks were installed. However, in some cases where the component construction material exhibited severe corrosion or pitting attack, the sample of the same alloy on the test

TABLE 2—*Nominal chemical compositions.*

| Alloy | Weight Percent | | | | | |
	Cr	Mo	Ni	Fe	C	Other
B-2	1.0^a	27.0	bal	2.0^a	0.01^a	
C-22	22.0	13.0	bal	4.0	0.015^a	2.0 W
C-276	15.5	16.0	bal	5.5	0.01^a	4.0 W
C-4	16.0	15.5	bal	3.0^a	0.01^a	0.7 Ti
625	21.5	9.0	bal	5.0^a	0.10^a	3.5 Cb + Ta
						0.4 Al, 0.4 Ti
H	22.0	9.0	bal	18.0	0.01^a	2.0 W
G-30	29.0	5.0	bal	15.0	0.03^a	0.7 Cb, 1.7 Cu
						2.5 W
G-3	22.0	7.0	bal	19.5	0.015^a	0.5 Cb + Ta,
						2.0 Cu
G	22.0	6.5	bal	19.5	0.05^a	2.0 Cb + Ta,
						2.0 Cu
718	19.0	3.0	bal	18.5	0.05^a	1.0 Ti, 0.1 Ta
825	21.5	3.0	bal	23.0	0.05^a	1.0 Ti, 2.0 Cu
400	...	...	bal	1.5	0.15^a	30.0 Cu
800	21.0	...	32.0	bal	0.10^a	0.4 Al, 0.4 Ti
M-532	22.0	5.0	26.0	bal	0.025	
700	20.0	4.5	25.0	bal	0.03	
904L	19.5	4.2	25.0	bal	0.02	1.5 Cu
28	27.0	3.5	31.0	bal	0.02^a	1.0 Cu
Al-6X	20.0	6.25	24.5	bal	0.025	
20CB-3®	20.0	2.5	34.0	bal	0.06^a	3.5 Cu
255	25.5	3.0	5.5	bal	0.04^a	0.17 N, 1.75 Cu
29-4-2	29.5	4.0	2.0	bal	0.006	
317LM	19.0	4.25	14.0	bal	0.03^a	
317L	19.0	3.5	13.0	bal	0.03^a	
316L	17.0	2.5	12.0	bal	0.03^a	
304L	19.0	...	10.0	bal	0.03^a	
Ti	...	...	...	...	0.05	bal Ti
Zr702	...	...	...	...	0.05	bal Zr

^{a}Maximum.

rack showed no apparent corrosion. A number of reasons may be given for this. Corrosion often occurs during conditions of upsets when the corrosivity of environments is more severe than at ordinary times. In addition, operating procedures for given equipment may also change with time, changing the corrosivity of the environment.

Although a large number of corrosion racks were examined, results from only 60 racks are reported in this examination. In the evaluation of localized corrosion, two applications were primarily considered; flue gas desulfurization (FGD) systems in coal-fired utilities, and various applications in the pulp and paper industry. In the evaluation of uniform corrosion, the field test racks were exposed to typical conditions for a large variety of industries.

Results and Discussion

Table 3 shows the relative ranking of alloys in four laboratory pitting environments. With few exceptions, the relative rankings of alloys are the same. Typically, those alloys with higher chromium, molybdenum, tungsten, and nitrogen content show the best resistance to localized corrosion. The corrosivity of all environments was accelerated by the presence of ferric ions. The original selection of these environments had been as "unbuffered" redox systems. For example, the presence of corrosion products that introduce ferrous ions into the system will change the redox potential of the solution significantly [1]. Consequently, the redox potential will drift with time as corrosion occurs. Alloys containing large amounts of iron, which are more susceptible to general corrosion, will experience a less severe environment because the corrosion products contain ferrous ions, which reduce the redox potential of the solution.

Field test results cannot provide a critical pitting temperature because temperatures cannot be varied at successively increasing levels. Field test coupons only provide information on whether a particular alloy has pitted. However, this information can be valuable in rating alloys in the field if a large number of test environments are considered [2-4]. Tables 4 to 7 demonstrate the incidence of pitting in 48 different environments, either FGD systems or pulp and paper industry environments. From the tables, a rating of the alloys can be established by comparing the fraction of environments where pitting has not occurred. If this is done, then Tables 4 to 7 will provide a relative ranking for the incidence of localized corrosion.

TABLE 3—*Relative ranking of alloys in laboratory environments, critical pitting temperature* °C.

Alloy	Mod. Yellow Death	Ferric Chloride	Yellow Death	Green Death
C-22	...	>102	>150	120
C-276	...	>102	150	110
C-4	...	>102	140	90
625	...	...	101	75
G-3	...	...	80	50
M-532	55	...	60	...
255	55	45	50	...
904L	55	...	45	...
317LM	35	...	35	...
20CB-3⊗ᵃ	30	<20	20	<20
825	25	...	25	<20
317L	25	...	...	<20
316L	20	15	20	<20
304L	15	<15	...	<20

ᵃ20CB-3 is a registered trademark of Carpenter Technology Corporation.

TABLE 4—The incidence of pitting corrosion in FGD[a] field test exposure environments.

	Lab Pitting	Field Test No.																															
	Test Rating	1	2	3	4	5	6	7	8	9	10	11	12	13	14	15	16	17	18	19	20	21	22	23	24	25	26	27	28	29	30	31	32
C-22	—	—	n	n	—	—	—	—	—	—	—	—	—	—	—	—	—	—	—	—	—	—	—	—	—	—	—	—	—	—	—	—	—
C-276	n	n	n	n	n	n	n	n	n	n	n	n	n	n	n	n	n	n	n	n	n	n	n	n	n	—	n	n	n	n	n	n	n
C-4	y	y	—	—	—	—	—	—	—	n	—	n	n	n	n	—	—	—	n	n	n	n	n	—	n	—	n	n	n	n	n	n	n
625	y	y	y	y	y	y	y	y	y	y	n	n	n	n	n	—	—	—	n	n	n	n	n	—	n	n	n	n	n	n	n	n	n
G-3	y	y	y	y	y	y	y	y	y	y	y	y	—	—	—	—	—	—	—	n	—	—	—	—	—	—	—	—	—	—	—	—	—
G	y	y	—	—	—	y	y	y	y	y	y	—	y	y	y	y	n	n	n	n	n	n	n	n	n	n	n	n	n	n	n	n	n
M-532	y	y	y	—	y	y	—	y	y	y	y	y	y	y	y	y	y	y	y	y	y	y	y	—	n	—	n	n	n	n	n	n	n
255	y	y	y	y	y	y	—	y	y	y	y	y	y	y	y	y	y	y	n	y	y	y	n	y	y	—	n	n	n	n	n	n	n
700	y	y	—	—	—	—	—	y	y	y	y	y	y	n	y	—	—	—	y	n	n	y	y	y	y	y	y	y	n	n	n	n	n
904L	y	y	y	y	—	—	y	y	y	—	—	—	y	n	y	—	—	—	n	y	y	y	n	—	y	—	n	y	y	n	n	n	n
317LM	—	—	y	y	—	—	—	—	—	—	—	—	y	y	y	—	—	—	y	—	y	y	y	—	n	y	n	y	n	n	n	n	n
825	—	—	y	y	—	—	y	—	—	y	y	—	y	y	y	—	—	y	y	y	n	y	y	y	y	n	y	y	y	y	y	y	y
317L	y	y	y	y	—	—	y	y	y	y	y	y	y	y	y	y	y	y	y	y	y	y	y	y	y	—	y	y	y	y	y	y	y
316L	y	y	y	y	y	y	y	y	—	y	—	—	y	y	y	—	—	y	y	y	y	y	y	y	y	y	y	y	y	y	y	y	y
304L	y	y	—	—	—	—	—	—	—	—	—	—	y	y	y	—	—	y	y	y	y	y	y	y	y	—	y	y	y	y	y	y	y

[a] No. 4 was run in the scrubber on a copper smelter. All others were conducted in scrubbers on electric utility coal-fired boilers.
[b] y = pitted, n = not pitted, — = not tested.

TABLE 5—*Description of FGD field tests.*

Test	Location of Test	Test Conditions
1	stack bottom, wet limestone process.	approximately 14 months 29% of time on full bypass at 149°C, balance of time scrubbed gas was exhausted at 155°F.
2 to 3	outlet duct, gas mixing zone, wet limestone process.	11 months, ~93°C, scrubbed gas with bypass gas reheat.
4	copper smelter flue gas scrubber.	wet flue gas, 19% SO_3, 0.5% SO_2, trace H_2SO_4, low Cl^-, some H_2O, balance N_2, 149°C, 72 days.
5	outlet duct, dual alkali system.	wet flue gas, 38% of time on scrubbed gas 71°C, balance of time on full bypass 149°C, 4% S coal, 800 ppm Cl^-, 223 days.
6	presaturator (inlet) wet limestone system.	168 days, wet/dry interface area, 66 to 149°C.
7	stack breaching duct, wet limestone system.	12 months, 71°C, scrubbed wet gas.
8	outlet duct, gas mixing zone, wet limestone system.	17 months, 93 to 104°C, scrubbed gas with bypass gas reheat.
9	same as No. 5 except in test 337 days in a different duct with scrubbed gas 25% of the time and full bypass the remainder.	. . .
10 to 11	outlet ducts turning vanes, wet lime process system.	15 months, 10–20% bypass gas is added to the scrubbed gas for reheat, 54–93°C, slurry pH 1 to 2, 3000 ppm Cl^-.
12	absorber outlet above mist eliminators, sodium carbonate system.	saturated wet scrubbed gas, 440 days, 49°C.
13	presaturate, 3 ft (0.91 m) down-stream from inlet damper, sodium carbonate system.	recycle liquor liquid spray, 440 days, 116°C, pH 7.3.
14	outlet duct plenum near stack entrance, sodium carbonate system.	wet scrubbed gases, 237 days, 54°C, coal analysis: 0.56% S, 100 ppm Cl^-.
15	entrance to venturi gas stream, Wellman-Lord system.	flue gas with some liquid spray, 127 days, 135°C, 2500 ppm SO_2, pH 1.5 to 2, 1200 ppm Cl^-.
16	inlet duct to venturi Wellman-Lord system.	dry flue gas, 127 days, 135°C 75% N_2, 12% CO_2, 5% O_2, 7% H_2O plus SO_2 and SO_3.
17	after mist eliminator, Wellman-Lord system.	scrubbed flue gas saturated with water, 127 days, 52°C, pH 1.5 to 2, 1200 ppm Cl^-.
18	same as No. 14, except exposure time was 193 days.	. . .
19	outlet duct between reheat area and stack, Wellman-Lord system.	27.6 months, scrubbed flue gas with 100–700 ppm SO_2, without reheat >50% of time (52°C), with reheat >50% of time (79°C). Coal contained 0.8% S.
20	same as No. 13, except pH not reported.	
21	same as No. 13, except pH 7.4.	
22	same as No. 12.	
23	outlet plenum floor near stack entrance, sodium carbonate system.	wet scrubbed gases, 193 days, coal analysis: 0.56% S, 100 ppm Cl^-.
24	same as No. 18.	
25	same as No. 23, except at a greater distance from the stack entrance or immediately in front of the absorber outlet.	
26	same as No. 12, except the pH was reported as 1.5 to 4.	
27	same as No. 21.	

TABLE 5—*Continued.*

Test	Location of Test	Test Conditions
28	presaturator, 15 ft (4.6 m) downstream of inlet damper, sodium carbonate system.	440 days, 93°C, in recycle liquor, constant liquid spray.
29	same as No. 28	
30	same as No. 12.	
31	same as No. 12.	
32	same as No. 28, except pH reported as 7.8.	

TABLE 6—*Pitting in pulp and paper environments (environments detailed in Table 7).*

Lab Pitting	Field Test No.[a]															
Test Rating	1	2	3	4	5	6	7	8	9	10	11	12	13	14	15	16
C-22	−	−	n	−	−	−	−	−	−	−	−	−	−	−	−	−
C-276	y	n	n	−	n	n	n	n	n	n	n	n	n	−	−	n
C-4	−	−	n	−	−	−	−	−	−	−	−	−	−	−	−	−
625	−	−	n	n	−	−	−	−	n	n	−	−	−	−	−	n
G-3	y	y	y	−	n	n	n	n	−	−	n	n	n	−	−	−
G	−	−	−	−	−	−	−	−	n	n	−	−	−	−	−	n
M-532	−	−	−	−	−	−	−	−	−	−	−	−	n	n	n	−
255	y	y	y	y	y	y	y	y	−	−	n	n	n	n	n	−
700	−	−	−	−	−	−	−	−	y	y	−	−	−	−	−	n
904L	y	−	y	−	y	y	y	−	−	−	y	y	−	−	−	−
317LM	−	−	−	n	−	−	−	y	−	−	−	−	−	−	−	−
825	−	−	−	−	−	−	−	−	n	n	−	−	−	−	−	n
317L	y	−	−	n	y	y	y	−	−	−	y	y	y	y	y	−
316L	y	y	−	n	y	y	y	−	y	y	y	y	y	−	−	y
304L	y	−	−	y	y	y	y	−	−	−	y	y	−	−	−	−

[a]Refer to Table 4 for test descriptions.
[b]y = pitted, n = not pitted, − = not tested.

Notice that HASTELLOY alloys C-22 and C-276 have not shown localized corrosion in any of the environments, except for pulp and paper environment No. 1. As alloys lower on the lists are considered, the localized corrosion resistance of these alloys in laboratory tests is decreased, and pitting corrosion in field tests occurs in more and more environments. There are a few exceptions, but generally the ranking of localized corrosion in the field test environments parallels the ranking of the alloys in the laboratory pitting tests. This occurs in spite of the large variation in redox potentials anticipated in the field environments. Environments in the pulp and paper industry contain redox systems that are as noble as the hypochlorite redox reaction occurring at almost 1 V positive to the saturated calomel electrode. The list also contains redox systems that are controlled by the presence of oxygen or even environments with very low redox potentials. Thus, the ranking of alloys, as determined by critical pitting temperatures in ferric chloride solutions, shows an excellent correlation with field tests in the two industries demonstrated in Tables 4 to 7.

The relative general corrosion resistance of alloys in a number of laboratory environments is shown in Tables 8 to 11. Notice that the relative corrosion resistance of alloys changes consider-

TABLE 7—*Description of pulp and paper field tests.*

Test	Location of Test	Test Conditions
1	H-stage bleach washer	CaOCl bleach solution, 438 days, 100°C, pH 8.9, 300 to 700 ppm Cl^-.
2	inlet duct of a gas cooling tower.	2% SO_2, 204 to 260°C, 86 days, on dry side of a wet/dry interface.
3	blow tower, sodium sulfite mill	sodium sulfite waste liquid plus ~ 10% H_2SO_4, 45 to 82°C ~ 1000 h.
4	Kraft Mill head box sampler.	white water, pH 7.8, 37°C.
5	Kraft Mill, inside a bleached stone ground-wood furnish tank.	sodium hydrosulfite bleach, (high sodium thiosulfate) low pH, 258 days.
6	Kraft Mill, in a Bleach Kraft furnish tank.	CaOCl, with high Cl^- 258 days.
7	Kraft Mill, in a white water chest.	White water, low pH, high Cl^-, high sodium thiosulfate, 258 days.
8	Kraft recovery boiler precipitator.	190 days, environment unknown.
9	vapor dome of a hemidegrader.	7.3% NaOH, pH 14, 170 psi, 185°C, 235 days.
10	red stock washer headbox.	ammonia base spent sulfite liquor, pH 1.8 to 2.0, 200 ppm Cl^-, 65 to 70°C, 246 days.
11 and 12	run in the same plant as Test No. 9, except in different units.	. . .
13	liquid phase H-stage bleach washer.	NaOCl plus trace C/O_2, 52°C, pH 8.8, 964 ppm Cl^-, 158-day test.
14 and 15	red stock washer headbox.	ammonia base spent sulfite liquor, pH 1.5 to 2.2, 200 to 300 ppm Cl^-, 70 to 80°C, 275 days.
16	first stage bleach washer.	Cl, H_2SO_4, HCl, pH 1.7 to 2.0, 35°C, 258 days.

TABLE 8—*Laboratory corrosion test results ASTM G 28 test.*

Alloy	Corrosion Rate, mpy/mm/y
Zr702	1/0.03
G-30	7/0.18
800	8/0.20
20CB-3	10/0.25
G-3	11/0.28
29-4-2	13/0.33
718	14/0.36
Al6X	17/0.43
255	23/0.58
304	25/0.64
316	28/0.71
625	29/0.74
C-22	39/0.99
C-276	250/6.35

ably with different environments and no single rating was found, as in the case for localized corrosion resistance in chloride environments. Increased molybdenum content generally results in increased resistance in environments such as hydrochloric acid and sulfuric acid, while increased chromium content may play a significant role in corrosion resistance in phosphoric acid or other very highly oxidizing acids, such as in the ASTM Method for Detecting Susceptibility to

TABLE 9—*Laboratory corrosion tests results in boiling 10% H_2SO_4.*

Alloy	Corrosion Rate, mpy/mm/y
B-2	2/0.05
C-22	12/0.30
C-276	14/0.36
G-3	23/0.58
255	30/0.77
G-30	32/0.81
625	45/1.14
20CB-3	51/1.30
H	54/1.37
718	350/8.9
316	600/15.2
304	2300/58.4

TABLE 10—*Laboratory corrosion test results, boiling.*

Alloy	85% H_2PO_4 Boiling Corrosion Rate mpy/mm/y	55% H_3PO_4 Boiling Corrosion Rate mpy/mm/y
Ta	0.3/0.01	—/—[a]
B-2	10/0.25	4/0.11
G-3	25/0.64	5/0.13
C-276	35/0.89	6/0.15
20CB-3	35/0.89	0.1/nil
G-30	35/0.89	—/—
825	45/1.14	6/0.15
28	70/1.78	—/—
904L	75/1.91	—/—
625	100/2.54	7.5/0.19
316L	750/1.91	16/0.41
255	1000/25.4	—/—
304	10 000/254	470/11.9

[a]— is not tested.

Intergranular Attack in Wrought Nickel-Rich, Chromium-Bearing Alloys (G 28) test. Table 12 presents the relative corrosion rankings of a number of alloys in samples retrieved from field test programs. An attempt was made to compare the laboratory and field test corrosion data.

In the first two environments of Table 12, which are pickling environments, no clear trends in relative corrosion resistance as related to laboratory tests in H_2SO_4 can be established. For example, alloys which show the best performance in laboratory H_2SO_4 or H_2SO_4 with ferric ions (oxidizing conditions) do not show the best performance in the pickling solutions in the field. Pickling solutions in the field are alternately oxidizing and nonoxidizing. For example, during metal pickling, the environment becomes nonoxidizing. When material is not being

TABLE 11—*Laboratory corrosion test results,*
1% HCl boiling.

Alloy	Corrosion Rate mpy/mm/y
B-2	0.7/0.02
G-30	0.4/0.01
625	1/0.03
C-22	2/0.05
C-276	11/0.28
400	40/1.02
Ti	70/1.78
20CB-3	90/2.29
825	135/3.43
904L	250/6.35
316	920/23.4

pickled, aeration of the environment makes these solutions highly oxidizing. The presence of cupric or ferric ions increases the corrosion rates. Results of field test programs show that FERRALIUM® alloy 255 and HASTELLOY alloy C-276 provided the best performance in copper pickling and steel pickling, respectively.

Five other sulfur-bearing environments are shown in Table 12. These include a sulfur dioxide (SO_2) cooler in the pulp and paper industry, SO_2 condensate in the FGD system, a sodium sulfide evaporator, and a prewasher in an FGD system. Sulfuric acid laboratory tests cannot be used to predict relative corrosion resistance in these environments. Therefore, laboratory uniform corrosion tests should be done with as much information as possible before making alloy recommendations.

The example provided for relative corrosion resistance of alloys in a fungicide process containing HCl does rate alloys relative to laboratory test data. Field experience shows that plant equipment made of HASTELLOY alloy B-2 in HCl service without oxidants and of HASTELLOY alloys C-276 and C-22 performs very well in HCl. The cases provided in Table 11 show that HASTELLOY alloys C-276 and B-2 perform best in HCl-related applications.

The relative ranking of alloys tested in laboratory-grade phosphoric acid and in field test racks in phosphoric acid generally do not show a good correlation. Impurities affect the corrosion rates in phosphoric acid manufacturing, and field test programs are necessary for proper materials selection. A comparison between test racks in Table 12 in various phosphoric acid manufacturing locations and those in Table 10 confirms the above observations.

Conclusions

The relative localized corrosion resistance of stainless steels and nickel-base alloys can generally be predicted using laboratory tests. Pitting temperatures obtained in various environments containing ferric ions correlate well with those obtained in actual field performance for localized corrosion in the FGD systems and pulp and paper mills. Conversely, the uniform corrosion resistance of alloys cannot be easily predicted from laboratory tests and generally, additional information is required before accurate predictions can be made of the relative ranking of alloys in the field. Field test results are especially valuable for proper materials selection in the latter case.

TABLE 12—*Field corrosion test results in various environments, mpy/mm/y.*

Environment	B-2	C-276	625	G/G-3	20CB-3	255	316	304
Steel pickling solution 9% H_2SO_4, 13 day exposure, ~71°C welded	120 / 3.05	1.8 / 0.05	5.0 / 0.13	6.0 / 0.15	25 / 0.64	90 / 2.29	… / …	340 / 8.6
Copper alloy pickling solution, 40% H_2SO_4, 13-day exposure, 110°C	33 / 0.84	56 / 1.41	15 / 0.38	35 / 0.89	… / …	7.0 / 0.18	230 / 5.8	… / …
Coal-fired FGD scrubber outlet duct SO_2/SO_3 condensate, 12 months, 121 to 177°C	… / …	0.1 / nil	2.0 / 0.05	4.5 / 0.11	… / …	20 / 0.51	42 / 1.07	… / …
SO_2 cooler, pulp and paper industry, 12 months	… / …	2.2 / 0.06	0.8 / …	1.6 / 0.02	… / …	21 / 0.53	dissolved	… / …
Phosphate rock digester, 30% P_2O_5, 5 weeks, 93°C	… / …	2.8 / 0.07	0.7 / 0.02	1.0 / 0.03	2.7 / 0.07	0.6 / 0.02	5.5 / 0.14	… / …
Na_2S evaporator 20% Na_2S, 6 months	12 / 0.31	… / …	… / …	35 / 0.89	… / …	30 / 0.76	… / …	… / …
HCl, fungicide environment, 64 days, 115°C, outlet duct SO_2/SO_3 welded samples	<1.0 / <0.03	<1.0 / <0.03	… / …	<4.0 / 0.10	… / …	27 / 0.69	… / …	60 / 1.52
Na polysulfide pH = 7, 5 months, 121 to 177°C	… / …	64 / 1.63	5.8 / 0.15	5.7 / 0.14	… / …	0.6 / 0.02	… / …	3.2 / 0.08
72% P_2O_5 evaporator, 31 days, 182°C	… / …	… / …	… / …	1.3 / 0.03	2.1 / 0.05	2.1 / 0.05	4.9 / 0.13	… / …
HF scrubber, 12% HF, 68% H_2SO_4 160 days, 77°C	… / …	54 / 1.37	74 / 1.88	76 / 1.93	… / …	… / …	… / …	… / …
Coal-fired FGD scrubber presaturator, 24 weeks, 66 to 149°C	… / …	1.7 / 0.04	7.2 / 0.18	3.7 / 0.09	… / …	… / …	63 / 1.60	… / …
Herbicide/insecticide	74 / 1.88	50 / 1.27	92 / 2.34	175 / 4.45	… / …	… / …	… / …	… / …
incinerator scrubber Cl/S componds, 60 days, 88°C	7 / 0.18	14 / 0.36	… / …	25 / 0.64	… / …	20 / 0.51	3000 / 76.2	… / …
Reboiler Xylene, HF, BF_3, 120 h, 175°C	… / …	2.9 / 0.07	1.7 / 0.04	2.5 / 0.06	5.0 / 0.13	… / …	65 / 1.65	… / …
Storage tank for 2 to 30% P_2O_5, 30 days, RT-88°C	… / …	… / …	4.3 / 0.11	4.7 / 0.12	20 / 0.51	… / …	… / …	… / …
H_3PO_4 evaporator 42% P_2O_5, 1600 h, welded samples	… / …	… / …	… / …	… / …	… / …	… / …	200 / 5.08	… / …

References

[1] Kolts, J. and Sridhar, N., "Temperature Effects in Localized Corrosion," *Corrosion of Nickel Base Alloys,* R. C. Scarbarry, Ed., American Society of Metals, 1984, p. 195.

[2] Silence, W. L. and Manning, P. E., "Laboratory and Field Corrosion Test Results Related to Flue Gas Desulfurization Systems," *CORROSION/83,* Paper No. 185, National Association of Corrosion Engineers (NACE) 18–22 April, Anaheim, CA.

[3] Manning, P. E., Tuff, W. F., Zordan, R. D., and Schuur, P. D., "Laboratory and Field Evaluation of High Performance Alloys in the Pulp and Paper Industry," Fourth International Symposium on Corrosion in the Pulp and Paper Industry, June 1984, Stockholm, Sweden.

[4] Manning, P. E., "Maintenance Materials Selection Through Field Corrosion Testing," *CORROSION/ 84,* Paper No. 243, NACE, 2–6 April 1984, New Orleans, LA.

R. N. Parkins[1]

Synthetic Solutions and Environment Sensitive Fracture

REFERENCE: Parkins, R. N., "**Synthetic Solutions and Environment Sensitive Fracture,**" *The Use of Synthetic Environments for Corrosion Testing, ASTM STP 970*, P. E. Francis and T. S. Lee, Eds., American Society for Testing and Materials, Philadelphia, 1988, pp. 132–141.

ABSTRACT: The stress corrosion cracking of alloys is markedly sensitive to the environment to which they are exposed, so that an alloy developed to have low susceptibility to such failure in a particular environment will not necessarily show such behavior in a different environment, as illustrated by a number of examples. It is also possible for results from tests using standard environments to lead to the exclusion of the use of materials that may behave satisfactorily in service. While it is obvious that tests should be conducted, whenever possible, in environments relevant to particular service conditions, these are not necessarily known when alloy development programs are being undertaken in the laboratory, hence the need for standardized or synthetic solutions. Evidence for the cracking domains for ferritic steels and α-brasses suggests that stress corrosion cracking in a range of environments occurs at potentials and pH's where the lower oxide of the relevant metal can form. This suggests that laboratory tests should involve a single standardized environment, and that variation from test to test of potential and pH, guided by the potential pH diagram, may provide a more reliable guide to stress corrosion cracking propensities than that obtained from tests in a single solution.

KEY WORDS: environment sensitive fracture, stress corrosion, corrosion fatigue, ferritic steels, α-brasses, potential, pH effects, standard solutions, environments

This brief review is concerned with the use of synthetic solutions in the context of environment sensitive fracture, and like laboratory corrosion tests generally, the objective is essentially the acquisition of data in relatively short times compared to service lifetimes, that data then being used, empirically or otherwise, to predict service behavior. It would appear obvious that such tests should involve environments relevant to particular service conditions, and especially so when alloy selection is to be based upon laboratory data, or an empirical relationship must be available to relate laboratory data involving some standard environment to service behavior. Stress corrosion cracking of some materials is well known to be highly specific to particular environments. Thus, low-strength steels are well known to display such failure in aqueous solutions of nitrates, hydroxides, and bicarbonates, for example, but are not susceptible to stress corrosion in chlorides or sulfates. On the other hand, high-strength steels can be caused to fail at appropriate potentials in most aqueous environments, and indeed some nonaqueous ones, so that the often-used 3.5% sodium chloride (NaCl) solution may be simulative of many environments in relation to high-strength steels, because the primary requirement is probably simply a source of hydrogen to induce embrittlement. However, the greater specificity of environments in relation to the stress corrosion cracking of low-strength steels suggests that a universal test environment, such as 3.5% NaCl solution, is unlikely to be appropriate. Even when laboratory tests involve an environment known to promote cracking in a particular type of alloy the relative

[1]Head, Department of Metallurgy and Engineering Materials, The University, Newcastle upon Tyne, NE1 7RU England.

resistances or susceptibilities of a range of similar alloys may not be the same in a different potent environment, as may be illustrated by various examples.

Low Alloy Ferritic Steels

These will display stress corrosion cracking over specific potential ranges in various environments, as already indicated. To illustrate the point that a steel developed to have good resistance to cracking when exposed to one environment may not show similar behavior in another environment, the effects of molybdenum additions upon cracking by hydroxide and carbonate-bicarbonate solutions may be cited [1]. Figure 1 shows the results from slow strain rate tests at various controlled potentials in these two environments for a steel with and without 5% molybdenum present. With the carbonate-bicarbonate environment, the molybdenum addition clearly has a beneficial effect in that the area bounded by the curve for that steel is appreciably smaller than for the carbon steel. However, quite the reverse result is obtained with the hydroxide solution, cracking being more severe at the most susceptible potential and the potential range for cracking being markedly extended as the result of the molybdenum addition to the steel.

Similar differences in cracking response may be observed in relation to nickel additions to low-strength ferritic steels. Thus, when almost 6% nickel is present, the resistance to cracking at various potentials by both hydroxide and carbonate-bicarbonate solutions is good, and considerably better than without the presence of nickel [1], as is apparent from Fig. 2. However, in boiling 42% magnesium chloride ($MgCl_2$) solution the nickel-containing steel cracks very readily [2] yet the unalloyed steel shows no propensity for failure in such a solution.

High-Strength Pre-Stressing Steel

The importance of laboratory test environments showing relevance to service conditions if alloy selection is to be reliable is equally important where standard test solutions are used for the acceptance or rejection of particular heats of a material. An ammonium thiocyanate solution is often used for stress corrosion tests upon pre-stressing steels for use in concrete [3], such an environment facilitating hydrogen entry into the steel whereby it fractures if appropriately

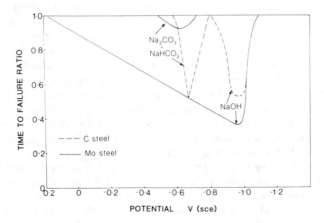

FIG. 1—*Effects of applied potential upon the time to failure ratio in slow strain rate tests of C-Mn steel, with and without a 5% molybdenum addition, in boiling 8 M NaOH and 1 M NaHCO$_3$ + 0.5 M Na$_2$CO$_3$ at 75°C.*

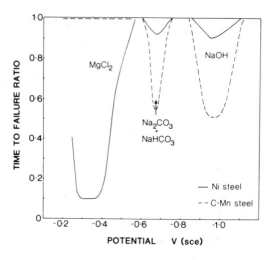

FIG. 2—*Effects of applied potential upon the time to failure ratio in slow strain rate tests of C-Mn steel, with and without a 6% nickel addition, in boiling 8 M NaOH, 1 M NaHCO$_3$ + 0.5 M Na$_2$CO$_3$ at 75°C, and boiling 4.4 M MgCl$_2$.*

stressed. When pre-stressing steel is completely surrounded by concrete, it is unlikely to suffer corrosion, but the ingress of aggressive ions at defects may result in breakdown of the passivity and departure from the relatively high pH (~ 12.5) environment that would otherwise exist. Thus, the ingress of chloride, for example, leads to pitting and a local fall in pH because of hydrolysis. Figure 3 shows the effects of applied potential upon the cracking of cold-drawn and heat-treated pre-stressing steels in 1-g/L calcium hydroxide (Ca(OH)$_2$) solution at adjusted pHs of 6 to 8.5, that is, similar to environments that can exist in the vicinity of pre-stressed bars in concrete. There are two regimes of potential within which environment sensitive fracture occurs, separated by a domain, in the vicinity of −0.7 V (saturated calomel electrode [SCE]), within which ductile fracture is displayed. At potentials below about −0.9 V, failure is by hydrogen ingress, but at higher potentials, above about −0.6 V, dissolution controls crack growth and is associated with the pitting potential being exceeded [4]. Figure 3 shows that tests at the free corrosion potential (fcp) are associated with the upper regime of potentials and measurements of the potentials of pre-stressing steels in contact with concrete show similar or higher values. Since the ammonium thiocyanate test applied to pre-stressing steels induces hydrogen assisted cracking, a query arises as to its realism in view of the results shown in Fig. 3, and the values of open circuit potentials observed with concrete-clad steels. To select steels on the basis of tests in thiocyanate may not result in the use of the most appropriate material, unless it is shown that susceptible and resistant steels have the same rankings where hydrogen related cracking is not involved. That point may have particular significance in relation to the use of the heat-treated variety of pre-stressing steels, which appear to be less favored than cold-drawn materials. That may be the result of the poorer performance in the thiocyanate test, since Fig. 3 shows that, at low potentials conducive to the ingress of hydrogen, the heat-treated steel is somewhat more susceptible than the cold-drawn steel. However, that figure also shows that, at potentials more likely to be met in service, the heat-treated material can perform as well as the cold-drawn steel.

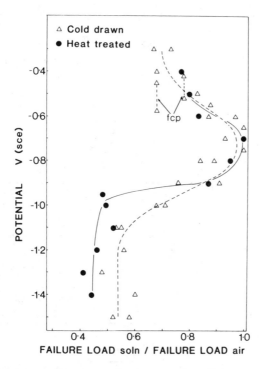

FIG. 3—*Effects of applied potential upon the failure load ratio of notched specimens of a cold drawn and a heat-treated pre-stressing steel in 1 g/L Ca(OH)₂ containing HCl, pH 6 to 8.5, in slow strain rate tests. The ranges of potential observed in tests at the free corrosion potential (fcp) are shown as bands.*

70/30 Brass

It is well known that small amounts of certain elements in specific alloys may markedly influence their stress corrosion cracking behavior, the effects of small amounts of carbon in 18Cr8Ni steels being perhaps the most extensively studied. Some recent work [5] on the susceptibilities of two brasses to the same specification may be used to illustrate similar differences arising from the presence of a trace element, but which extends to show variations in behavior in different environments. The two brasses differed most significantly in that one contained 0.032% arsenic, while the other contained only 0.002% arsenic. In an acid sulfate solution, the latter displayed no tendency for stress corrosion cracking, only general dissolution at higher potentials, whereas the 0.032% arsenic brass displayed stress corrosion in the same environment, as shown by the slow strain rate results in Fig. 4. Table 1 lists the cracking responses of the two brasses to cracking in various environments and makes essentially the same point as for the steels mentioned earlier, that the relative cracking tendencies depend upon the environments in which those susceptibilities are determined.

Importance of Surface Conditions

Figures 1 through 4 show for various systems that stress corrosion cracking only occurs within reasonably well defined ranges of potential, and it appears vital that any testing program should reflect that fact. This is primarily because open circuit potentials in operating plant can be

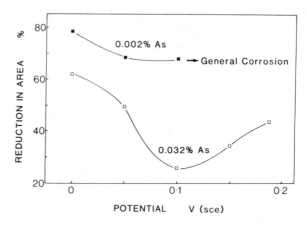

FIG. 4—*Reductions in area to failure in slow strain rate tests on two α-brasses in 0.5 M Na₂SO₄ — 0.005 M H₂SO₄ as a function of potential* [5].

TABLE 1—*Comparison of stress corrosion cracking susceptibilities of the two brasses in various environments* [5].

Environments in Which 0.032% Arsenic Brass Cracked, but not 0.002% Arsenic Brass	Environments in Which Both Brasses Cracked	Environments in Which 0.002% Arsenic Brass Cracked, but not 0.032% Arsenic Brass
Acid SO_4 solution	SO_2 vapors	acid SO_4 solution with $KAsO_2$ added
Brackish water	acid SO_4 vapors distilled water deionized water water vapour	...

appreciably different from those observed in laboratory tests. Thus, in the latter it is usual to use test specimens that are machined or polished, surface conditions that are rarely met in service. Yet surface conditions can significantly influence potential, as indeed can other factors, such as the extents to which oxygen has access to a system, the steel surface is creviced or is coupled to other materials. Mill scale, rust, and other oxidation products, so frequently present on surfaces in plant constructed of the lower alloyed steels in particular, may also result in open circuit potentials significantly different from those taken up by machined or polished surfaces. Since open circuit potentials in plant are not invariably easy to determine, and indeed unknown in the case of plant at the design stage, it appears necessary for laboratory tests to be conducted over a range of potentials, related to values likely to be experienced in a given environment, for a reliable assessment of cracking propensity.

Alternatively, and probably preferably, tests involving surface conditions similar to those likely to exist in plant may be involved, because the significance of realistic surfaces may extend beyond those mentioned in relation to potential. Figure 5 shows the results from some tests upon a controlled rolled pipeline steel immersed in a carbonate-bicarbonate solution and indicates that mill scaled surfaces are appreciably more susceptible to the initiation and propagation of stress corrosion cracks than ground surfaces [6]. The reasons for the differences in cracking responses between these two surfaces are not known, but may relate to composition differences in the steel in the near-surface region of the as-rolled material, which would be

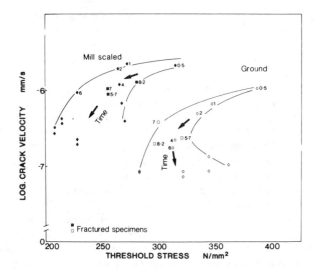

FIG. 5—*Trends in crack velocity and threshold stress with increasing test time for mill scaled and ground surfaces of a pipeline steel in 1 M NaHCO₃ + 0.5 M Na₂CO₃ at 75°C and −0.65 V (SCE) (points without associated numbers, 14-day tests; others show test times in days)* [6].

removed when the surfaces were ground. It is also known [7] that the presence of mill scale on pipeline surfaces plays a critical role in the achievement of cracking potentials in the presence of cathodic current densities of appropriate magnitude derived from the cathodic protection system.

However, even if laboratory tests involve more realistic surfaces than the carefully prepared and cleaned surfaces very frequently involved, it still appears desirable to conduct tests over an appropriate range of potentials in view of the likelihood that potentials achieved in plant may be different from those taken up in laboratory tests at open circuit.

Potential-pH Diagrams

While the dangers inherent in using standardized environments in relation to environment sensitive fracture are readily indicated by the examples quoted and many others not quoted, there remains a problem in relation to alloy development when possible service environments may not always be identifiable at the time of the development program. In such circumstances it appears desirable that an alloy should be assessed in a range of environments rather than a single one with the possible implications already indicated. Even then it is necessary for realism to be injected into the program if an excessively large number of test environments are not to be involved.

The significance of cracking only occurring in restricted ranges of potential must be related to specific potential-dependent reactions. Thus, the cracking domains shown in Fig. 1 probably occur where they do because in hydroxides dissolution of iron as the ferroate ion is possible and in carbonate-bicarbonate solutions iron dissolves as ferrous ions or as a soluble complex ion, $Fe(CO_3)_2^{2-}$, as favored by Davies and Burstein [8]. In addition, the coincidence of such cracking domains with active-passive transitions [9] implies a role for filming reactions in such dissolution-related cracking. On the other hand, where the failure mechanism is related to the ingress of hydrogen into the metal then cracking will only occur if the potential is below that at which hydrogen discharge can occur. Thus, in Fig. 3 the lower potential regime of cracking begins at about -0.7 V (SCE) which, for neutral solutions, approximates the potential at which hydrogen release is feasible.

Whether the crack growth mechanism is dissolution or hydrogen related the potential dependence of cracking will usually show a pH dependence, because of the influence of pH upon the potentials at which the various reactions are possible. It may be expected then that if the regimes of potential for cracking are plotted on a potential-pH diagram some correlation with reactions involved in the cracking mechanism may be observed.

Figure 6 shows the potential and pH range for the stress corrosion cracking of ferritic steels in various environments, together with the pH-dependent equilibrium potentials for various reactions at 90°C, representing an average temperature for the various systems involved [10]. Clearly, with the exception of the acetate environment, each cracking domain is crossed by the boundary between the ferrosoferric oxide (Fe_3O_4) and ferric oxide (Fe_2O_3) domains. Even the acetate cracking domain is close to the Fe_3O_4/Fe_2O_3 line and cracking by this environment is associated with the formation of Fe_3O_4 films. Although Fe_3O_4 is frequently associated with other phases, for example, ferrous carbonate ($FeCO_3$) in the case of cracking by carbonate-bicarbonate solutions and ferrous phosphate ($Fe_3(PO_4)_2$) for cracking by phosphate solutions, cracking in all of the environments shown in Fig. 6 is associated with the formation of Fe_3O_4 films. Moreover, for most of the systems shown in Fig. 6 only ductile failures occur in slow strain rate tests carried out at potentials high enough to form Fe_2O_3 alone, that is, at about 0.25 V above the calculated equilibrium potentials for reaction between Fe_3O_4 and Fe_2O_3. In those environments (nitrates and high-temperature water) where cracking is associated with those relatively high potentials where Fe_2O_3 is the stable phase, the cracks grow from pits wherein local acidification and potential changes from those external to the pits and cracks may account for the presence of Fe_3O_4 within the cracks but not on the external surfaces.

All of the cracking domains indicated in Fig. 6 are appreciably above the equilibrium potential for hydrogen discharge, and so it is unlikely, especially for the buffered or more alkaline solutions where pH reductions within pits or cracks do not occur, that cracking involves a hydrogen-related mechanism. However, some of the environments shown in Fig. 6 promote second cracking domains at lower potentials than those shown in Fig. 6, that is, two separated regimes of cracking are identifiable, as with the pre-stressing steels in Fig. 3. Figure 7 shows a

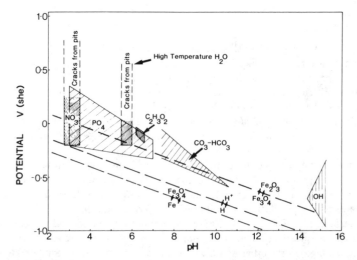

FIG. 6—*Potential and pH ranges for the stress corrosion cracking of ferritic steels in various environments, together with the pH-dependent equilibrium potentials for various reactions* [10].

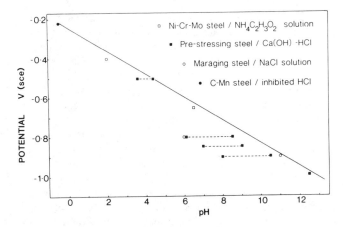

FIG. 7—*Highest potentials at which hydrogen-induced cracking was observed in various ferritic steels exposed to different solutions and subjected to slow strain rate tests. The line represents the equilibrium potential for hydrogen discharge as a function of pH.*

plot of the highest potentials at which hydrogen-induced cracking was observed in various ferritic steels exposed to different solutions from which it is clear that there is reasonable agreement between these potentials and those calculated from the hydrogen discharge reaction as a function of pH.

Although less systematically studied than the ferritic steels from this viewpoint, cracking domains for α-brasses in various nonammoniacal environments are shown superimposed upon the Cu-H$_2$O potential-pH diagram in Fig. 8. As with the ferritic steels, where cracking was largely associated with the formation of the lower oxide (Fe$_3$O$_4$) there is a tendency for the cracking domains for α-brasses to be concentrated in regions where cuprous oxide (Cu$_2$O) is likely to form.

While Figs. 6 and 8 show the cracking domains superimposed upon the relevant metal-water potential-pH diagrams this is not meant to imply that species other than those present in water do not play a role in the cracking mechanisms. In some cases the added species may do little

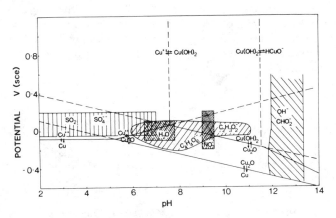

FIG. 8—*Potential and pH ranges for the stress corrosion cracking of α-brass in various environments, together with the equilibrium potentials for various reactions.*

more than change the pH of water, and hence the potential range for cracking, but other additions must be involved in more than this secondary role. Moreover, metallurgical influences in cracking are not adequately represented in Figs. 6 and 8, since, for example, very pure iron does not crack in some, if any, of the environments shown in Fig. 6. Similarly, pure copper, which can be caused to display stress corrosion cracking in nitrite solutions, does not crack in waters and probably also not in some of the other environments to which Fig. 8 refers. Nevertheless, from the point of view of standard solutions for alloy development work in the laboratory, Figs. 6 and 8 suggest that cracking is likely to occur in widely varying solutions providing that the lower oxides can form. This suggestion is in accord with the observation over the last decade or so that the list of environments that will promote stress corrosion cracking of ferritic steels and brasses has increased considerably. Thus, it is not long since it was thought that α-brasses only failed in ammoniacal environments and that low-strength ferritic steels only displayed stress corrosion cracking in nitrate and hydroxide solutions. The extensive growth in the number of potent solutions promoting cracking in both of these materials appears best rationalized in terms of pH and potential effects. This in turn suggests that it would be valuable in designing laboratory test programs to involve solutions of different pH's and to explore the cracking propensity as a function of potential guided by the indications of Figs. 6 through 8, so far as ferritic steels and brasses are concerned.

It also seems probable that similar relationships to those of Figs. 6 through 8 may hold in relation to the stress corrosion cracking of other materials, but corrosion fatigue presents a different problem. Although it is now established for a number of metal-environment systems that display static-load stress corrosion cracking that cyclic loading will enhance such cracking, the solution specificity associated with dissolution-related stress corrosion cracking is not a limitation in relation to corrosion fatigue. This is probably because with static or monotonically increasing stress, as in stress corrosion, the retention of crack sharpness requires properties of the metal-environment combination that will allow dissolution at the crack tip but passivation of the crack sides, hence the significance of oxide formation in stress corrosion. There is no such requirement with cyclic loading, because the crack is maintained sharp by mechanical processes [11]. Consequently while cyclic loading will promote cracking under the same environmental conditions as static loading, it will also promote cracking under conditions where the filming requirements for static-load stress corrosion are not met. Hence corrosion fatigue of C-Mn steels will occur in the presence of nitrates, hydroxides, bicarbonates, and so forth that promote stress corrosion, but will also occur in chloride or sulfate solutions that do not promote stress corrosion. This implies that the restrictions on stress corrosion cracking shown in Figs. 6 and 8 may not apply under cyclic loading conditions, although where the crack growth mechanism involves the ingress of hydrogen the implications of Fig. 7 are still applicable. The choice of laboratory test environments for corrosion fatigue is therefore less restricted, although solution pH and potential are still of relevance, and the extent and nature of the species present will have implications for crack growth rates through their influence upon dissolution rates or the hydrogen discharge reaction.

References

[1] Parkins, R. N., Slattery, P. W., and Poulson, B. S., Corrosion, Vol. 37, 1981, pp. 650–664.
[2] Poulson, B. S. and Parkins, R. N., Corrosion, Vol. 29, 1973, pp. 414–422.
[3] Federation Internationale de las Precontrainte Symposium 1978, "Stress Corrosion Cracking Resistance Test for Prestressing Tendons," Technical Report 5, Warren Springs Laboratory, Slough, United Kingdom.
[4] Parkins, R. N., Elices, M., Sanchez-Galvez, V., and Caballero, L., Corrosion Science, Vol. 22, 1982, pp. 379–405.
[5] Parkins, R. N., Rangel, C. M., and Yu, J., Metallurgical Transactions, A, Vol. 16A, 1985, pp. 1671–1681.

[6] Parkins, R. N. and Fessler, R. R., "Line Pipe Stress Corrosion Cracking—Mechanisms and Remedies," *Corrosion '86*, Paper 320, National Association of Corrosion Engineers, Houston.
[7] Parkins, R. N., O'Dell, C. S., and Fessler, R. R., *Corrosion Science*, Vol. 24, 1984, pp. 343-374.
[8] Davies, D. H. and Burstein, G. T., *Corrosion*, Vol. 36, 1980, pp. 416-422.
[9] Parkins, R. N., *Corrosion Science*, Vol. 20, 1980, pp. 147-166.
[10] Congleton, J., Shoji, T., and Parkins, R. N. loc. cit, Vol. 25, 1985, pp. 633-650.
[11] Knott, J. F., *Fundamentals of Fracture Mechanics*, Butterworths, London, 1973, p. 238.

DISCUSSION

E. Heitz[1] *(written discussion)*—Are there systems other than the carbonate/bicarbonate system that show during CERT experiments a maximum SCC susceptibility at critical strain rates?

R. N. Parkins (author's closure)—Yes. For example, low-strength steels in air-contaminated ammonia, Mg-Al alloys in chromate-chloride solutions and Ti-6Al-V alloy in NaCl solutions, all of which are mentioned in *Stress Corrosion Cracking—The Slow Strain Rate Techniques, STP 665*, G. M. Ugiansky and J. H. Payer, Eds., American Society for Testing and Materials, Philadelphia, 1979.

Dale McIntyre[2] *(written discussion)*—What strain rate would you recommend for testing austenitic stainless steels in hot dilute chloride solutions using the slow strain rate method?

Have you had success predicting potential ranges for cracking using your fast scan-slow scan potentio-dynamic polarization technique on ferritic steels in monoethanolamine solutions?

R. N. Parkins (author's closure)—There is an International Standards Organization (ISO) standard on slow strain rate testing that is about to be published, which makes recommendations about the choice of strain rate. In my laboratory we have caused cracking in austenitic stainless steel immersed in hot dilute (10 to 1000 ppm) chloride solutions by straining at strain rates from about 10^{-6} to 10^{-4}/s. I would expect cracking to occur also at lower strain rates, but a useful strain rate would be about 5×10^{-6}/s for most tests in this system.

Some results on prediction of potential ranges for cracking are given in a paper in *Materials Performance*, Vol. 25, No. 10, 1986, pp. 20-27. This particular system is addressed in more detail in this respect in Paper 188, *Corrosion 87*, NACE.

[1]Dechema-Institut, Theodor-Hauss-Allee 25, D-6000 Frankfurt, West Germany.
[2]Cortest Laboratories Inc., 11115 Mills Rd., Suite 102, Cypress, TX 77089.

Dereck R. Johns[1]

Intergranular Corrosion of Austenitic Stainless Steels—An Electrochemical, Potentiokinetic Test Method

REFERENCE: Johns, D. R., "Intergranular Corrosion of Austenitic Stainless Steels—An Electrochemical, Potentiokinetic Test Method," *The Use of Synthetic Environments for Corrosion Testing, ASTM STP 970*, P. E. Francis and T. S. Lee, Eds., American Society for Testing and Materials, Philadelphia, 1988, pp. 142-151.

ABSTRACT: An anodic polarization test in a perchloric acid-sodium chloride electrolyte has been developed to determine the intergranular corrosion susceptibility of three austenitic stainless steels, American Iron and Steel Institute (AISI) 302, 304, and 316. A region of secondary activity is present in the polarization curves, and the magnitude and potential of the maximum of this secondary activity is related to susceptibility to intergranular corrosion. The initial work involved polarization testing of the main surface of strip and plate material in the solution-treated condition and after heat treatment at 948 K (675°C) for up to 7 days. The correlation between the penetration rate determined by ASTM C 262 and the electrochemical activity is excellent for some materials but poor for others. In the latter, it was clear that end and edge grain attack were dominant in the nitric acid test. However, if the electrochemical activity of all three orthogonal faces of the materials is measured, and a weighted sum determined to reflect the relative areas of each face exposed to nitric acid, then a consistently good correlation with penetration rate is obtained. For material heat treated to provide a range of penetration rates (0.08–14 mm/year) as measured by ASTM C 262, a relationship of mm/year = A (activity)$^{0.5}$ is found. For low penetration rates (<0.5 mm/year), the power index becomes approximately 0.25. This indicates that the nitric acid test is less able to differentiate between highly resistant materials that still exhibit measurably different electro-chemical activities.

KEY WORDS: austenitic stainless steels, intergranular corrosion, sensitizing, electrochemistry, nitric acid, corrosion tests

Austenitic stainless steels offer the most satisfactory combination of corrosion resistance and mechanical properties for many applications in the chemical, petroleum, and power generation industries. The construction of plant manufacturing them, however, frequently involves fabrication techniques that subject the steels to heating cycles in the temperature range at which the precipitation of chromium carbides, or the segregation of elements to grain boundaries, may result in susceptibility to intergranular corrosion. In very severe environments, significant grain boundary corrosion may occur in nominally unsensitized material.

In applications in which such problems are anticipated, the steel supplier may be required to prove the quality of the product using screening tests such as those included in the ASTM Recommended Practices (A262-75, Practices A–E). However, these procedures are costly and time consuming.

Therefore, alternatives have been sought and electrochemical methods may offer a rapid and quantitative assessment technique. Such methods permit the study of the passive film formed

---

[1]Investigator, British Steel Corp., Swinden Laboratories, Moorgate, Rotherham, England S60 3AR.

on stainless steels in the presence of ionic species and at oxidation levels governed by the choice of electrolyte and potential. Investigating intergranular corrosion requires the establishment of a test procedure that is capable of measuring the susceptibility of grain boundaries to corrosion when destabilization of the passive film at grain boundaries occurs.

The electrochemical reactivation method (EPR) has been the subject of detailed investigation [1,2]. It measures the activity associated with the selective reactivation of grain boundaries susceptible to intergranular corrosion. This measurement is made during a rapid potential sweep in a suitable electrolyte in the cathodic direction from the passive potential region to the corrosion potential. The procedure is complicated by the need for metallographic preparation of the specimens and subsequent photomicrography. Electrochemically, it is also a nonequilibrium procedure.

A second electrochemical technique has been discussed in the literature, both as a potentiostatic [3] and a potentiokinetic [4] method, which is the virtual antithesis of the EPR test. The method involves the quasiequilibrium, low scan rate, anodic polarization of the specimen in a perchloric acid-sodium chloride electrolyte, and measures the degree to which the passivation of the grain boundaries is retarded relative to the bulk material. This retardation is manifested as a secondary active peak in the polarization curve.

This paper presents the results of an investigation comparing the response of austenitic stainless steels, with a wide range of resistance to intergranular corrosion, in the perchloric acid (PCA) test and the standard boiling nitric acid test, A262-75, Practice C. The electrochemical procedures adopted aimed only limit the specimen preparation to that required in Practice C— i.e., wet grinding only. Also, the generation of a quantitative parameter should require minimum mathematical manipulation and should be capable of direct comparison with the penetration rate obtained from the chemical test.

Experimental Procedures

Materials

Six commercial or experimental casts of AISI 304H (0.1% C), 304L, and 316L were initially investigated, in both the solution treated condition and after heat treatment at 948 K (675°C) for from 1 min to 7 days. Testing was then extended to 20 coils of commercial AISI 302, 304, and 304L strip in the as-received condition (bright annealed or softened and descaled). Of these samples, 18 were also evaluated after a 30 min heat treatment at 948 K (675°C).

Electrochemical and Chemical Testing

The procedures and equipment employed were in general agreement with ASTM Practice for Standard Reference Method for Making Potentiostatic and Potentiodynamic Anodic Polarization Measurements G 5-82, but used approximately 4 cm $\times$ 2 cm specimens with an electrical contact wire soft-soldered to the top end cut. The specimen and contact area were masked with an acid-resistant tar-vinyl paint to leave approximately 1 cm^2 exposed to the electrolyte. This area was prepared by wet grinding to a 400-grit silicon carbide finish. The polarization tests were performed in a standard three-electrode cell using a 1 M perchloric acid ($HClO_4$) + 0.2 M sodium chloride (NaCl) electrolyte, deaerated with argon and maintained at 293 K (20°C). The response of the specimen was recorded on an XY-t logarithmic chart recorder with range options of 1 μA–100 mA or 0.1 μA–10 mA. The potential sweeps were conducted at 1.2 V/h relative to the saturated calomel electrode (SCE) from a start potential of -600 mV. The boiling nitric acid (BNA) testing was carried out in accordance with ASTM 262-75 Practice C.

Results

The initial polarization measurements were restricted to the main surface of the flat products but poor correlation was obtained with the BNA results due to major end or edge attack on the BNA test specimens. Polarization tests were then carried out based on the principle illustrated in Fig. 1. Electrochemical data were obtained on the three grain orientations exposed to the nitric acid in the BNA specimen. An example of the influence of grain orientation on the secondary peak current density obtained is shown in Fig. 2. From a knowledge of the dimensions and geometry of each BNA specimen, the specimen current density, i_{spec}, of the individual specimens was calculated using the equation defined in Fig. 1. The resulting values of specimen current density were then compared with those from the BNA test.

The form of the polarization curves produced is illustrated in Figs. 3 to 6. The high carbon Type 304H material (Fig. 3) gave no secondary activity in the solution treated condition, but a very clear secondary peak was found after heating for 1 min at 948 K (675°C). The secondary peak current density increased markedly with time at this temperature. In less susceptible material, such as solution-treated Type 304L, a range of curve forms was obtained, which included a clearly separated secondary peak (confirmed as anodic), or a peak separated by a distinct minimum (Fig. 4). These curve forms accounted for some 70% of the low susceptibility materials tested (BNA penetration rate <0.5 mm/year). Of the remainder of these materials, 15% gave a clear plateau region adjacent to the primary peak (Fig. 5) and 5% merely showed a current oscillation in the relevant potential range. Ten percent of the samples produced a discernible secondary activity, but this was not followed by a rapid fall to passive current levels, but instead by an approximately constant activity (Fig. 6). We assumed this latter curve form was

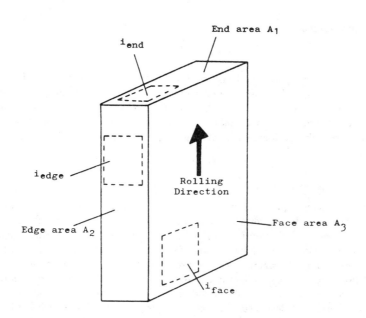

$$i_{specimen} = 2(A_1 \cdot i_{end} + A_2 \cdot i_{edge} + A_3 \cdot i_{face})/A_4$$

FIG. 1—*Test areas for derivation of specimen activity.*

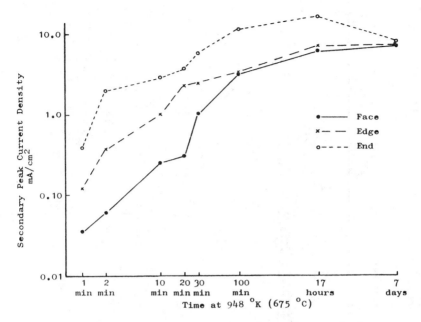

FIG. 2—*Effect of grain orientation on secondary peak current density.*

the result of crevice formation at the junction of the exposed specimen and the paint film. When repeated, the tests produced more normal curve forms.

Table 1 presents an example of the secondary peak current density values obtained from the three orientations of the strip and plate, the specimen current density (i_{spec}, mA/cm^2) derived from the corresponding BNA specimen geometry, and the actual BNA penetration rate (P, mm/year) recorded. Figure 7 illustrates the relationship between the latter parameters. Regression analysis of the results for AISI 302/304 provides the power equation $P = 3.91\, i_{spec}^{0.53}$, which, over the penetration rate range 0.1 to 13.64 mm/year, has a correlation coefficient, r, of 0.944 and a significance of correlation, $C = r(N - 1)^{1/2} = 8.01$ for a population N of 73 or 219 polarizations.

In the penetration rate range of prime interest, i.e. <0.5 mm/year, analysis of the individual orientation activity values and the derived specimen values produces the following equations from 35 specimens:

$$P = 0.91\, i_{face}^{0.26}, r = 0.68 \tag{1}$$

$$P = 0.48\, i_{edge}^{0.18}, r = 0.62 \tag{2}$$

$$P = 0.45\, i_{end}^{0.25}, r = 0.65 \tag{3}$$

$$P = 0.96\, i_{spec}^{0.29}, r = 0.83 \tag{4}$$

Each of these expressions is highly significant, with a C value >3.5. The high correlation coefficient obtained using the specimen activity, i_{spec}, shows the validity of using this as a measure of the susceptibility of the material to intergranular corrosion.

Also indicated in Fig. 7 are the relationships found by analysis of the AISI 316L results. While the strip material behaves similarly to the AISI 302/304 alloys, the plate form produces a

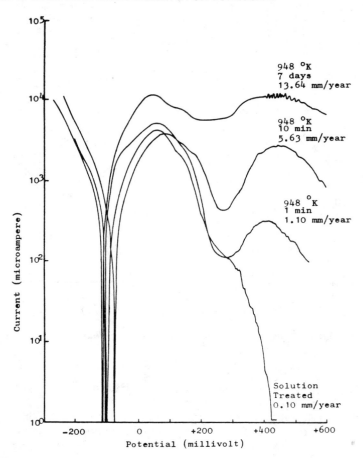

FIG. 3—*Influence of sensitization time on secondary activity.*

markedly different response. Subsequent examination shows that the plate material used was from an experimental cast containing significant amounts of sigma phase because of its process history. However, the presence of sigma phase does not contribute to secondary activity during the polarization test.

A limited metallographic examination of PCA tested specimens has been undertaken. The mode of attack in material highly susceptible to intergranular attack is confined to a very narrow grain boundary zone, but the etching effect of the primary active zone obscures the effect of the secondary activity in materials that are less susceptible. Potentiostatic experiments at the potential of the secondary peak are required, and those few carried out on specimens highly susceptible to intergranular attack exhibit visible grain detachment within the polarization cell after only minutes at potential.

Discussion

The results presented above on commercial and experimental materials, deliberately heat treated and processed so that they would have a wide range of corrosion properties, demonstrate that the electrochemical test provides data that can be correlated with those obtained with the

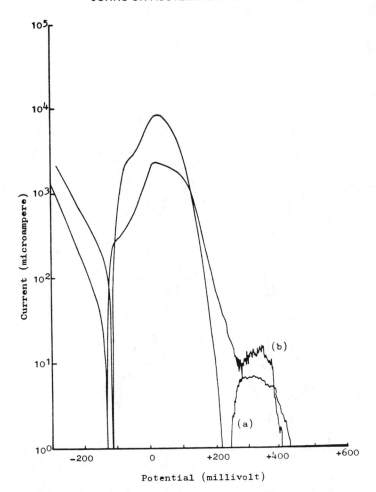

FIG. 4—*Well-formed secondary activity at low sensitization level:* (a) *separated secondary peak.* (b) *distinct minimum between peaks.*

standard chemical test, ASTM 262-75 Practice C. Testing each grain orientation exposed to the chemical test to derive an equivalent specimen activity appears sound and the availability of computer controlled, multichannel potentiostats will negate the increased test load incurred. The response of the PCA test does not appear to diminish at high or low sensitization levels. The data indicate that the Practice C procedure is less discriminating at low penetration rates, resulting in a marked reduction in the power index relating the penetration rate to the secondary current density when the analysis is restricted to <0.5 mm/year.

The work so far has shown that the PCA test merits further investigation as a release test method and as a technique capable of investigating the process of intergranular corrosion. The technique is sensitive to materials in which precipitation of chromium-rich carbides has occurred, but whether this is due to chromium depletion or a direct dissolution of carbides requires clarification. In high carbon steels or steels deliberately heat treated to precipitate grain boundary carbide, the presence of such phases resulted in a marked increase in electrochemical

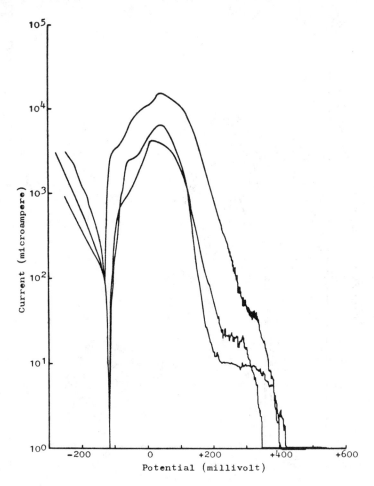

FIG. 5—*Three examples of plateau type secondary activity.*

activity. In comparison, the presence of sigma phase in the experimental AISI 316L melt did not contribute to the electrochemical activity. The measurement of significant secondary activity in specimens where grain boundary precipitation is minimal suggests that segregation effects may be affecting intergranular corrosion. The effect of the grain size of the materials tested has been ignored in the present work. However, because both tests assess the sensitivity to grain boundary corrosion, the influence of grain size requires evaluation. The present data have been generated on flat products, although many product forms are used for applications involving nitric acid. The application of correlation equations derived on flat products to other products, such as tubes and forgings, has not been tested but should be possible.

Conclusions

Potentiokinetic polarization in a perchloric acid-sodium chloride electrolyte offers a quantitative and rapid method of assessing the state of sensitization to intergranular corrosion of austenitic stainless steels.

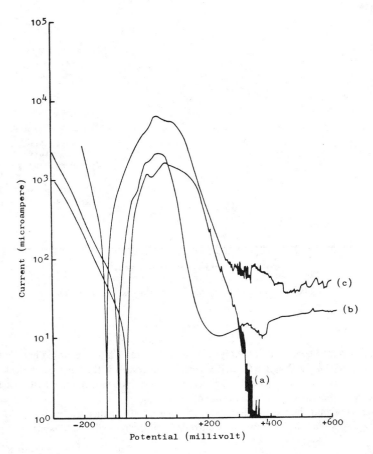

FIG. 6—*Uncertain secondary activity:* (a) *Current oscillation.* (b), (c) *High passive current.*

TABLE 1—*BSC Type 304L strip (0.73 mm).*

Time at 948 K	i_{face}, mA/cm^2	i_{edge}, mA/cm^2	i_{end}, mA/cm^2	BNA Penetration, mm/year	PCA i_{spec}, mA/cm^2
0 min	0.0058	0.0090	0.063	0.225	0.006
2 min	0.049	0.010	0.118	0.490	0.053
10 min	0.067	0.120	0.136	0.581	0.071
20 min	0.025	0.029	0.091	0.311	0.027
30 min	0.038	0.054	0.090	0.361	0.040
100 min	0.089	0.117	0.190	0.990	0.093
17 h	0.217	1.348	1.205	1.830	0.293
7 days	0.073	0.222	0.291	1.045	0.085

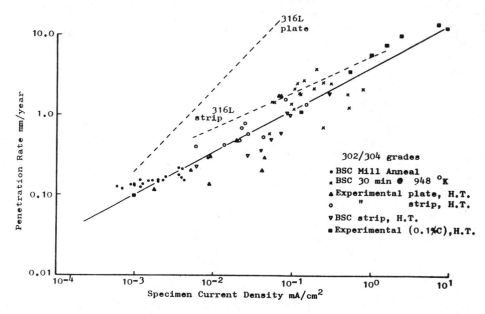

FIG. 7—*Comparison of electrochemical and chemical data.*

The data obtained correlate well with the penetration rate values recorded in a boiling nitric acid test, ASTM 262-75 Practice C. Better correlation results if the electrochemical activity of each grain orientation exposed to the nitric acid is determined to give an activity normalized to the geometry of the nitric acid test specimen.

The data show that the electrochemical technique merits further study, a process made less onerous by the availability of computer-controlled, multichannel potentiostats.

Acknowledgments

The author would like to thank Dr. R. Baker, Director of Research and Development, British Steel Corporation for permission to publish this paper.

References

[1] Clarke, W. L., Romero, V. M., and Danko, J. C., "Detection of Sensitization in Stainless Steel Using Electrochemical Techniques," NACE Corrosion Conference, Paper No. 180, San Francisco, CA, March 1977.

[2] Clarke, W. L., Cowan, R. L., and Walker, W. L., *Intergranular Corrosion of Stainless Alloys, ASTM STP 656*, R. F. Steigerwald, Ed., American Society for Testing and Materials, Philadelphia, 1978, pp. 99–132.

[3] Aaltonen, P., Aho-Mantila, I., and Hanninen, H., *Corrosion Science,* Vol. 23, No. 4, 1983, pp. 431–440.

[4] Chung, P. and Szklarska-Smialowska, S., *Corrosion-NACE,* Vol. 37, No. 1, January 1981.

DISCUSSION

B. Singh[1] (discusser's questions)—What are the effects of roughness of the specimen on such polarization data, and the directional aspects of roughness?

D. R. Johns (author's responses)—Work was done on roughness values corresponding to 80 grit, 240 grit, and 400 grit, with no discernible differences. In fact, the secondary peaks were much the same. All the tests were subsequently done at 400 grit for comparison purposes.

J. M. Sykes[2] (discusser's question)—Do you have a good reason for using a positive-going potential sweep? Have you tried negative-going sweeps, as in the EPR test? If you begin with a passive surface, the anodic activity of the grain boundaries might be masked less by the active peak.

D. R. Johns (author's response)—The PCA test measures the retardation of passivation of grain boundaries previously activated in the primary anodic loop. In all but the most sensitive specimens we find full passive behavior in the secondary potential range during negative-going sweeps.

D. McIntyre[3] (discusser's question)—What correlation would you expect between your test and the ASTM A 262 practice test?

D. R. Johns (author's response)—Where the IGC was due to chromium carbide precipitation, I would expect a good fit between the tests. Perhaps step structures would be seen after oxalic testing in materials giving an i_{spec} below some value, a ditch above some higher value, and dual structures in between.

In low carbon alloys where the PCA activity is due to some other mechanism, I imagine there would be little correlation.

[1]Yard Ltd. Charing Cross Tower, Glasgow, United Kingdom.
[2]Department of Metallurgy and Science of Materials, University of Oxford, Parks Rd., Oxford, OX1 3PH.
[3]Cortest Laboratories, 11115 Mills Rd., Suite 102, Cypress, TX 77429.

Einar Mattsson,[1] Rolf Holm,[2] and Lars Hassel[2]

Ammonia Test for Stress Corrosion Resistance of Copper Alloys

REFERENCE: Mattsson, E., Holm, R., and Hassel, L., **"Ammonia Test for Stress Corrosion Resistance of Copper Alloys,"** *The Use of Synthetic Environments for Corrosion Testing, ASTM STP 970*, P. E. Francis and T. S. Lee, Eds., American Society for Testing and Materials, Philadelphia, 1988, pp. 152–164.

ABSTRACT: Within the ISO Committee TC 26/SC 2, Wrought Copper and Copper Alloys, a Working Group, WG4, has been set up to propose an ammonia test to evaluate the stress corrosion resistance of copper alloy products. To create a basis for such a proposal, the Swedish Corrosion Institute has initiated a test program, which has been carried out by Metallverken in Sweden. The proposal is based on earlier experience with ammonia tests at Metallverken, including a method originating from F. Aebi and used in Switzerland, Great Britain, and the Federal Republic of Germany. The proposed test procedure involves a 24-h exposure of the test material at room temperature to an ammoniacal atmosphere in equilibrium with an ammonium chloride solution. The vapor pressure of ammonia in equilibrium with the solution depends on the pH value. So, by changing the pH value, the severity of the test can be regulated. For comparison, a selection of materials has been subjected to the proposed test, to other accelerated tests based on ammoniacal corrodants, to the mercurous nitrate test, and to a four-year field test in the urban atmosphere of Stockholm. The results have shown that the proposed method has advantages over other accelerated SCC tests hitherto used for production and acceptance control. The test has been described in ISO standard 6957.

KEY WORDS: copper base alloys, stress corrosion tests, accelerated tests, field corrosion test, acceptance tests, standardization

Within the ISO Committee TC 26/SC 2, Wrought Copper and Copper Alloys, a Working Group, WG4, has been set up to propose an ammonia test to evaluate the stress corrosion resistance of copper alloy products. To create a basis for such a proposal, the Swedish Corrosion Institute has initiated a test program, which has been carried out by Metallverken in Sweden. The following members of the Working Group have actively contributed to the planning and pursuing of the project: Prof. E. Mattsson (convener), Swedish Corrosion Institute, Sweden; Mr. O. Claesson (secretary), Metallnormcentralen, Sweden; Dr. K. P. Camenisch, Metallwerke AG Dornach, Switzerland; Mr. R. Holm, Metallverken AB, Sweden; Mr. A. van Ooijen, De Vitrite Fabriek, The Netherlands; Mr. M. Schmidt, A/S NKT, Denmark; Dr. H. H. Sick, Wieland Werke AG, Federal Republic of Germany; Mr. G. Wildsmith, IMI, Yorkshire Copper Tube Ltd, Great Britain.

Previously Used Tests

For stress corrosion testing of copper alloys mainly the following methods have been used.

[1]Professor, Swedish Corrosion Institute, Box 5607, S-114 86 Stockholm, Sweden.
[2]Senior research fellow and assistant researcher, respectively, Metallverken, S-721 88 Västerås, Sweden.

Test in Mercurous Nitrate Solution

This test is described in ISO/R196-1961 and in the Swedish standard SS 11 71 01. The test is considered moderately severe, i.e., somewhat too mild to represent common outdoor exposure in an urban or rural atmosphere, but somewhat too severe to represent indoor storage conditions. The main drawbacks with the test are that mercurous nitrate very rarely occurs as a corrodant in practice, and that it should not be used from an environmental point of view.

Test in Ammoniacal Copper Sulfate Solution According to Mattsson

This test is described in SS 11 71 02. In this test, a solution with a pH value of 11.2 is used. A variant with a solution at pH 7.2 is used only for *binary brasses*. This test is covered by ASTM Test Method for Use of Mattsson's Solution of pH 7.2 to Evaluate the Stress-Corrosion Cracking Susceptibility of Copper-Zinc Alloys (G-37-85). Because of surface roughening during exposure, difficulties may arise on inspection. These difficulties can be overcome, however, by slight deformation of the specimen before inspection, opening up the cracks.

Test in Ammoniacal Atmosphere According to Eichhorn

This test is described in DIN 50916: Part 1, and in SS 11 71 03. The test is too severe to be representative of common outdoor exposure conditions, but applicable in cases in which exceptional safety requirements have to be met.

Test in Ammoniacal Atmosphere According to Aebi

A way to obtain an ammoniacal atmosphere of a lower concentration for the SCC testing of copper alloys without encountering the inconvenience mentioned has been advised by F. Aebi [1]. The method has been described in BS 2871: Part 3:1972 as an alternative to the mercurous nitrate test for revealing detrimental internal tensile stress in heat exchanger tubes of Al- or Sn-alloyed special brasses. For the same purpose, the test method has also been introduced by the Swiss, as reported by M. Germann at an ISO/TC 26/SC 2 meeting in Berlin in 1976. Both test practices prescribe exposure of the specimens in the ammoniacal atmosphere in equilibrium with a 2-M ammonium chloride water solution (i.e., 107 g/L NH$_4$Cl) at a solution pH value of 10 obtained by adding sodium hydroxide. The exposure duration is set to 24 h at room temperature. According to Aebi, the partial pressure of ammonia and, therefore, the severity of the test, can be regulated by varying the pH value of the ammonium chloride solution, for practical purposes, within a range of a few units above 8. The partial pressure of the water vapor is not affected in this range.

In September 1985, a new German Standard, DIN 50916: Part 2, has also been issued for SCC resistance testing of construction parts of copper alloys using the Aebi solution to provide the ammoniacal test atmosphere. The severity and selectivity of the test according to the German Standard is regulated by the variation of the exposure time, i.e., a time between 4 h and 7 days at pH 10 or between 16 h and 28 days at pH 9 (at fixed room temperature). This standard, however, seems too elaborate and time-consuming for production or acceptance control.

Plan of Investigation

The test outlined by Aebi was chosen as a basis for the accelerated laboratory test. The investigation comprised the following parts:

(a) studies of the various testing parameters and proposal of a detailed test procedure,

(b) comparison of the results of the proposed test and other laboratory tests, and
(c) comparison of the results of the proposed test and a field test.

Experimental Procedure

Materials Tested

The materials tested are given in Tables 1–3. Their nominal composition is specified by the ISO designation. The specimen forms used in the tests are shown in Fig. 1. Strip materials were tested in the form of cups and some rod materials in the form of C-rings. The remaining rod materials and all the tube materials were tested in an unchanged form.

Specimens were prepared to represent different levels of residual or applied tensile stress. For cup, tube, and rod specimens, the stress levels were achieved by annealing the cold-worked specimens at different temperatures and lengths of time. For C-ring specimens, the different levels of elastic tensile stress were achieved by appropriate compression of the open rings in accordance with the ASTM Standard Practices for Making and Using C-Ring Stress Corrosion Cracking Test Specimen G-38-73 (1984). The rings were manufactured from rod and stress-relieve annealed before compression. The specimens exposed in the field test were replicates of specimens in the laboratory tests.

The average stress levels were determined by Dr. Camenisch, Metallwerke Dornach AG, Switzerland, for some of the brass extrusion variants on the basis of deflections after slitting the rod ends. The results are given in Table 2.

Procedure of the Proposed Test Method

Initially the various testing parameters were studied experimentally. Taking into account the results then obtained in addition to those earlier published by Aebi [1] as well as the prescriptions in the related specifications, the authors arrived at the following testing conditions.

Preparation of Testing Medium—The preparation of the solution producing the ammoniacal atmosphere was started by dissolving 107 ± 0.1 g ammonium chloride (NH_4Cl) in 500 ml of deionized water. Then a fairly concentrated sodium hydroxide solution was slowly added (e.g., 50–75 ml of an approx. 50% by weight NaOH solution) to give the ammonium chloride solution the desired pH value, measured by a pH meter with an accuracy of ± 0.05 units. Finally, the test solution was filled up to 1.0 L with de-ionized water and the solution was kept in a closed vessel. The sodium hydroxide additions were varied to give solution pH values of 9.3, 9.5, 10.0, and 10.5 for different test series. The relationship between the ammonia pressure in the atmosphere and the pH value of the ammonium chloride solution is described by Aebi [1] in Fig. 2.

Pretreatment of Specimens—The specimens were degreased in trichloroethene vapor, pickled in 5% b.v. sulfuric acid solution, rinsed first in running cold water, then in hot water, and finally dried in hot air.

Exposure—The clean, dry specimens were stored in a closed, dry vessel. This vessel and the test exposure vessel, a 10 L desiccator with 2 L of freshly prepared, pH-adjusted ammonium chloride solution, were both thermostated to take the same temperature of $25 \pm 1°C$. Then the specimens were transferred to the test vessel and exposed there for generally 24 h, with the temperature maintained at $25 \pm 1°C$. The importance of adequate temperature control and of thermostating the specimens to the test vessel temperature before exposure is shown in Table 1. Comparative tests showed improved reproducibility of results by ascertaining the access of a sufficient amount of corrodant to the exposed specimen surfaces. This was achieved by using a test solution volume of at least 20% of the test vessel volume, and by providing at least 100 mL of test solution per dm^2 of exposed specimen surface.

Duplicate or triplicate specimens of each material variant were examined. Specimens removed for examination were not returned for continued exposure.

TABLE 1—*Influence of specimen temperature and humidity conditions on SCC results. Exposure of C-rings of CuZn39Pb3 at different levels of tensile stress in the atmosphere above a 2 M NH₄Cl-solution with pH 10.0 (see Exposure section).*

Tensile Stress in C-rings (Fig. 1b), N/mm²	Humidification Before Exposure	Temperature (°C)		Results of Inspection After Indicated Exposure							
		Before Exposure	During Exposure	1 h	2 h	3 h	4 h	6 h	8 h	16 h	24 h
300	…	25	25						+	−	
200									+	−	
100									+	+	−
50									+	+	+
300	…	20	25						−		
200									−		
100									+	−	
50									+	+	+
300	in spray cabinet 0.5 h, 25°C	25	25	+	+	−					
200				+	+	+	−				
100				+	+	+	+	−			
50				+	+	+	+	+	+		

NOTE: +, no cracks present; −, cracks present.

TABLE 2—Results from the SCC testing of copper alloys; comparison of laboratory test results for rod.

Materials, ISO Designation	Form (Fig. 1)	Annealing Conditions (Temperature, Time)	Calculated According to the Slitting and Deflection Method, Average Value, N/mm²[b]	0.5 h in HgNO₃-Solution According to ISO R 196[b,c]	Ammoniacal Atmosphere According to Eichhorn DIN 50916:1[c,d]	Ammoniacal Copper Sulfate Solution According to Mattsson SS 117102[d]	pH 9.5	pH 10.5
CuZn39Pb3	rod, round	nonannealed	130	+	+	+	+	+
		200°C, 1 h	65	+	+	+	+	+
		300°C, 1 h	33	+	+	+	+	+
	rod, hexagonal	nonannealed	284	−	−	−	−	−
		200°C, 1 h	125	+	−	−	+	−
		300°C, 1 h	51	+	+	+	+	+
	rod, asymetric	nonannealed	195	−	−	−	+	−
		200°C, 1 h	104	+	−	−	+	+
		300°C, 1 h	45	+	+	+	+	+
CuZn32Pb2AlSnAs	rod, round	nonannealed	157	+	+	+	+	+
		200°C, 1 h	136	+	+	+	+	+
		300°C, 1 h	73	+	+	+	+	+
CuZn38Sn1	rod, hexagonal	nonannealed	397	−	−	−	−	−
		200°C, 1 h	180	−	−	−	−	−
		300°C, 1 h	73	+	−	−	+	+

[a] +, no cracks present; −, cracks present.
[b] Test carried out by Dr. K. Camenisch, Metallwerke AG Dornach.
[c] Test carried out by Dr. H. Sick, Wieland-Werke AG.
[d] Test carried out by the authors.

TABLE 3—Results from the SCC testing of copper alloys: comparison of laboratory and field test results.

Results of Inspection after Indicated Exposure Time[a]

Material, ISO Designation	Form (Fig. 1)	Tensile Stress Level by Annealing Conditions (Temperature, Time) or by Compression of C-rings	Ammoniacal Atmosphere above 2 M NH₄Cl-solution with pH Values of: 9.3	9.5	10.0 Regular	10.0 Premoistened spec.	24 h in: Ammoniacal Atmosphere According to Eichhorn DIN 50916:1	Ammoniacal Copper Sulfate Solution According to Mattsson SS 117102	Four Years Outdoors in Stockholm under Rain Shelter
CuZn37	cups	nonannealed	−	−	−	−	−	−	−[c]
		200°C, 1 h	−	−	−	−	−	−	−[c]
		250°C, 1 h	−	+	−	−	−	−	+[c]
		300°C, 1 h	+	+	−	−	−	−	+[c]
CuZn24Ni12	cups	nonannealed	+	+	+	+	−	−	+
		250°C, 1 h	+	+	+	+	−	−	+
		300°C, 1 h	+	+	+	+	−	−	+
		350°C, 1 h	+	+	+	+	−	−	+
CuZn20Al2As	cups	nonannealed	−	−	−	−	−	−	−
		300°C, 5 min	−	−	−	−	−	−	−
		400°C, 5 min	+	+	+	+	+	+	+
		500°C, 5 min	+	+	+	+	+	+	+
	tubes	nonannealed	−	−	−	−	−	−	−
		400°C, 5 min	+	+	+	+	+	+[b]	+
		500°C, 5 min	+	+	+	+	+	+[b]	+
		600°C, 5 min	+	+	+	+	+	+[b]	+
CuZn39Pb3	C-rings	300 N/mm²	+	+	−	−	−	+[b]	+[c]
		200 N/mm²	+	+	−	−	−	+[b]	+[c]
		100 N/mm²	+	+	−	−	−	+[b]	+[c]
		50 N/mm²	+	+	+	−	−	+[b]	+[c]

TABLE 3—Continued.

| Specimens | | | Results of Inspection After Indicated Exposure Time[a] | | | | | | |
| | | | Ammoniacal Atmosphere above 2 M NH$_4$Cl-solution with pH Values of: | | | | 24 h in | | Four Years Outdoors in Stockholm under Rain Shelter |
Material, ISO Designation	Form (Fig. 1)	Tensile Stress Level by Annealing Conditions (Temperature, Time) or by Compression of C-rings	9.3	9.5	10.0 Regular	10.0 Premoistened spec.	Ammoniacal Atmosphere According to Eichhorn DIN 50916:1	Ammoniacal Copper Sulfate Solution According to Mattsson SS 117102	
CuZn32Pb2AlSnAs	C-rings	280 N/mm²	+	−	−	⋮	⋮	⋮	⋮
		180 N/mm²	+	−	−	⋮	⋮	⋮	⋮
		80 N/mm²	+	+	+	⋮	⋮	⋮	⋮
		30 N/mm²	−	+	+	⋮	⋮	⋮	⋮
CuAl10Fe5Ni5	C-rings	420 N/mm²	−	−	−	⋮	⋮	⋮	⋮
		320 N/mm²	+	−	−	⋮	⋮	⋮	⋮
		220 N/mm²	+	+	+	⋮	⋮	⋮	⋮
		170 N/mm²	+	+	+	⋮	⋮	⋮	⋮
CuNi10Fe1Mn	cups	nonannealed	+	+	+	⋮	⋮	⋮	⋮
	tubes	nonannealed	+	+	+	⋮	⋮	⋮	⋮
CuSn6	cups	nonannealed	+	+	+	⋮	⋮	⋮	⋮

ᵃ +, no cracks; −, cracks present.
ᵇ superficial intergranular corrosion.
ᶜ superficial dezincification.

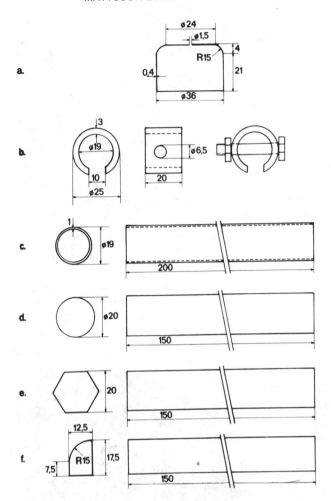

FIG. 1—*Specimen forms:* (a) *Cup, deep-drawn from soft-annealed strip and annealed to different levels of tensile stress,* (b) *C-ring, soft annealed after machining from rod and compressed by bolts to different levels of elastic tensile stress,* (c) *Tube,* (d) *Rod, round,* (e) *Rod, hexagonal, and* (f) *Rod, asymmetric (specimens cut from hard-drawn material and annealed to different levels of tensile stress).*

Evaluation—After exposure, the specimens were removed from the desiccator and immediately pickled clean in the sulphuric acid solution, containing 2–5% b.v. concentrated hydrogen peroxide. The specimens were rinsed, then dried in hot air. The surfaces were examined for possible cracking with a binocular at a magnification of ×15. Tube and rod specimens were subjected to slight deformation before inspection to open up possible cracks and make them more easily visible. For heavy sections, a surface zone was sliced off and bent before inspection, as shown in Fig. 3. When cracks were detected in any of the replicate specimens, this variant was considered to have failed (denoted "−" in Tables 1 and 2).

Effect of Premoistening—For some specimens of each variant, a further acceleration of the test was shown possible by premoistening the specimens immediately before exposure. The premoistening was done by exposing the specimens for 30 min to a mist of deionized water in a spray cabinet [2] at the same temperature as in the SCC test vessel.

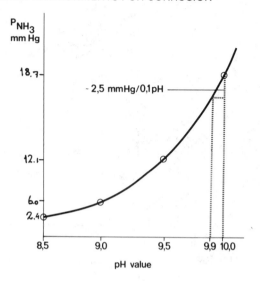

FIG. 2—*Relation between the partial pressure of ammonia in the atmosphere and the pH value of the ammonium chloride solution at 20°C* [1].

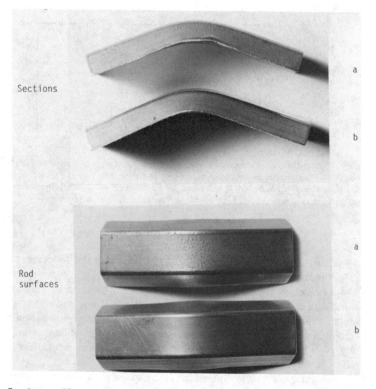

FIG. 3—*Specimens of hexagonal rod (CuZn39Pb3) exposed for 24 h to the atmosphere above the test solution with pH 10.5. The figure shows sliced-off surface zones deformed for opening-up of cracks:* (a) *rod annealed at 200°C for 1 h, and* (b) *rod annealed at 300°C for 1 h.*

Procedures of the Other Laboratory Tests

For comparison, tests were made with the mercurous nitrate test, the Mattsson solution test, and the severe Eichhorn atmosphere test, using a test material identical to that used in the test method proposed in the section on procedure.

The pretreatment of the specimens, the cleaning after exposure, and the evaluation of results were identical with those used in the test with the proposed method. The other test conditions used agree with those given in the test standards.

Field Test

An exposure test was carried out at the urban test site in Stockholm. There, the mean value of NH_3-N in the air is about 6 $\mu g/m^3$ and that of S-compounds, about 100 μg S/m^3 [3]. Earlier SCC field tests performed by Landegren and Mattsson [4] showed this site represents a moderately severe stress-corrosion environment.

Specimens were exposed under a rain shelter mounted on racks. The test was restricted to cup specimens of the alloys CuZn37 and CuZn24Ni12, cup and tube specimens of CuZn20Al2As, and C-ring specimens of CuZn39Pb3 in different tensile states. Every material variant was represented by twenty specimens. Replicates were tested by accelerated ammonia methods in the laboratory.

The exposure was terminated after four years (in May, 1985). Then the specimens were brought in for cleaning and inspection according to the same procedures as applied in the laboratory tests. When cracks were detected in any of the replicate specimens, this variant was considered to have failed (denoted "−" in Table 3).

Results

Comparison of Different Laboratory Tests

The results obtained with the different laboratory methods are shown in Table 2. These tests are restricted to certain variants of brass extrusions and have been carried out as pass/fail tests using an exposure time of 30 min for the mercurous nitrate test and 24 h for all the ammonia tests.

Comparison of the Field Test and Ammonia Tests in the Laboratory

Results from the field test and the ammonia tests of cup, C-ring, and tube specimens are given in Table 3. Also in this case, the ammonia tests have been carried out as pass/fail tests using an exposure time of 24 h. The crack paths determined by metallographic examination and cracking patterns visually observed on cup specimens are shown in Fig. 4.

Discussion

Comparison of the Proposed Test and the Other Laboratory Tests

From Tables 2 and 3, the following conclusions can be drawn about the correlation between the proposed test and the other tests investigated:

• The proposed test with pH 9.5 is somewhat less severe than the mercurous nitrate test, while the proposed test with pH 10.5 is slightly more severe. Against this background the proposed test with pH 10.0 appears equivalent to the mercurous nitrate test when applied to the materials studied here.

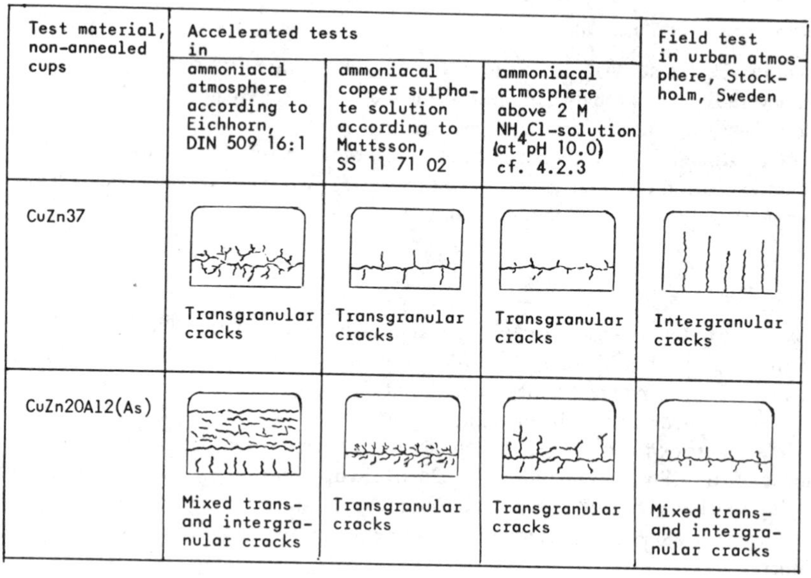

Test material, non-annealed cups	Accelerated tests in			Field test in urban atmosphere, Stockholm, Sweden
	ammoniacal atmosphere according to Eichhorn, DIN 509 16:1	ammoniacal copper sulphate solution according to Mattsson, SS 11 71 02	ammoniacal atmosphere above 2 M NH_4Cl-solution (at pH 10.0) cf. 4.2.3	
CuZn37				
	Transgranular cracks	Transgranular cracks	Transgranular cracks	Intergranular cracks
CuZn20Al2(As)				
	Mixed trans- and intergranular cracks	Transgranular cracks	Transgranular cracks	Mixed trans- and intergranular cracks

FIG. 4—*Cracking patterns and paths on cup specimens after exposure.*

• The Eichhorn test is the most severe test. It is more severe than the proposed test with pH 10.5, but premoistening the specimens in the proposed test with pH 10.0 made this test generally equivalent with the Eichhorn test, except for nickel silver (CuZn24Ni12).

• The Mattsson solution test was generally found more severe than the proposed test, even with pH 10.5. Yet, exceptions were noted for C-ring specimens of leaded brass (CuZn39Pb3) and for tube specimens of aluminum brass (CuZn20Al2As). It should also be noted that the Mattsson solution caused superficial intergranular corrosion on the alloys mentioned.

• There is a correlation between the average residual tensile stress calculated (Table 2) and the results from the accelerated tests investigated, as only specimens with the highest stress levels failed. It can further be noted that a certain calculated stress level was more detrimental in a hexagonal rod and an asymmetric rod than in a round rod, probably because of the concentration of stress at edges. This was confirmed by the fact that the cracks were located at the edges.

Comparison of the Proposed Test and the Field Test

Table 3 shows that there is a good correlation between the results from the field test and the proposed test with pH 9.5, although the latter is generally a little more severe. An exception, however, exists for nonannealed tube specimens of aluminum brass (CuZn20Al2As), which cracked in the field test, but did not crack in the proposed test until the pH value was raised to 10.0. The other ammonia tests were considerably more severe than the field test. Another exception was noted for the highest stress level of leaded brass, which cracked in the proposed test, but did not crack in the field exposure. The cause was probably superficial dezincification in the latter case.

As for crack path and cracking pattern, the proposed test and the field test differ as shown in Fig. 4. The difference in crack path probably results because the pH value of the corrosive agent is higher in the proposed test as well as in the other accelerated ammonia tests than it is on exposure in the urban atmosphere of Stockholm; such pH dependence of the crack path has been shown earlier [5]. The difference in cracking pattern may be related to a nonuniform stress distribution in the section of the specimen. As stress corrosion is probably initiated at different penetration depths in the two cases, different cracking patterns may result. It does not seem possible, however, to design one accelerated test that is representative of all exposure conditions occurring in service.

Parameters for the Proposed Test

pH Value—As seen in Tables 2 and 3 the pH value of the test solution is crucial to the severity of the test. A pH value of 9.5 is representative of the urban atmosphere in Stockholm; pH 10.0 provides a test approximately equivalent to the mercurous nitrate test, and pH 10.5 gives a test nearly as severe as the Eichhorn test.

It is important to use a test solution with sufficient capacity to maintain the desired ammonia content in the atmosphere throughout the exposure period. In this respect, an ammonium chloride solution, with its comparatively high buffer capacity, is preferable to an ammonium hydroxide solution.

Temperature—As shown in DIN 50916: Part 2, the temperature has to be kept constant within ±1 to 2°C during the exposure to avoid dew condensation on the specimens.

The results in Table 1 show the importance of temperature conditioning the specimens before bringing them into the testing vessel. Dew condensation taking place on the surface of a specimen with a lower temperature than the testing atmosphere will accelerate the test somewhat. The same effect was obtained by premoistening the specimens (Tables 1 and 3).

Exposure Time—For a pass/fail at-shop or acceptance test, it is desirable to keep the exposure time as short as possible. An exposure time of 7 to 15 h, however, may cause difficulties in working hours; 24 h is appropriate. The effect of occasional variations in dew condensation during the exposure will then level out (Tables 1 and 3, and the following section), which results in good reproducibility.

In a pass/fail test of this type, it is also appropriate to work with a fixed exposure time. Using a varying exposure time would require either a great number of specimens or reexposure of specimens after cleaning and examination, which is not considered acceptable.

Premoistening of Specimens—The results in Table 1 show that premoistening of the specimens before exposure shortens the time to cracking, i.e., it accelerates the test. The likely cause is that a solution film is required on the metal surface for the stress corrosion process to start. After long exposure, however, the effect of premoistening largely disappears (Table 3).

Deformation of Specimens before Inspection—When testing heavy sections using the proposed method and others, the cracks may be very fine and difficult to detect, especially with great surface roughness present. When deforming the exposed specimen, e.g., by bending, the cracks open up and become easily visible as shown in Fig. 3. For very heavy sections, when deformation of the whole specimen is difficult, a thin surface zone may be sliced off for deformation.

Specimens exposed to a test solution, e.g., the Mattsson solution, or exposed for a long time to an outdoor atmosphere, may have suffered intergranular corrosion or dezincification. Such attack may have the appearance of a network of very fine superficial cracks after deformation, and should be distinguished from the more distinct, often deep cracks formed by stress corrosion. This difficulty has not been encountered, however, using the proposed method.

Conclusions

The proposed stress corrosion test for copper alloys has shown the following features favoring its use for R & D as well as for production and acceptance control:

• Its severity can be regulated by changing the pH value of the test solution to comply with different requirements in practice. Thus, a test with pH 9.5 seems representative of the urban atmosphere; pH 10.0 provides a test approximately equivalent to the well-known mercurous nitrate test; and pH 10.5 gives a test nearly as severe as the Eichhorn test, used when an extreme safety margin is required.

- The exposure time is 24 h, which generally is convenient for practical use.
- The equipment and the procedure for testing are simple and also suitable for shop application.
- The method does not involve handling poisonous chemicals.
- A similar test has already been used in some countries, e.g., Switzerland, Great Britain, and the Federal Republic of Germany.

The test has been approved for ISO standardization (ISO 6957).

References

[1] Aebi, F., *Zeitschrift für Metallkunde,* Vol. 49, No. 2, 1958, pp. 63–68.
[2] Spray cabinet Type 450 032/033, H. Kühler KG, Lippstadt, FRG.
[3] Mattsson, E., Lindgren, S., Rask, S., and Wennström, G., *Proceedings of the 4th Scandinavian Corrosion Congress,* Helsinki, 1964, pp. 171–180.
[4] Landegren, W. and Mattsson, E., *British Corrosion Journal,* Vol. 11, No. 2, 1976, pp. 80–85.
[5] Mattsson, E., *Electrochimica Acta,* Vol. 3, No. 4, 1961, pp. 279–291.

Paul D. Goodman,[1] *Victor F. Lucey,*[1] *and Carlo R. Maselkowski*[1]

A Synthetic Environment to Simulate the Pitting Corrosion of Copper in Potable Waters

REFERENCE: Goodman, P. D., Lucey, V. F., and Maselkowski, C. R., "**A Synthetic Environment to Simulate the Pitting Corrosion of Copper in Potable Waters,**" *The Use of Synthetic Environments for Corrosion Testing, ASTM STP 970,* P. E. Francis and T. S. Lee, Eds., American Society for Testing and Materials, Philadelphia, 1988, pp. 165-173.

ABSTRACT: The BNF rapid test to assess the ability of water to support the pitting corrosion of copper is described. The test cell provides an environment that simulates conditions that, in service, lead to the pitting of copper. The mode of operation is explained in terms of copper corrosion mechanisms elucidated by investigations carried out at BNF. The development of a synthetic water capable of supporting the pitting corrosion of copper is also described.

KEY WORDS: copper, local corrosion, pitting, crevice corrosion, electrochemical techniques

Copper is widely used for pipe work in potable water plumbing systems in many countries throughout the world, and copper hot storage vessels are commonly used in the United Kingdom, New Zealand, and other countries where low pressure hot water systems are employed. Failures are few, but those that do occur are usually the result of pitting corrosion. Various types of fresh water pitting of copper have been reported in the literature [1-5]. In the United Kingdom, the United States, Belgium, and Holland, the majority of pitting failures are due to what is now designated as "Type 1" pitting corrosion [5].

A number of factors determine the occurrence and severity of Type 1 pitting. It is principally a cold or warm water phenomenon. Pitting in hot water can occur, but this proceeds by an apparently different mechanism and is known as "Type 2" pitting [5]. The initiation of Type 1 pitting also requires periods in which the copper is in contact with stagnant or relatively slow moving water. Type 1 pitting in copper tubes is frequently associated with the presence of carbon films in the bore resulting from manufacturing processes[2,3] [6-12].

It is well established that pitting occurs only in waters of certain compositions. Campbell [15] has shown that surface derived waters containing organic matter would not cause Type 1 pitting corrosion of copper. Waters from underground sources, however, are usually very low in organics, and in such waters, the occurrence of pitting depends on the balance of the inorganic constituents.

A number of investigators in various countries [14-16] carried out statistical analyses to determine the relationship between the composition of waters and the known incidence of pitting

[1]Materials engineer, principal materials engineer, and materials engineer, respectively, BNF Metals Technology Center, Grove Laboratory, Wantage, Oxfordshire, OX12 9BJ, England. Mr. Goodman is now Senior Research investigator, Cookson Group PLC, 7 Wadsworth Rd., Perivale, Greenford, Middlesex UB6 7JQ, England.
[2]Devroey, P. and Depommier, C. Visseries et Trefileries Reunies S. A., Machelen, Belgium, (unpublished report), 1962.
[3]Borsma, H. J., Elzenga, C. H. J., and Nijolt, H., KIWA, Holland, (unpublished work).

occurring in the areas to which the waters were supplied. Nomograms were developed to predict the pitting propensity of a water of a given composition. Although there was some agreement about the most important factors, each investigation found a different combination of variables to be most significant. This is not surprising, since many water composition variables are inter-related by various equilibria. Furthermore, the number of variables which need to be considered is large, and the available data relatively limited. In addition, the range of water compositions was somewhat different in each investigation. In Lucey's work [16] the principal determining factors were the concentration of chloride and sulfate ions, pH, and dissolved oxygen, with sodium and nitrate content as secondary but still important variables. In other work, the sulfate to chloride ratio [14] and dissolved carbon dioxide (CO_2) content (related to pH) [15] were among other significant factors. Within a limited range of compositions, nomograms have proved useful in predicting the likely aggressivity of supply waters. However, more accurate and generally applicable tests were required to determine water aggressivity.

Pourbaix [17] and Lucey [18] have noted that measurements of the electrochemical potential of copper in water show that tube samples in which no pitting is occurring tend to establish rest potential values below about 50 mV SCE. Potentials above 100 mV are usually associated with tubes undergoing pitting corrosion. Thus, monitoring the electrode potential of copper tubes in test rigs simulating normal domestic service can assess the pitting propensity of water supplies. However, although very reliable, these tests take a long time (usually 6 to 12 months or more), and consume large quantities of water. Consequently the procedure is not suitable for the routine assessment of the pitting propensities of waters or, for example, the effects of blending together waters from different supplies.

The need for a reliable, simple, inexpensive, and rapid method of assessing the propensity of a supply water to cause Type 1 pitting has resulted in the development of the electrochemical technique described below. Work is also underway to develop a simple test to assess the degree of carbon film contamination in copper tubes. A synthetic solution designed to simulate a natural water and used for this test is described.

Background of the Test

Description of Test Method

The electrochemical cell (sometimes referred to as the rapid pitting cell) was developed over a number of years by Goodman and Lucey [19] following the initial work of Lucey and Shaw [20]. The principle of the test is to pass a small current from an external source through a multiple copper anode in a cell containing the water being tested. One of the anode elements, which are connected in parallel, is provided with a diffusion barrier at its surface. This promotes pitting corrosion in aggressive waters, but has no effect in waters that do not support Type 1 pitting corrosion. The depolarization that occurs in the occluded electrode in an aggressive water results in a higher proportion of the total current being carried by that element. The degree of nonuniformity of the current distribution at the end of the test is used to measure the pitting corrosion propensity of the water.

The construction of the cell is illustrated in Fig. 1. The anode base consists of 1.2-mm copper wires sealed into a rectangular acrylic block. Twelve wires are arranged in a circular configuration with an additional central anode. Each anode wire is connected via a 510 Ω resistor to a common point. The cell body is a section of 32-mm acrylic tube. A cathode consisting of a copper wire, and water inlet and outlet tubes are built into a cell top. The hemispherical end of a vertically positioned glass rod is allowed to rest on the central anode, providing the diffusion barrier. The anode block is prepared by abrading the surface with "wet and dry" paper down to 1200 grit. It is washed with distilled water and dried with an air blast, and the test cell is then assembled.

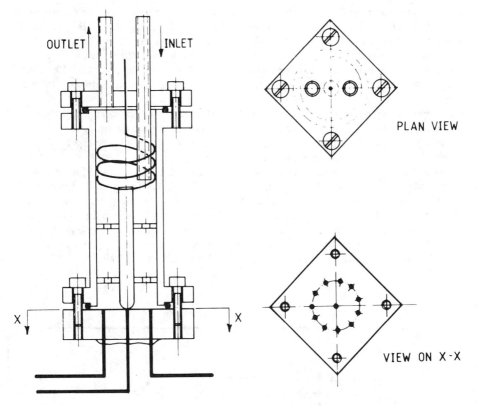

OUTLET   INLET

PLAN VIEW

X   X

VIEW ON X-X

FIG. 1—*The electrochemical test cell.*

Water for the test is contained in a closed vessel, e.g., a neoprene rubber air bag, to avoid loss or gain of dissolved gases. About 4 L of water are normally used for the test. The water is allowed to flow slowly through the cell and out for a period of 100 h, during which time a constant applied current of 15 μA is passed through it. A flow rate of 40 mL/h is usually employed. However, varying the flow rate between 20 mL/h and 2 L/min does not significantly affect the results. The current passed by each anode element is determined by measuring the potential drop across its series resistor.

Use of the Electrochemical Pitting Cell to Assess the Aggressivity of Supply Waters

If a water which does not support Type 1 pitting is tested, a protective oxide film is formed over all the anodes. With an extremely aggressive water, up to 100% of the total current at the end of the test may be carried by the central anode. Most waters lie somewhere between these two extremes. Practical experience with the rapid pitting test cell indicates that the aggressivity of a supply water towards copper plumbing components is related to the proportion of the total current carried by the central anode of the cell at the end of the test period.

Twenty-nine different supply waters were tested in the rapid pitting cell, and the current carried by the central anode was compared with the calculated pitting propensity rating (PPR) of these waters [16]. The PPR of a supply water is calculated by means of a nomogram based on its chemical composition, and is directly related to the expectation of failure of copper plumbing components by Type 1 pitting corrosion in a water of that composition within a given period of time [16]. The PPR rating system was developed by Lucey [16], who studied the relationship between time to failure and water composition for plumbing components in 120 supply waters. Increasingly positive PPR's correspond to increasingly rapid expected times to failure. PPR values of 0, 2, 4, and 6 indicate most probable times to failure of approximately 30, 10, 3, and 1 years, respectively. A PPR value of zero is taken as the practical boundary between pitting and nonpitting waters.

The relationship between the test-cell central anode current and the calculated PPR rating for the twenty nine supply waters is shown in Fig. 2. This provides a quantitative comparison between the statistically expected aggressivity of these supply waters and the rapid-pitting cell results. A good general correlation is observed.

BNF has used the rapid pitting cell to test more than 60 waters that were suspected (e.g., by Water Authorities, copper tube manufacturers, and other interested parties) of causing the failure of copper plumbing components. Based on examination of service failures most of these waters were associated with Type 1 pitting corrosion in the past. The rapid-pitting cell gave results that were always consistent with the available service experience. On the basis of the data compiled in Fig. 2 and the above-mentioned service comparisons, general guidelines have been drawn for rating the aggressivity of supply waters. These are shown in Table 1.

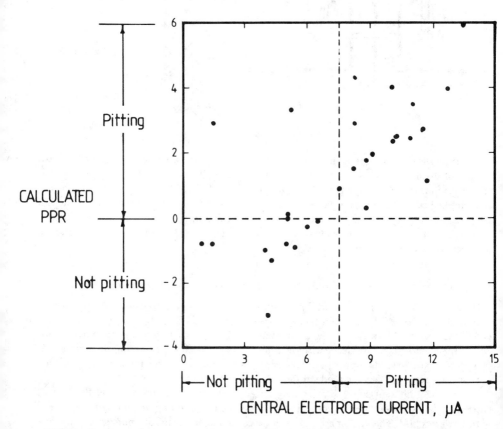

FIG. 2—*Correlation between rapid cell results and calculated PPR ratings for twenty seven supply waters.*

TABLE 1—*Guidelines for rating the aggressivity of supply waters.*

Percentage of Total Current Passed by Central Element	Nature of Water
<20%	nonpitting water
20%-50%	borderline case (failures not expected within realistic times)
50%-70%	fairly aggressive pitting water
>70%	very aggressive pitting water

The broken lines in Fig. 2 also indicate the demarcation between waters that would and would not be expected to cause the failure of copper plumbing components by Type 1 pitting corrosion on the basis of both the nomogram and the electrochemical test cell.

In addition to the generalized results described above, several specific examples of the use of the rapid-pitting cell illustrate its effectiveness. An apparent anomaly occurred in testing water supplied to Hatfield, Hertfordshire. This water has caused many pitting failures in the past. However, the results of recent tests carried out on four separate occasions indicated that the water should not support Type 1 pitting. The water was subsequently tested using a long-term copper-tube test rig. This confirmed that the water was now no longer capable of supporting Type 1 pitting.

The cell has been demonstrated on a number of occasions to produce more reliable results than the Lucey nomogram. Some examples are as follows:

1. The nomogram predicted that water from Weymouth in Dorset should be nonaggressive, ascribing to it a PPR of −1. Tests carried out using the electrochemical cell on three separate occasions indicated that the water varied from borderline to very aggressive. Since in terms of pitting corrosion, no service experience was available, a copper-tube test rig was installed. The results obtained confirmed that Weymouth water is indeed fairly aggressive.

2. Water from the Egham pumping station has a PPR of 4. This corresponds to a most probable time to failure of copper tubes and cylinders of about 3 years [16]. However, pitting problems have not been reported from this area. The electrochemical cell gave a central anode current of about 50%, indicating a slightly aggressive water. The electrochemical cell result is thus more in line with service experience.

3. Five water samples with sulfate levels less than 10 ppm were tested. Center anode element current values for all five waters were much higher than would be expected from the calculated PPR values using the Lucey nomogram. Reexamination of the service data from which the nomogram was derived showed that as a result of statistical averaging, the nomogram systematically underestimated the aggressivity of very low sulfate waters. Thus, for this type of water, the electrochemical test cell gives a better indication of the expected service life.

4. The electrochemical cell has been used to investigate the effect of base exchange softening on the aggressivity of supply waters. It is known from service experience that this form of water treatment increases the tendency for a water to cause pitting. This would be predicted by the nomogram on the basis of the increased sodium ion content. Two water samples from Germany were examined, one taken before and the other after base exchange softening. The respective PPR values were calculated as −1.3 and +10.3, i.e., there was a change from a nonpitting water to one in which failures would be expected in less than 3 months [16]. On the other hand, the electrochemical cell indicated that the base-exchange softened water had only slightly more aggressive characteristics. This is a much more realistic assessment of the effect of base exchange softening.

Explanation of Electrochemical Cell Operation

Mechanism of Pit Initiation

Various aspects of the mechanism of Type 1 pitting have been studied by Lucey [18,21-24]. The establishment of a membrane electrode is an essential step in initiating Type 1 pitting. Cuprous ions are produced at anodic sites when copper is initially immersed in aerated water.

$$Cu \rightarrow Cu^+ + \epsilon \tag{1}$$

There will be a corresponding reaction at cathodic sites.

$$O_2 + 2H_2O + 4\epsilon \rightarrow 4OH^- \tag{2}$$

Cuprous oxide is formed according to the equilibrium reaction:

$$2Cu^+ + H_2O \rightleftharpoons Cu_2O + 2H^+ \tag{3}$$

Initially, the equilibrium will go to the right, since the H^+ ions produced will diffuse rapidly into the adjacent solution. Cuprous oxide is porous and electrically conducting. Further oxide formation will occur principally at the interface between the metal and the oxide. Reactions (1) and (3) result in the formation of positively charged species. Charge neutrality will be maintained by the migration of anion species from the bulk solution into the reaction interface, with the diffusion of cations (Cu^+ and H^+) into the bulk solution. The only stable cuprous salt that can be formed from the anions normally present in supply waters is cuprous chloride [25]. This is stable at pH values less than 3 or 4. At less acidic pH values, cuprous chloride is eventually hydrolyzed, to form cuprous oxide. In the absence of chloride, any free cuprous ions are unstable and will be converted to the cupric form. (Chloride ions are in fact believed necessary for the occurrence of Type 1 pitting).

The equilibrium of reaction (3) will go completely to the right if the rate of removal of H^+ ions is greater than the rate of increase in Cu^+ ion concentration in the reaction region. However, if diffusion through the oxide layer is sufficiently restricted, and if chloride ions are present in the water, the species in Eq (3) will co-exist in equilibrium. This means that the cuprous oxide layer will be separated from the metal below by a layer of solution containing cuprous, hydrogen, and chloride ions. Solid cuprous chloride may or may not be present as a layer on the copper metal, depending on the pH and cuprous ion concentration. This arrangement constitutes a membrane cell.

Once the membrane cell has been established, pitting may proceed. The process is illustrated schematically in Fig. 3. In a membrane cell, the underside of the membrane supports the anodic oxidation of cuprous ions to cupric ions.

$$Cu^+ \rightarrow Cu^{2+} + \epsilon \tag{4}$$

The cupric ions produced may either remain in the electrolyte under the membrane solution of diffuse through the membrane. In the former case, the cupric ions may diffuse to the nearby copper metal, reacting with it to generate more cuprous ions.

$$Cu^{2+} + Cu \rightarrow 2Cu^+ \tag{5}$$

The cupric ions passing through the membrane are either precipitated, pass into the bulk of the water, or react at the upper surface of the membrane to regenerate cuprous ions.

$$Cu^{2+} + \epsilon \rightarrow Cu^+ \tag{6}$$

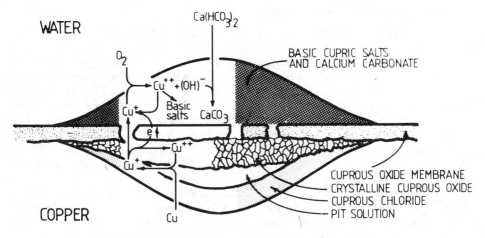

FIG. 3—*Mechanism of pitting corrosion of copper in potable waters.*

The cuprous ions in the solution above the membrane (these also include cuprous ions that have diffused through the membrane) will be oxidized to cupric by the oxygen in the water.

$$4Cu^+ + O_2 + 2H_2O \rightarrow 4Cu^{2+} + 4OH^- \tag{7}$$

This results in the precipitation of relatively insoluble cupric compounds (e.g., $CuCO_3 \cdot Cu(OH)_2$) and the reduction of cupric ions to cuprous ions on the upper surface of the membrane according to Eq 6. The pH increase in the region above the membrane also results in the precipitation of calcium carbonate if the water is hard.

$$HCO_3^- + OH^- \rightleftharpoons CO_3^{2-} + H_2O \tag{8}$$

The overall process results in pitting, with copper transported from the metal through the membrane into the mound above it and the bulk solution.

Pit Initiation in the Electrochemical Cell

The principles outlined above may be used to explain the operation of the electrochemical cell. The passage of anodic current results initially in the formation of cuprous oxide (Eqs 1 and 3) on all electrode elements. At the central anode, however, cuprous and H^+ ions diffusing into the solution will tend to build up in the most occluded region. This will occur after oxygen initially present in this region has been consumed in the conversion of cuprous ions to cupric ions.

The accumulation of the Cu^+ and H^+ ions reduces the net diffusional flux of these species away from the reaction interface into the adjacent local solution. This facilitates the formation of a membrane cell as previously outlined. The central electrode subsequently begins to draw proportionately more current than the other electrodes. This is because the passage of anodic current on the other electrodes causes the protective oxide film to thicken, thus offering more resistance to the passage of current than in a membrane cell containing the concentrated pit solution.

The Development of an Artificial Pitting Solution

In waters which support Type 1 pitting, the occurrence of copper tube failures is usually associated with the presence of residual carbon films in the tube bore. Manufacturers of copper tube take measures to remove residual films, but a technique for assessing the degree of any remaining carbon contamination is still needed. BNF is currently engaged in developing such a test, for which a laboratory-prepared solution capable of supporting Type 1 pitting corrosion is required. Goodman [26] has prepared such a solution, consisting of calcium bicarbonate with sodium chloride and sodium sulfate additions. The calcium bicarbonate is prepared by bubbling CO_2 gas through a suspension of calcium carbonate. Air is bubbled through the resulting solution until a pH of 6 to 7 is attained. The solution composition is Cl^-: 30 ppm, sulfate ion (SO_4^{2-}): 90 ppm, Na^+: 63 ppm, Ca^{2+}: 120 ppm, Total hardness: 300 ppm, Total alkalinity: 300 ppm.

According to the rapid pitting test, the solution is moderately aggressive. Attempts to make pitting solutions from sodium bicarbonate, calcium chloride and calcium sulfate were unsuccessful. In spite of positive PPR values, these solutions did not result in pitting in electrochemical cell tests [26]. The reasons for this apparent anomaly are not understood.

Conclusions

An electrochemical test has been developed that was extremely successful in predicting the aggressivity of supply waters for copper plumbing components. The technique represents a considerable improvement over previously established nomogram methods. It is rapid and inexpensive, and requires only a small volume of the water under test.

References

[1] Campbell, H. S., Water Treatment and Examination, Vol. 20, 1971, p. 11.
[2] BNF Miscellaneous Publication No. 568, February 1972.
[3] Linder, M. and Lindman, E. K., Proceedings of the 9th Secondary Corrosion Conference, Vol. 2, September 1983.
[4] Suzuki, I., Ishikawa, Y., and Hisamatsu, Y., Corrosion Science, Vol. 23, 1983, p. 1095.
[5] Campbell, H. S., BNF Miscellaneous Publication No. 574, August 1982.
[6] Campbell, H. S., Journal of the Institute of Metals, Vol. 77, 1950, p. 345.
[7] Von Franque, O., Werkstoffe und Korrosion, Vol. 5, 1968, p. 377.
[8] Schafer, G. J., Australasian Corrosion Engineering, Vol. 6, 1962, p. 15.
[9] Nielsen, K., Dansk V. V.S. (Copenhagen), Vol. 8, 1971, p. 367.
[10] Einarsson, A. and Elisson, G., Idnadamal (Reykjavik), 1971.
[11] Campbell, H. S., Proceedings of the 2nd International Congress on Metallic Corrosion, National Association or Corrosion Engineers, New York, 1966, p. 237.
[12] Campbell, H. S., Published discussion of P. T. Gilbert, Proceedings of the Society of Water Treatment and Examination, Vol. 15, 1966, p. 180.
[13] Campbell, H. S., Journal of Applied Chemistry, (London), Vol. 4, 1954, p. 663.
[14] Depommier, C. T., INCRA Project No. 317, 2nd Progress Report, July 1981.
[15] Diegle, R. B. and Berry, W. E., INCRA Project No. 272, Final Report, April 1980.
[16] Lucey, V. F., BNF Research Report RRA 1838, BNF Metals Technology Centre, Oxfordshire, England, December 1972.
[17] Pourbaix, M., Corrosion, Vol. 25, No. 6, 1969, p. 267.
[18] Lucey, V. F., BNF Research Report RRA 1866, BNF Metals Research Centre, Oxfordshire, England, May 1984.
[19] Goodman, P. D. and Lucey, V. F., BNF Report R392/13, BNF Metals Research Centre, Oxfordshire, England, July 1984.
[20] Shaw, D. and Lucey, V. F., BNF Research Report RRA 1956, BNF Metals Research Centre, Oxfordshire, England, June 1979.
[21] Lucey, V. F., BNF Research Report A 1484, BNF Metals Research Centre, Oxfordshire, England, July 1964.

[22] Lucey, V. F., *BNF Research Report A 1552*, BNF Metals Research Centre, Oxfordshire, England, September 1965.
[23] Lucey, V. F., *British Corrosion Journal*, Vol. 2, 1967, p. 166.
[24] Lucey, V. F., *British Corrosion Journal*, Vol. 7, 1972, p. 36.
[25] Cotton, F. A. and Wilkinson, G., *Advanced Inorganic Chemistry*, John Wiley Interscience, 1962, p. 749.
[26] Goodman, P. D., *BNF Paper S392.9*, BNF Metals Research Centre, Oxfordshire, England, June 1981.

James E. Castle,[1] *Anthony H. L. Chamberlain,*[1] *Bradley Garner,*[1]
M. Sadegh Parvizi,[1] *and Abraham Aladjem*[1]

The Use of Synthetic or Natural Seawater in Studies of the Corrosion of Copper Alloys

REFERENCE: Castle, J. E., Chamberlain, A. H. L., Garner, B., Parvizi, M. S., and Aladjem, A.,
"The Use of Synthetic or Natural Seawater in Studies of the Corrosion of Copper Alloys," *The Use
of Synthetic Environments for Corrosion Testing, ASTM STP 970,* P. E. Francis and T. S. Lee,
Eds., American Society for Testing and Materials, Philadelphia, 1988 pp. 174–189.

ABSTRACT: The inorganic components of seawater are well known and easily simulated by mixtures of the appropriate salts. The adoption of such mixes has the advantage that the buffering ability of seawater is simulated to some extent but leaves open the question of the role of the organic component of seawater, which is not simulated in any of the normal recipes available. It is always open to the experimenter to avoid the problem of simulating seawater by carrying out the test exposures in the sea itself. However seawater has such a variability that any exposure site or time of year will be inadequate in some unknown regard.

This paper describes our experience in holding water in the laboratory and of work relating to the influence of organic natural products on the corrosion of Kunife 10 (alloy CA706, 10Ni1.5FeCu alloy). To attempt to create a "modus operandi" for waters containing organic species, it is essential to first envisage their possible actions, for example, chelating action, inhibition, stimulation of transport, a reduction in ionic activity in solution leading to a change in the Pourbaix diagram, modification of crystal habits, or the formation of weak boundary layers. These aspects will be discussed in the paper.

A detailed study of the micro-organisms, which develop in tanks containing seawater over the period of their use as a stock of test medium and of the concomitent changes in water chemistry, has been made and will be reported. A more direct method of examining the effect of the organic component is to remove it from the water in order that it may be added back to synthetic mixes or used to augment the seawater itself. A system, which permits the removal and retention of high molecular weight compounds, will be described.

It is concluded that the organic material may have less influence on the corrosion of the Kunife alloy than on some other systems but that the techniques established to deal with organic components of seawater have been successful. In particular an influence of organic molecules on the cathodic process at low oxygen levels has been demonstrated.

KEY WORDS: corrosion, copper alloys, seawater, laboratory tests

The inorganic components of seawater are well known and easily simulated by mixtures of the appropriate salts. Thus several standard recipes exist [1,2] and are widely used for corrosion studies. The adoption of such mixes has the advantage that the buffering ability of seawater is simulated to some extent, which is not so when merely the ionicity of seawater is simulated, for example, by the use of 3.4% sodium chloride solution [3]. Inorganic mixes, however, leave open the whole question of biological activity and interactions including the role of the organic com-

[1]Professor, lecturer, research officer, research officer, and visiting scientist, respectively, Departments of Materials Science and Engineering and Microbiology, University of Surrey GU2 5XH, United Kingdom.

ponent of seawater, which is not simulated in any of the normal recipes available. Copper alloys are known to show intriguing differences in corrosion properties according to the season of first exposure, and we have already shown that the products formed on aluminum brass are quite different if they are exposed to natural seawater than if they are exposed to synthetic waters [4,5]. It seems likely that some of this variability stems from the organic component of the seawater, which is complex and enormous in its range of molecular species (Fig. 1) [6,7]. This

Scheme 1. Biologically active metabolites from seaweeds, algae and crustaceans.

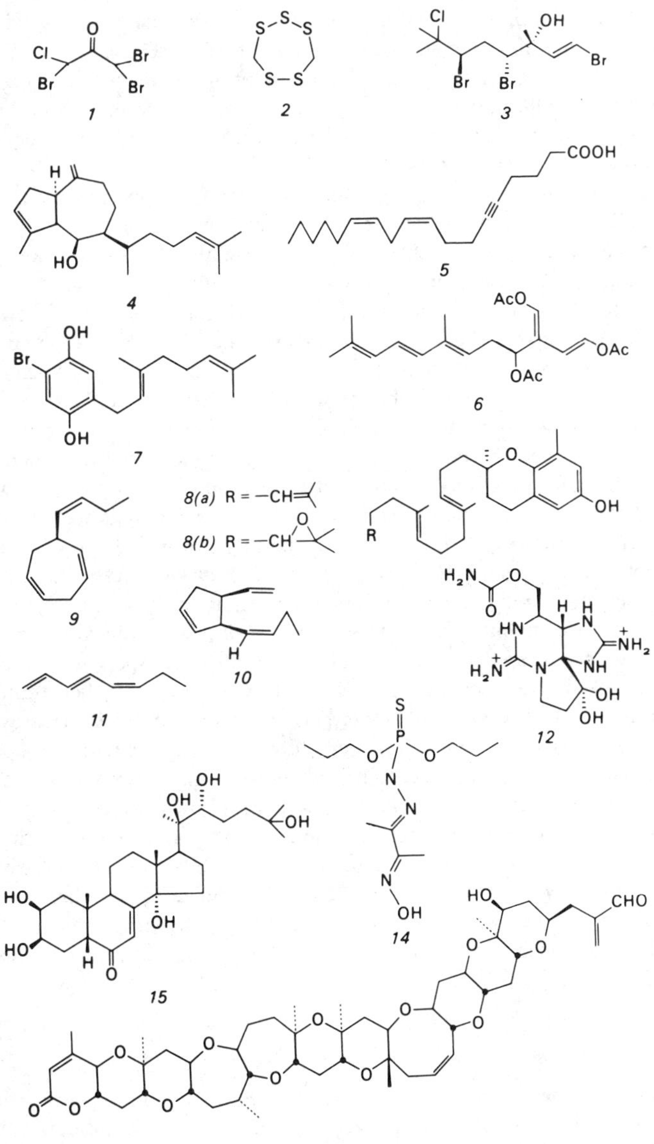

FIG. 1—*A selection of the smaller molecules associated with microorganisms as illustrated by Naylor* [6]. *The numbers enable the molecules to be identified in Naylor's paper.*

paper describes our experience in holding water in the laboratory and of work relating to the influence of organic natural products on the corrosion of Kunife 10 (Alloy Ca706, 10Ni1.5FeCu alloy).

The Possible Roles for Organic Compounds

It is always open to the experimenter to avoid the problem of simulating seawater by carrying out the test exposures in the sea itself. Such exposure may test the influence of the components not included in the synthetic mixtures but is difficult to standardize because seawater has such a notorious variability that any exposure site or time of year may be inadequate in some unknown regard. Thus data based only on exposure trials poses impossible problems for those charged with the task of investigating some future corrosion outbreaks. For example, did corrosion ensue because of the presence of some natural accelerator or indeed because of the absence of a natural inhibitor? The alternative of accounting in laboratory experiments for the possible roles played by organic molecules is a complex problem. To attempt to create a "modus operandi," it is essential to first envisage their possible actions.

Many organic molecules have a strong chelating action for copper, we think for example of the inhibiting effect of molecules, such as benzotriazole and its analogues. In some cases the chelate may be easily reversible, so that it actually stimulates the transport of copper, as does cystein [8]. Some molecules may be more specific for a given oxidation state than is benzotriazole (BTA), so that copper is held in a single valence state; we have, for example, already reported that Cu(I) appears to have an exceptional stability on surfaces exposed to natural seawater [5]. The reduction is activity of ions of copper in the aqueous phase when a stable chelate is formed may be such as to require a change in the boundaries of the Pourbaix diagram applicable to the system. Such a decrease in activity on germination of the copper tolerant alga *Ectocarpus* has been described by Hall and Baker [9], and we have used this phenomenon as the basis for explaining the selective dissolution of copper from aluminum brass with the consequential enrichment of aluminum on the surface. This occurs only in natural seawater and is contrary to thermodynamics as represented by the normal Pourbaix diagram. The adsorption of organic compounds on the surface may interfere with electrode reactions and may themselves undergo electrochemical reduction or oxidation, according to the prevailing oxygen potential. Alternatively organic compounds act as brighteners in electrocrystallization, modifying crystal habits, or indeed preventing the growth on surfaces of particular species. Such compounds are of course widely used as antiscaling agents in heat exchangers. Another example of organic action is that of providing weak boundary layers. Such layers are recognized by the adhesives and coatings industries as responsible for premature failure; Blunn [10] has shown that bacterial colonization of corrosion products can provide zones for the easy separation of the corrosion products, perhaps the origin of the sloughing action [11], which keeps copper alloys free of macrofouling layers. One final example of organic action may be found in the generation of waste products, such as ammonia or sulfides. Strictly these fall in the ambience of the inorganic components yet their level is determined by the activity of microbes in relation to the biological oxygen demand of the system. Recently Newman [12] has drawn attention to the generation from sulfides of the thiosulfate ion, and we have found this to influence corrosion of copper alloys at the ppm level.

Experimental Procedures

The collection of seawater for laboratory use initiates changes in the ecology of the environment contained within it, which we can barely control but certainly not oppose. Providing the water is aerated and provided with light then metabolism of the organisms present in the original stock will proceed. Because of the differing abilities of these organisms to cope with the situation and because of the removal of some threats of predation, the system will begin to diverge more and more from the original state. It has been our policy to channel this divergence into three representative seawaters. The procedure is illustrated in Fig. 2. Seawater is collected from the marine biology station of the Central Electricity Generating Board (CEGB, Great Britain) at Fawley Power Station and transported the 40 miles (64 km) to Guildford. It is divided and charged into lit and aerated tanks on the day of collection. One part is filtered to 0.2 μm, the second to 20 μm, and the third is unfiltered. These differing levels of filtration permit different classes of microorganisms to remain in the water. The lighting is supplied from "Trulite" tubes supported above the tanks and operated on a 12-12 h light-cycle, the unfiltered tank receiving the greatest intensity. The temperature of the tanks remained between 17 and 22°C and the pH value between 8.00 and 8.15; nutrients were not supplied and the water volume (104 L) supported optimum diaton populations for about ten weeks. The development of free floating planktonic populations was precluded by the inclusion of airlifts acting as undergravel filtration systems. In the partly filtered tanks yellow-brown stains of diatom colonies always appeared on the walls after a few days.

A detailed survey of the micro-organisms, which develop over the period of their use as a stock of test medium and of the concomitant changes in water chemistry has been made. The tanks were examined over a 78-day period corresponding to the length of run normally used during the earlier corrosion studies. The following parameters were measured initially on a daily basis: pH, temperature, conductivity, and dissolved oxygen levels. Bacterial counts, fungal counts, nitrate, phosphate, and silicate were measured on a weekly basis and samples for estimation of all other micro-organisms were taken periodically and subjected to light and scanning electron microscopy. These samples were obtained from arrays of vertically placed coverslips suspended just above the gravel base.

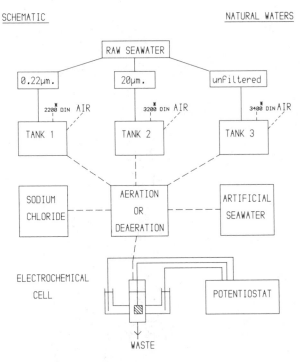

FIG. 2—*Schematic outline of the seawater pathway in these studies of corrosion.*

The results of the daily routine analyses showed all tanks were subject to the same tempera-ture variations within 1°C and that the pH was level at pH 8.2. This indicated that even where tanks had relatively high populations of surface-dwelling organisms their photosynthetic pro-cesses were not sufficient to cause a pH excursion to more alkaline values. Similarly, although fluctuating very slightly, conductivity and dissolved oxygen levels were steady throughout the study period indicating that there were no losses caused by evaporation or tendencies to anaero-biosis, respectively. Indeed the aeration of the waters all oscillated about the value for full satu-ration of water by oxygen with respect to atmospheric levels. Again this indicated that the levels of photosynthesis compensated for the respiratory processes in all the tanks. Planktonic bacte-rial counts were also relatively constant around 2×10^3 organisms per millilitre. At no time was a value obtained for suspended chlorophyll. This was probably because of the bottom filtration caused by the undergravel filters operated by the airlift systems.

The analyses of silicate in Tanks 1 and 2 (Fig. 3) and phosphate in all three tanks were inter-esting as they showed considerable increases over the levels found in the initial seawater sam-ples. This suggested leaching of these ions from the heat sterilized gravel, and in all three tanks the levels of phosphorous as phosphate and of silicon as silicate were well above any limiting threshold value. This was not true however of the nitrate levels, which although they increased to a variable extent in the three tanks, all became undetectable and therefore limiting to contin-ued algal growth. This occurred first in Tank 2 at 52 days, then Tank 3 at 58 days and finally at

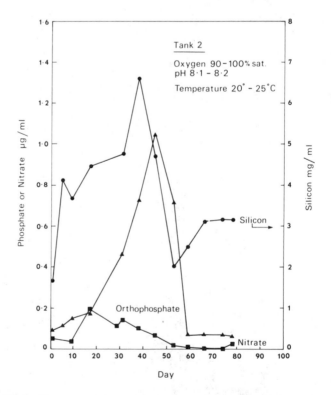

FIG. 3—*The concentration of key nutrients in Tank 2 as a function of time.*

Tank 1 at 65 days. Correlated with these events the silicate levels, which had been falling sharply because of explosive growth of surface dwelling diatoms, inflected and began to increase again. The phosphate levels, which had also fallen markedly during this high growth period did not undergo a sudden increase but stabilized at the initial fresh seawater levels, presumably because of consumption by the other nonphotosynthetic organisms, which however would not require silicate.

The very stable bacterial populations were apparently maintained by large populations of choanoflagellates in the initial stages up to four to five weeks, and their subsequent replacement by high numbers of grazing ciliate and flagellate protozoa.

The diatom populations reached peak levels successively in Tanks 2, 3, and 1, and on each occasion the total depletion of nitrogen in the form of nitrate resulted in a crash in the populations. This stressing of the diatoms and their subsequent death was enhanced by the attack of a colonial parasitic protozoan *Labyrinthula,* which swarmed over the diatom film, penetrating the cells and causing their lysis. The fact that bacterial numbers did not increase but if anything declined was also due to an increase in population of the bacterial-feeding protozoans.

Thus, in conclusion, although undergoing different pre-treatments the three tanks followed an identical course of events and showed a typical succession of micro-organisms with the biological stability of the tanks finally destroyed by the total depletion of one key nutrient, nitrate. This study showed that the tanks followed the same path, albeit on a slightly different time scale. The purpose of having comparative waters was thus frustrated. However the source of the similarity between the highly filtered tank and the others was subsequently found to be cross-contamination by airborne spray. When the tanks were tightly covered the most highly filtered tank remained free of any algal growths for months on end, and subsequent comparisons used this water as the Tank 1 standard condition.

These tanks provide water that would be expected to contain different types and concentrations of metabolic products. This is well illustrated by the X-ray Photoemission Spectroscopy (XPS) spectra of samples of slime removed from the glass walls (Fig. 4). These spectra show a steady increase in oxygen content of the film material with a decreasing degree of filtration of the tank water. The carbon 1-s line indicates a parallel increase in polar carbon group and in nitrogen content. Thus a comparison of corrosion behavior of test pieces exposed in each tank should be revealed when a significant influence is being exerted by the molecules and organisms contributing to the slime film. This is the way in which the tanks are normally used, water being syphoned from each tank to pass over a sample contained in an electrochemical cell.

A similar system enables either isotonic sodium chloride or synthetic seawater to be drawn down through the cell. In all cases differences in behavior of the samples relative to each other are sought.

A more direct method of examining the effect of the organic component is to remove it from the water in order that it may be added back to synthetic mixes or used to augment the seawater itself. This in the long term could be the basis for the preparation of standard synthetic seawaters containing an organic fraction. A system that permits the removal and retention of high molecular weight compounds is shown in Fig. 5. Seawater that has already been filtered to the level used in the highly filtered tank is used to fill and top-up a recirculating system containing a high pressure ultrafiltration cell. The water is processed at $4°C$, and the concentrate is stored at $-25°C$. By repeated passes through the cell a final product is produced, 1 L of which contains the organic molecules (RMM [root mean molecular weight] > 1000) from 500 L. At its peak period the level of dissolved organic material (DOM) may exceed 1 mg/L. However the levels may be many times less than this at certain seasons, for example, in late autumn it can sink to 10% of the peak value. We have used such extracts to study the adsorption of relevant organic material by copper alloys, as might occur in the sea. Clearly there is variability in the components extracted at different seasons, but by storing these, comparative trials can be undertaken under laboratory conditions.

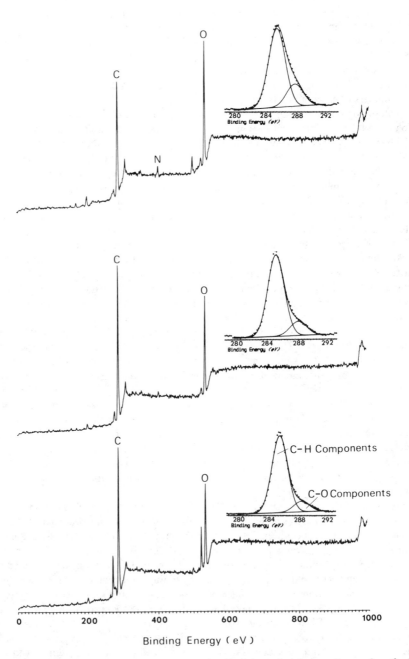

FIG. 4—*Spectra of slime removed from the internal walls of the tanks: the inset figures show the carbon spectrum.*

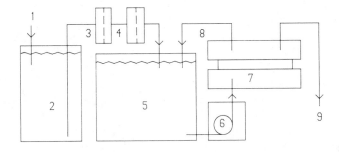

1) GAS LINE (OXYGEN FREE NITROGEN) MAINTAINS PRESSURE HEAD
   FOR RESERVOIR CONTAINING FRESH UNFILTERED SEAWATER.

2) RAW SEAWATER DRIVEN TO PRIMARY FILTRATION UNIT.

3) PREFILTRATION REMOVES PARTICULATE MATERIAL.

4) SUB-MICRON FILTER (0.22um.) REMOVES BACTERIA.

5) GLASS ASPIRATOR CONTAINS CONCENTRATED SEAWATER.

6) PUMP SYSTEM DELIVERS SEAWATER TO MAIN FILTER UNIT.

7) ULTRAFILTRATION CASSETTE SEPARATES DISSOLVED ORGANICS.

8) SEAWATER CONTAINING HIGH MOLECULAR WEIGHT ORGANICS IS
   RETURNED TO THE ASPIRATOR.

9) ULTRAFILTRATE EFFLUENT IS DISCARDED.

FINAL VOLUME - ONE LITRE FROM 500 LITRES OF SEAWATER.

PROCESS TEMPERATURE 4°C. STORAGE AT -25°C.

FIG. 5—*Schematic outline of the extraction and concentration of dissolved organic materials.*

Results

Adsorption of Organic Compounds

Strong organic adsorption accompanies the corrosion process under natural conditions as shown by the data in Fig. 6. Nitrogen is again a good indicator of organic uptake and some values are set out in Table 1. Samples were also exposed to differing concentrations of the organic extracts in 20 mL of fully filtered seawater for a period of up to 90 h. XPS observation of the nitrogen peak showed that this reached a saturation level of between 10 and 14% nitrogen for both with highly enriched water. At the highest concentration a thick film could be seen to be formed. Note the strong similarity between the spectrum for this alloy surface that is fully saturated with absorbed molecules and that from a tube exposed naturally in a tube rig in North Carolina (Fig. 7).

When metal samples are exposed, the surface analytical measurements show the net effect of absorption and the formation of new corrosion products. A clearer idea of the nature of adsorption in the absence of corrosion has been obtained from studies of the adsorption of albumin on separate copper(I) and copper(II) oxides [13]. The conclusions were that both oxides adsorb strongly and rapidly, effective saturation was reached after only 10 s, but only some 60% of the surfaces were covered (Tables 2 and 3). The use of selective eluents for the surface showed that

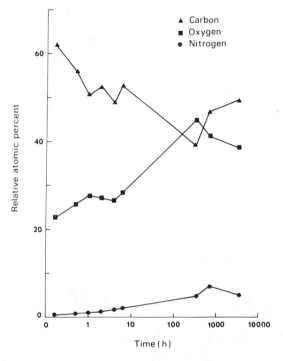

FIG. 6—*The levels of CNO arising from adsorption of DOM during marine exposure.*

TABLE 1—*Surface concentration of nitrogen on Kunife alloy after different times in the sea.*

Time, h	Nitrogen, %
0.25	0.4
0.5	0.7
1	1.0
2	1.2
4	1.6
6	2.0
336	4.8
672	7.0
3408	5.0
5280	6.0

adsorption occurred by the formation of strong ligands only broken by citric acid. Further, when the cross-linkages with the protein were broken by use of iso-propyl alcohol, the surface coverage increased dramatically (Tables 2 and 3) to nearly 100%.

In view of this strong adsorption it is not surprising that DOM is picked up by the test piece during the brief passage of a given seawater through the electrochemical cell. Typical contrasting spectra obtained after exposure to sodium chloride (NaCl) and to Tank 1 water are shown in Fig. 8, illustrating the associated changes in inorganic composition of the film. These presumably stem from the action of organic components.

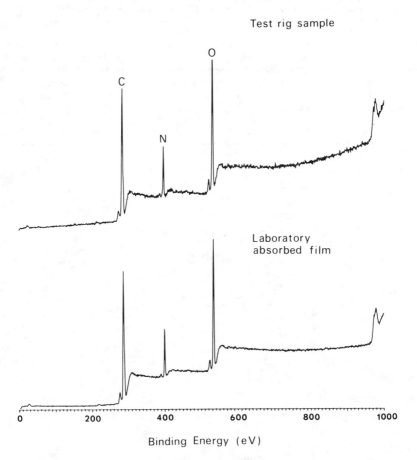

FIG. 7—*A comparison between the surface chemistry of a sample from the INCO seawater tube rig in North Carolina and a sample exposed to enhanced (500 mg/L) dissolved organic material.*

TABLE 2—*The interaction of the protein, bovine albumin, and copper (I) oxide. The nitrogen signal gives an indication of surface coverage during adsorption. Data obtained with copper (II) oxide was very similar (see Ref 13).*

Element	Surface Composition, atomic %					
	10 s	1 min	5 min	20 min	1 h	3 h
C	50.3	49.3	49.1	48.3	50.4	47.2
N	9.5	9.0	9.8	7.8	11.0	8.7
O	34.1	35.2	35.7	37.2	33.2	38.4
Cu	6.1	6.5	5.4	6.7	5.4	5.7

TABLE 3—*The interaction of the protein, bovine albumin, and copper (I) oxide. The nitrogen signal gives an indication of surface coverage under the influence of aggressive eluents. Data obtained with copper (II) oxide was very similar (see Ref 13).*

			Eluent		
Element	None	NaCl	Citric Acid	Methanol/Water	IPA/Water
		SURFACE COMPOSITION, ATOMIC %			
C	47.3	47.8	47.3	51.1	56.2
N	9.6	9.9	7.0	11.4	14.3
O	37.9	34.0	35.7	31.8	26.9
Cu	5.2	8.3	10.7	5.7	2.6

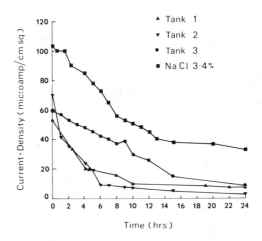

FIG. 8—*The passivation currents in sodium chloride and in seawater from the tanks.*

Inhibition or Promotion of Corrosion

A number of trials and comparisons have been carried out to assess the effect of adsorbed species on the normal corrosion rate. The comparative rates of passivation of the surface at an anodic potential in sodium chloride and in water drawn from each of the three tanks is given in Fig. 9. The current does fall more rapidly in seawater but the effect is a matter of degree only and is consistently so over many studies. Measurements of the corrosion current from the intersection of the Tafel lines also suggests very little difference between seawater and the single chloride (Table 4). This is affirmed by results of tests in which the alloy is pre-exposed to NaCl containing albumin and then examined by a potential sweep, they are virtually identical to those obtained in sodium chloride without the protein adsorbed. Finally samples have been pre-exposed to the organic extract and exposed both in the laboratory (Table 4) and in the sea. The laboratory data show again that there is very little effect on the corrosion rate, the anodic Tafel constant or the rest potential. The exposures to the sea (Tables 1 and 5) indicated that whatever the starting level, there was a tendency for the surfaces to become similar in nitrogen content. No measurements of corrosion rate were made on the marine samples, but determination of the Cu/Ni and Cu/Fe ratios by energy dispersive X-ray analysis (EDAX) giving the mean concentration of the outermost 1 μm of the surface showed that these also converged after about a weeks exposure. The marked enrichment in iron suggests that there is selective loss of copper

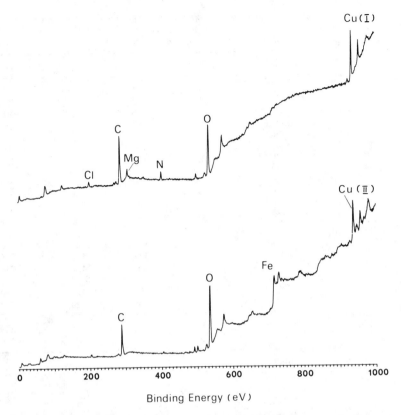

FIG. 9—*Surface analysis of samples after exposure to sodium chloride (lower curve) and seawater (upper curve) from the laboratory tanks. Note the higher levels of carbon and nitrogen and the completely different surface composition formed in seawater.*

and that this is almost identical for all specimens exposed. The preformed organic films did appear to have an influence on the degree of microfouling, however this was less marked on the samples carrying the highest concentrations of film material.

The cathodic reaction is of particular importance to the corrosion of Kunife since it is the ready inhibition of this reaction that gives the alloy many of its worthwhile properties. Throughout the sequence of trials the cathodic curves were also very similar, becoming highly polarized after exposure of about a day and, with one exception, whatever the initial level of organic material (Table 4). The exceptional case was for that sample exposed to 500-mg/L DOM and thus carrying a thick film. The cathodic Tafel slope (BC) for this sample remained as found for the unexposed samples.

Surface Composition

Much of the published work of this laboratory has dealt with the composition of the surface film as determined by the surface analytical technique of electron spectroscopy [14,15]. In this paper we shall only focus on some points of difference between seawater and isotonic (3.4%) sodium chloride. Firstly, analysis of the carbon line shape in the XPS spectrum confirms the findings based on the nitrogen peak; the components relating to polar groups increase with time in both natural and tank seawaters but are not observed in synthetic environments, and the

TABLE 4—*Electrochemical data obtained for a number of different "seawater-type" media.* B_A *and* B_C *are the Tafel constants for the anodic and cathodic processes, respectively.* I_{corr} *and* E_{corr} *are the corrosion current,* $\mu g/cm^2$*, and potential (mV).*

Medium	Pretreatment	B_A	B_C	I_{corr}	E_{corr}(SCE)
Tank 3	None	41	81	2.2	−225
NaCl	,,	58	88	3.6	−230
,,	,,	61	95	2.7	−229
,,	24 h at OCP	46	357	3.7	−198
Tank 3	24 h at OCP	45	203	4.3	−211
NaCl (3.4%)	24 h at −150mV	54	109	1.6	−217
,,	,,	66	121	2.6	−214
Seawater	,,	42	174	2.0	−181
Tank 1	,,	54	170	3.1	−211
Tank 3	,,	72	337	9.0	−218
,,	,,	74	189	4.0	−211
,,	,,	104	319	7.4	−218
0.1% Albumin in NaCl 3.4%	,,	45	177	3.0	−205
,,	,,	47	182	3.4	−206
,,	,,	45	203	4.0	−211
12-mg/L DOM in NaCl 3.4%	,,	45	199	3.3	−198
,,	,,	46	166	2.5	−204
500-mg/L DOM in NaCl 3.4%	120 h	63	75	2.3	−229
,,	,,	34	95	0.5	−144

NOTE: OCP is open circuit potential.

TABLE 5—*Surface concentration of nitrogen after 90-h exposure to kunife to seawater enriched with DOM (typical values for natural seawater, percent of nitrogen interpolated from Table 1.*

Concentration of Medium		Surface Concentration of Nitrogen
Protein, ng/L	Carbohydrate, $\mu g/L$	Nitrogen, %
0	0	<0.1
50	420	3.5
100	850	2.6
500	4 200	3.1
1500	400 000	14

organic material permeates the whole of the inorganic film [5]. Second, copper(I) compounds are in large excess throughout the films formed in seawater, natural or tank, but in synthetic solutions the outer surface quickly becomes covered with a layer of crystalline Copper(II) basic chloride Fig. 8. Nelson [16] has shown that organic films adsorb $CuCl_2^-$ very strongly, and this is one possible reason for the observation. This is in accord with the fact that the films often behave as if they are electrically insulating in the spectrometer and with the fact that in one instance the Cu(I) film was on top of the Cu(II) layer. Alternatively it may be difficult for crystalline material to nucleate on an organic slime film. Finally, there is a difference in the formation of an iron rich film in the corrosion products. This is easily formed at a slightly anodic potential in sodium chloride (Fig. 8) and also appears to form in the sea itself providing the temperature is not too low. However it was not possible to form this layer at normal temperatures in seawater passing from the tanks as a once-through stream. It did form on samples suspended directly in the tanks before this mode of exposure was abandoned because of its influence on the microbiology of the environment.

The Influence of Oxygen

One clear and dramatic difference between seawater containing organic material and that which does not, lies in the influence of low oxygen potential on cathodic inhibition. We have shown repeatedly that samples that have been aged in tank seawater or in synthetic solutions to which has been added either organic extract or albumin lose cathodic inhibition when the solution is rapidly deaerated. This can be seen in Fig. 10 by the increase in slope of the cathodic Tafel line after deaeration. We believe this to arise from the electrochemical reduction of the adsorbed molecules, perhaps to give sulfide ions [17]. The deaerations have now been repeated by decreasing oxygen to a controlled concentration (Table 6) and show that passivity is lost at about 80 ppb (1% of normal). In addition the rate of passivation at anodic potentials was much reduced at a low oxygen level. The work highlighted the great difficulty there is in defining oxygen level in the presence of a biological oxygen demand, that is, levels of oxygen that are reproducible and controllable in synthetic media are very difficult to control in natural waters.

By-Products of Microbes

The inorganic products of micro-organisms can be and frequently are added to test media. Sulfide ions have been the subject of much investigation in the context of corrosion of the present alloy. Ammonia too is known to stimulate the corrosion of copper alloys. To some extent

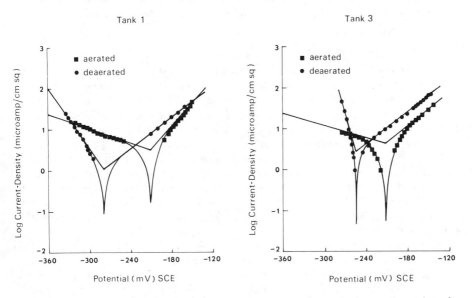

FIG. 10—*Tafel lines for samples held 24 h in Tanks 1 and 3. The increase in slope on deaeration arises from loss of cathodic inhibition and does not occur in sodium chloride solutions.*

TABLE 6—*The influence of oxygen concentration on the anodic current measured at* -150-*mV SCE.*

Oxygen Content/ ppb Range	Time, min								
	0	2	4	10	15	20	30	60	600
	ANODE CURRENT DENSITY, $\mu A/CM^2$								
7700	210	...	...	...	...	...	90	83	31
800 to 1000	118	61	50	46	43	...	...	39	28
320 to 400	180	...	...	...	...	158	...	152	93
0 to 80	180	...	...	...	...	176	...	166	158

these are considered aberrant conditions of seawater, important only in polluted or low oxygen waters. This loses sight of the fact that anaerobic cells of activity may occur under a thick biofilm and that these inorganic metabolites may transform to other products in well oxygenated environments away from their seat of production. In this way thiosulfate and nitrite ions will be present in seawater at particular times and locations; yet these appear to be hardly ever added to standard mixes.

There is however a much more complex interaction with the environment in the presence of organisms rather than just their metabolites. The development of an inorganic/organic film may provide a suitable habitat for a wide range of microorganisms, bacterial and algal. Many of these are able to function heterotrophically, that is, in the presence of light they will generate oxygen, while in its absence they will generate carbon dioxide. This, along with their other excretory materials, may lead to a diurnal rhythm in pH and complexing ability [18,19]. This dynamic interaction between the organisms present in natural water can never be simulated by additions to a recipe and indeed in the natural environment will follow an ever-changing pathway. In our laboratory work we have used a stream of water drawn from the tank to minimize this surface colonization; this will therefore always be a point of distinction from true exposure trials. Sufficient comparative exposures must be made in any given case to judge whether the laboratory water simulates all the important features of the corrosion process.

Conclusions

In the work reported here we have used essentially static conditions. Adsorption itself is unlikely to be influenced by flow rate, but other effects, such as corrosion product precipitation or sloughing of films may be. The work is a guide to what might be achieved, and points the way to the inclusion of organic species as in rig trials, which might otherwise be performed with inorganic solutions only. It has been shown that in the case of copper alloys, the organic constituents of seawater are strongly adsorbed to the surface. The rate of adsorption is rapid compared to the rate of passivation, and the evidence is that organic compounds may be found throughout the thickness of the inorganic film. The organic component may be separated by ultrafiltration and added to synthetic mixes, or alternatively seawater may be held in the laboratory providing the conditions necessary for living organisms are supplied. If this method is adopted we have shown how comparative ecologies can be established in order to judge the relative influences of different types of organic products.

In the specific case of Kunife 10 alloy there is little influence of organic material on corrosion rate in well aerated solutions, and some influence on the surface composition can be observed. The presence of a surface Cu(I) compound and the lack of a crystalline deposit of the basic chloride are deviations from the chemistry expected on the basis of purely inorganic solutions. There is a major difference of behavior in low-oxygen conditions, and this can be induced by the addition of organic extracts to sodium chloride or to seawater mixes.

Acknowledgment

The authors are pleased to thank the International Copper Research Association, INCRA, for their support of this work through Contract 254. Jane Newcombe and Julia Hoare undertook the microbiological and chemical assay of the seawater tanks.

References

[1] ASTM, Specification for Substitute Ocean Water, D 1141-52, *Manual of Industrial Water, STP 148A*, American Society for Testing and Materials, Philadelphia, 1954.
[2] Kester, D. R., Duedall, I. W., Connors, D. N., and Pytkowicz, R. M., *Limolnology and Oceanography*, Vol. 12, 1967, pp. 176–179.
[3] Uhlig, H. H., *The Corrosion Handbook*, Wiley, New York, 1948, p. 1121.
[4] Castle, J. E. and Epler, D. C., *Surface Science*, Vol. 53, 1975, pp. 286–296.
[5] Castle, J. E. and Epler, D. C., *Fouling*.
[6] Naylor, S., *Chemistry in Britain*, 1984, pp. 118–125.
[7] Duursma, E. V. and Dawson, R., Eds., *Marine Organic Chemistry*, Elsevier Scientific Publishing Co., Amsterdam, Oxford, New York, Elsevier Oceanography Series, Vol. 31, 1981.
[8] Rowlands, J. C., *Journal of Applied Chemistry*, Vol. 15, 1965, pp. 57–63.
[9] Hall, A. and Baker, A. J. M., *Journal of Materials Science*, Vol. 20, 1985, pp. 1111–1118.
[10] Blunn, G., *Proceedings of the 6th International Biodeterioration Symposium*, in press.
[11] Efird, K. D., *Materials Performance*, Vol. 15, 1976, pp. 16–25.
[12] Newman, R. C., *Corrosion Science*, Vol. 25, 1985, pp. 341–350; see also *Corrosion, NACE*, Vol. 38, 1982, pp. 261–265.
[13] Castle, J. E. and Paynter, R., Report on Contract 254, International Copper Research Association, 1981, submitted to *Journal of the Colloque Interface Chemistry*.
[14] Castle, J. E. and Parvizi, M. S., *Proceedings of the International Colloque Materials for Condenser Tubes*, Avignon, SFEN, France, 1982, pp. 93–102.
[15] Castle, J. E. and Parvizi, M. S., "Protective Surface Film Characteristics of Copper Alloys in Seawater," *Corrosion Prevention and Control*, 1986, pp. 1–17.
[16] Nelson, A., *Anales Chimica Acta*, Vol. 169, 1985, pp. 273–286.
[17] Castle, J. E., Parvizi, M. S., and Chamberlain, A. H. L., *Microbiology Corrosion Book 303*, Metals Society, London, 1983, pp. 36–45.
[18] Terry, L. A. and Edyvean, R. G. J., *Biodeterioration*, Vol. 5, T. A. Oxley and S. Barry, Eds., Wiley, New York, 1985.
[19] Wollmington, A. D. and Davenport, J., *Journal of Experimental Marine Biology Ecology*, Vol. 66, 1983, pp. 125–154.

Wing K. Cheung[1] and John G. N. Thomas[1]

The Effect of Dissolved Copper on the Erosion-Corrosion of Copper Alloys in Flowing ASTM Seawater

REFERENCE: Cheung, W. K. and Thomas, J. G. N., **"The Effect of Dissolved Copper on the Erosion-Corrosion of Copper Alloys in Flowing ASTM Seawater,"** *The Use of Synthetic Environments for Corrosion Testing, ASTM STP 970,* P. E. Francis and T. S. Lee, Eds., American Society for Testing and Materials, Philadelphia, 1988, pp. 190-204.

ABSTRACT: Erosion-corrosion experiments have been carried out on tubular copper alloys in ASTM artificial seawater, using two recirculating flow systems. The results showed that the order of resistance to erosion-corrosion of the alloys was correct, but the values of the critical breakaway flow velocities were considerably greater than those reported for once-through seawater cooling systems in practice. Analysis of the recirculating seawater showed that because of the high initial corrosion rates of the copper alloys, the concentration of dissolved metal ions increased rapidly to significant values, and supersaturation of copper ions was occurring. A technique has now been developed to remove dissolved copper and other heavy metal ions from the recirculating seawater in the flow systems by passage through a suitably prepared chelating ion-exchange resin. Under these conditions, the corrosion rates of tubes of 90/10 copper-nickel were considerably higher than those obtained previously in the presence of dissolved copper ions.

KEY WORDS: erosion-corrosion, seawater, copper alloys, corrosion rates, polarization resistance, critical breakaway velocity, supersaturation, chelating resin, dissolved copper

Seawater is used frequently as a process coolant in industrial plants, but the operation of heat exchangers with seawater as a coolant may be limited by corrosion of the heat exchanger tubes. Corrosion rates are affected by factors such as flow velocity, temperature, and tube materials. As the flow velocity increases, the corrosion rates of the tubes increase slowly until, at the critical breakaway velocity, rapid dissolution of the metal occurs because of erosion-corrosion, and this has a serious effect on tube life.

Reliable values of critical breakaway velocities for erosion-corrosion, which are essential in the proper design and operation of heat exchanger systems, must be obtained. The aim of the present work was to establish a reliable method using a laboratory environment to measure the breakaway flow velocities for copper-nickel alloys under practical conditions in heat exchangers using seawater coolant. The initial work was carried out to determine whether the polarization resistance technique [1–6] could be applied to the determination of erosion-corrosion rates. Breakaway flow velocities were obtained for three copper alloys, i.e., 90/10 copper-nickel, 70/30 copper-nickel, and aluminum brass. However, these values were considerably greater than those observed in "once-through" systems in practice. Since the seawater chemistry in a closed loop system may change because of the accumulation of soluble corrosion products which may influence the onset of erosion-corrosion, a method is described for reducing the contamina-

---

[1]Higher scientific officer and principal scientific officer, respectively, Division of Materials Applications, National Physical Laboratory, Teddington, Middlesex, England.

ion of the test solution in the system. Using this method, the effect of supersaturation of seawa-
er by dissolved copper ions on the corrosion rates of tubular 90/10 copper-nickel was investigated.

Experimental Procedure

Apparatus

Two flow systems have been developed for the present work; these are described below.

Flow System 'A'—Flow system 'A' is a small, bench-mounted flow loop. A schematic diagram is shown in Fig. 1. The system consists of a single-loop of mainly 32 mm internal diameter
pipe of ABS (acrylonitrile-butadiene-styrene) plastic containing an orifice plate and tapered-
tube flow meter F made from polyvinyl chloride (PVC) plastic. The test solution is circulated by
a PTFE (polytetrafluoroethylene) lined pump from a polypropylene tank approximately 4.0×10^4 cm^3 in capacity. A bypass return to the tank is provided, which includes a resin column
having a volume of about 9000 cm^3, constructed from pyrex glass. The flow through the bypass
can be separately controlled by a diaphragm valve D, and measured by means of a tapered-tube
flowmeter G made from PVC plastic. The rate of flow of solution through the column when
filled to a depth of about 75 mm with resin is about 100 cm^3 s^{-1}. The maximum flow velocity
which can be obtained through the 19-mm outer diameter specimen is about 8 ms^{-1}.

Specimen Mounting—The method of mounting the specimen is shown in Fig. 2. Two specimens, each about 64 mm long and 19 mm in outer diameter, are held together in a PTFE mount
using O-rings. The ends of the specimens are mounted in flanges with the upstream end slightly
projected to simulate the fixing of tubes into the end plates of heat exchangers. For polarization
studies, a counter electrode of platinum wire running down the center of the specimen(s) is
provided, as shown in Fig. 2. The platinum wire is supported at both ends on crosses of platinum wire suitably shaped to hold the wire in the central position. The symmetrical disposition

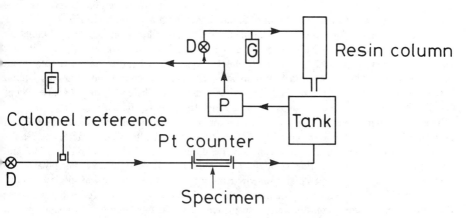

FIG. 1—*Schematic diagram of flow system A.*

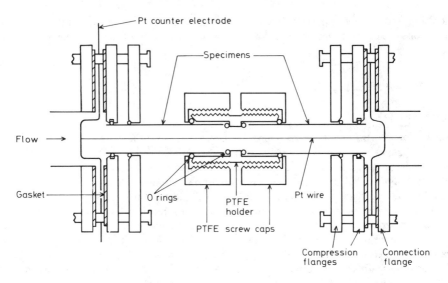

FIG. 2—*Diagram of mounting of twin specimen tubes.*

of the platinum counter electrode ensures uniform current density along the whole length of the specimen tube. The potential of the specimen may be measured using a calomel reference electrode, connected to the system through a PTFE tap and mounted in a sidearm upstream of the specimen.

Flow System 'B'—This flow system is a much larger flow loop. A schematic diagram is shown in Fig. 3. The test solution is circulated from a reservoir tank (approximately 3.2×10^5 cm^3 in capacity) by a pump, constructed from Type 316 stainless steel, through pipework of ABS plastic. A bypass return to the tank is provided in the system in which a resin column of about 3.9×10^4 cm^3 in volume, constructed from PVC plastic, is inserted. The flow through the bypass can be separately controlled by a diaphragm valve D, and measured by means of a tapered-tube flowmeter G made from PVC plastic. The maximum flow rate which can be achieved through the resin column when filled to a depth of about 75 mm with resin is about 450 cm^3 s^{-1}.

The flow system has four channels in parallel, and the flow velocity through each channel can be separately controlled by ball valves V, and measured by means of an orifice plate and tapered-tube flowmeter F made from PVC plastic. Each channel can contain a tubular metal specimen, about 127 mm long and 19 mm in outer diameter, the ends of which are mounted in flanges (as shown in Fig. 4) to simulate the fixing of tubes into the end plates of heat exchangers. The flow velocities through the specimens range from 2.6 to about 18 ms^{-1}.

The flow system consists mainly of 75-mm internal diameter pipe, and one channel is constructed entirely with this sized pipe. In the other three channels, the pipe diameter upstream of the specimens is 75 mm, but downstream the pipe diameters are 50 mm, 38 mm, and 32 mm respectively, to enable lower flow velocities to be measured more accurately. For polarization studies, a counter electrode of platinum wire running down the center of the specimen is provided in each channel, and the potential of the specimen may be measured using a calomel reference electrode connected to the system through a PTFE tap, mounted in a side arm upstream of the specimen.

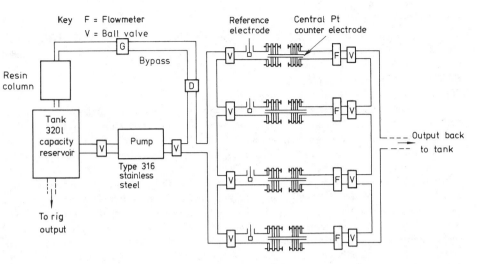

FIG. 3—*Schematic diagram of flow system B.*

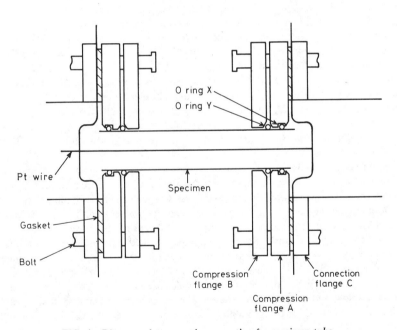

FIG. 4—*Diagram of compression mounting for specimen tube.*

Specimen Mounting—Each end of the specimen is held in a neoprene rubber O-ring, X, fitting into a groove in plastic flange A as shown in Fig. 4. The O-ring contacts the tube about 3 mm from its end, and around the projecting end of the tube, flange A is tapered to minimize the formation of a crevice. To prevent leakage of solution past O-ring, X, a second O-ring, Y, is compressed against the specimen by flange B. Flanges A and B are connected to the flow system by being bolted to flange C and sealed with a rubber gasket.

Materials and Test Procedure

The experiments have been carried out on tubular specimens of copper alloys. These alloys are 90/10 copper-nickel (Kunifer 10, nominal composition, 10-11% Ni, 1.5-1.8% Fe, 0.3-0.8% Mn, remainder Cu), 70/30 copper-nickel (Kunifer 30, nominal composition, 30.37% Ni, 0.85% Fe, 0.85% Mn, remainder Cu), and Yorcalbro aluminum brass (nominal composition, 76-78% Cu, 1.8-2.3% Al, 0.4% As, remainder Zn).

All the tubes were annealed and grit blasted in the bores to remove carbonaceous films on the surface. Before use, the specimens were degreased with trichloroethylene and cleaned in 10% sulfuric acid for one min, rinsed thoroughly in distilled water, dried in a stream of dry air, and then weighed. After exposure in the flow system, the specimens were rinsed in distilled water, dried as before, and reweighed.

The test solution of artificial seawater was prepared according to the ASTM Specification for Substitute Ocean Water D 1141-75 with addition of heavy metals. The flow velocities used ranged from 2.6 to around 12 ms^{-1} generally, and up to 18 ms^{-1} during one test. Before use, the pH of the seawater was adjusted to 8.2. Since the pH tended to increase during the course of the experiments, it was checked frequently and adjusted to 8.2 with hydrochloric acid when necessary. To compare the corrosion behavior of specimens in both ASTM artificial and natural seawater, an experiment was also carried out in which 90/10 copper-nickel was exposed in flowing natural seawater. The natural seawater was used within 48 h of collection. Air was bubbled continuously into the seawater during both storage and the experiment.

In general, the corrosion experiments were carried out over periods of 8 to 10 days with continuous exposure of the specimens to the flowing seawater. The temperature of the circulating seawater was maintained at 30 ± 1°C by passing chilled water through a coil of nylon tubing in the tank to balance the heat input from the pump.

In both flow systems A and B, the instantaneous corrosion rates of the specimens were measured as a function of time, using the polarization resistance technique [1-6]. The twin specimen mounting arrangement for flow system A (Fig. 2) was originally provided so that the downstream specimen could be used as a counter electrode, thus eliminating the need for an inert counter electrode. However, this arrangement was not very satisfactory because of IR drop effects.

In the present work, polarization resistance measurements were carried out by galvanostatic polarization using the central platinum counter electrode. Values of the reciprocal polarization ($1/R_p$) determined at intervals during the exposure of the specimens were plotted against time, and the areas under the curves integrated. These integrated values were then correlated with the total weight losses of the specimens to give the instantaneous corrosion rates in mm y^{-1}. These corrosion rates can only be considered approximate because the Stern-Geary constant B, determined from the total weight loss, was an average value for the total exposure time. Some variation of B with time is possible because of changes in surface condition and copper ion concentration. In experiments in which the copper ions had not been removed, the values of B increased slightly as the flow velocity increased. The range of values with increasing flow velocity were; 70/30 copper-nickel 0.039 to 0.052 V; 90/10 copper-nickel 0.026 to 0.039 V; aluminum brass 0.029 to 0.041 V.

The B value for the 90/10 copper-nickel at 5.5 ms^{-1} with copper ions removed was 0.030 V. A comparison of this value with that obtained at the same flow velocity when the copper ions had not been removed, i.e., 0.035 V, shows that removal of copper ions slightly decreases the B value, though the decrease may be within the range of experimental error. The use of instantaneous corrosion rates is very valuable in characterizing the general time variation of the corrosion process, since the latter cannot be determined from total weight loss.

The tests to determine corrosion behavior in seawater from which dissolved copper ions had been removed were conducted with 90/10 copper-nickel. In these tests, seawater was passed through the bypass loop containing a chelating resin (C-7901, Sigma Chemicals Co. Ltd). To ensure that the seawater would not be affected by the presence of the resin, the resin was converted to the calcium and magnesium forms by treatment with a solution containing magnesium and calcium chlorides in the same proportion as in seawater. A large volume of ASTM seawater was then passed through the resin to equilibrate it with seawater. After this treatment, the pH of the ASTM seawater used for the tests remained reasonably constant (pH 8.20 to 8.36) over the exposure period. During the test, the concentration of dissolved copper was analyzed spectrophotometrically at regular intervals, according to the method given in ASTM Test Methods for Copper in Water D 1688.

Results and Discussion

Initial Results

Figures 5 to 7 show the variation of the relative corrosion rates ($1/R_p$, k ohm^{-1} cm^{-2}) and mm y^{-1} with time of exposure to ASTM artificial seawater at four flow velocities for 70/30 copper-nickel, 90/10 copper-nickel, and aluminum brass specimens, respectively.

70/30 Copper-Nickel Alloy—The 70/30 copper-nickel specimen exhibited high initial corrosion rates, which decreased rapidly during the first 20 h of the test (Fig. 5). Beyond 20 h of exposure, little change in the corrosion rates was observed and no significant dependence of corrosion rates on flow velocity was noted throughout these tests. At the end of the test, the specimens were covered with a uniform thin greenish film. Clearly, the rapid passivation of the specimen and low corrosion rates at each flow rate resulted from this thin protective film.

90/10 Copper-Nickel Alloy—Initial corrosion rates for the 90/10 copper-nickel specimens were similar to those for 70/30 copper-nickel specimen. However, with the 90/10 alloy, passivation was not immediate and the corrosion rates at each flow rate increased at 20 h of exposure to

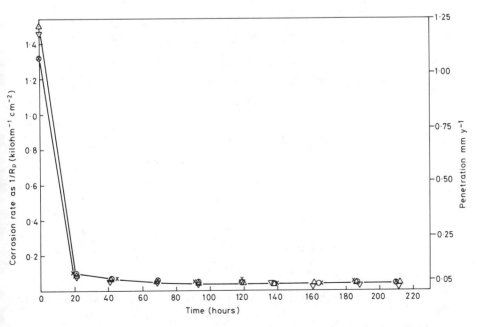

FIG. 5—*Variation with time of average corrosion rates of 70/30 copper-nickel tubes in artificial seawater flowing at 2.6 (X), 5.1 (○), 7.7 (△), 11.6 (▽) ms^{-1} (galvanostatic data).*

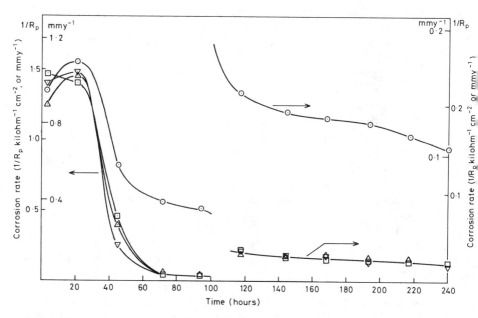

FIG. 6—*Variation with time of average corrosion rates of 90/10 copper-nickel tubes in artificial seawater flowing at 13–18 (○), 7.8 (□), 5.2 (△), 2.6 (▽) ms⁻¹ (experiment No. 1).*

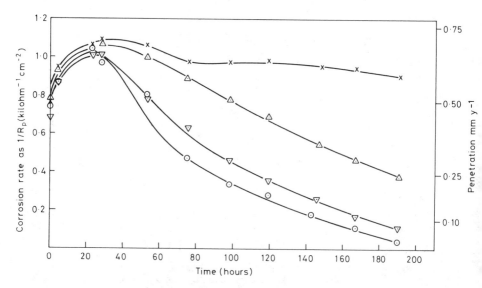

FIG. 7—*Variation with time of average corrosion rates of aluminum brass tubes in artificial seawater flowing at 2.7 (○), 5.2 (▽), 7.8 (△), 12.1 (X) ms⁻¹ (galvanostatic data).*

a maximum value before decreasing to lower values at subsequent time intervals. Apparently, passive films form more slowly on this alloy and this initial rise in corrosion rates may be due to removal of air-formed films on the alloy surface present at the start of the tests. The corrosion rates were largely independent of flow velocities at exposure periods up to 20 h, although the highest corrosion rates tended to occur for specimens exposed to the highest flow velocities. Beyond exposure times of 20 h, however, there was considerable dependence of corrosion rate at the highest flow velocity, and although the corrosion rates of the specimens at the three lowest velocities (viz 2.6, 5.2, and 7.8 ms^{-1}) decreased slowly to low values at exposure times of 70 h, the corrosion rate of the specimen exposed at 13-18 ms^{-1} remained substantially higher.

Examination of the 90/10 copper-nickel specimen at the end of the test revealed that the surfaces of the specimen exposed to the lower flow velocities were covered with a uniform greenish colored film. On the specimen exposed at 7.8 ms^{-1}, the film near the upstream end was thinner and darker in color. No localized corrosion was observed, however, thus validating the low corrosion rate observed for this specimen at longer exposure times during the test. At the highest flow velocity (13-18 ms^{-1}), the specimen carried the uniform green colored film at the downstream end, but at the upstream end there was a thin dark film extending about 5 mm from the tube inlet, followed by a region of bright metal extending for approximately 10 mm. These features are shown in Fig. 8. Clearly, erosion-corrosion is possible in this flow region.

Aluminum Brass—Corrosion rates were not markedly dependent on flow velocity over test periods up to 30 h. During the initial stages of the test, the behavior of this alloy was similar to that of the 90/10 copper-nickel alloy. The corrosion rates increased to a maximum value at 30 h of exposure time. Thereafter, the corrosion rates decreased gradually with time over the remainder of the exposure period, the extent of the decrease being greater with lower the flow velocity. The corrosion rate of the specimen exposed to a flow velocity of 12.1 ms^{-1} remained high over the entire exposure period and decreased only slightly with time.

At the end of the test, the internal surfaces of the specimens were in general covered with a light grey film. However, on the specimen exposed to the highest flow velocity of 12.1 ms^{-1}, a region of bright metal occurred similar to that described above for one of the 90/10 copper-nickel specimens.

For all the above alloys, the virtual independence of initial corrosion rates on flow velocity indicates that mass transfer effects in the solution are of secondary importance, and the rate determining factor is probably the activation controlled dissolution of copper or other constituents of the alloys.

Determination of Breakaway Flow Velocities

To determine the breakaway flow velocity with more accuracy, the final corrosion rates (after about 190 h) have been plotted against flow velocity (Fig. 9). Breaks in the slopes of the curves with evidence of enhanced corrosion occurred both for aluminum brass and 90/10 copper-nickel alloys, i.e., the alloys for which visual observations had indicated a bright metal region near the inlet of the tube. Similar results were obtained by plotting mass losses of the specimens against flow velocity (Fig. 10). With both sets of data, there was no evidence of erosion-corrosion on 70/30 copper-nickel specimens, in agreement with the visual evidence from examination of the specimens at the end of the tests. Mean values for the breakaway flow velocities were >7.8 ms^{-1} for the 90/10 copper-nickel alloy and 6.0 ms^{-1} for the aluminum brass alloy.

Taking data from the review of erosion-corrosion by Syrett [7], and maximum recommended flow velocities to avoid inlet erosion-corrosion of heat exchanger tubes in practice [8,9], maximum values observed in long-term tests or practical conditions for the breakaway flow velocities are 2.5 ms^{-1} for aluminum brass, 3.6 ms^{-1} for 90/10 copper-nickel, and 5.0 ms^{-1} for 70/30 copper-nickel. Thus, the order of the breakaway flow velocities in the present work for the three copper alloys is correct, but the values are considerably higher than those observed in practice in

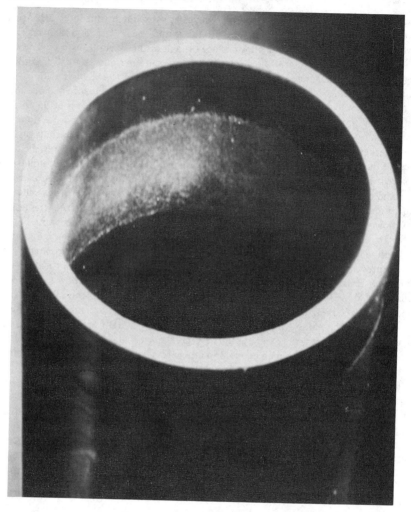

FIG. 8—*Erosion-corrosion effect at tube inlet for 90/10 copper-nickel: seawater flow velocity = 13–18 ms⁻¹.*

once-through seawater cooling systems. One possible reason for the discrepancy could be that the hydrodynamic conditions are more severe in practice.

The hydrodynamic conditions within the test pieces used in this study could not be readily modified, and furthermore, other aspects were more probable causes of the discrepancies. ASTM artificial seawater is possibly less aggressive than natural seawater. Also, changes in water chemistry resulting from the use of a recirculating flow system might be responsible, since significant concentration of extraneous ions might occur as a result of the high rates of dissolution of the specimens during the initial stages of the tests. These extraneous ions might stabilize protective films on the copper alloys. The presence of quantities of copper ions were thought to be the most significant. Experiments to test these two hypotheses are described below.

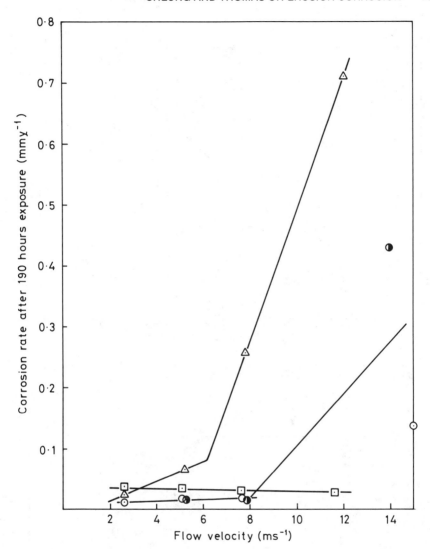

FIG. 9—*Variation with flow velocity of corrosion rates of copper alloy tubes after 190 h exposure in artificial seawater,* ○ ◑ *90/10 copper-nickel,* □ *70/30 copper-nickel,* △ *aluminum brass.*

Comparison of Artificial and Natural Seawater

Figure 11 shows the variation of relative corrosion rates with time of exposure for 90/10 copper-nickel in ASTM artificial and natural seawater. The results obtained using flow system A, showed that there was no significant difference in the corrosion behavior of 90/10 copper-nickel in ASTM and natural seawater. Therefore, the use of ASTM rather than natural seawater is not responsible for the difference in breakaway flow velocities obtained in a recirculated seawater system and those observed in practice.

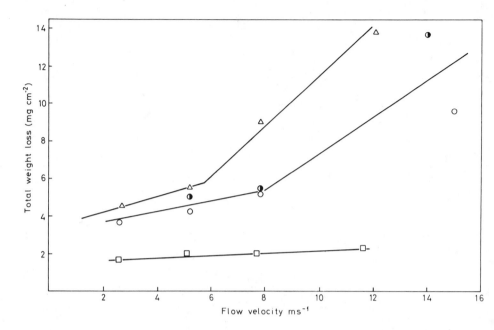

FIG. 10—*Variation of total weight losses of copper alloy tubes with flow velocity of artificial seawater,* ◯, ◑ *90/10 copper-nickel,* ☐ *70/30 copper-nickel,* △ *aluminum brass.*

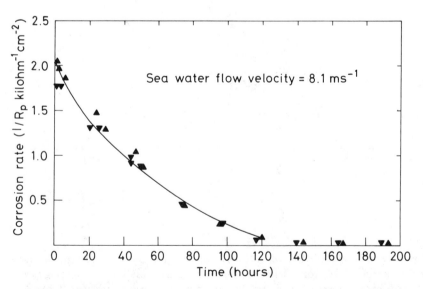

FIG. 11—*Variation with time of average corrosion rates of 90/10 copper-nickel tubes in flowing* ▲ *artificial or* ▼ *natural seawater (results obtained using flow rig A).*

Influence of Copper Ion Concentration

Figure 12 shows the variation of dissolved copper concentration with time of exposure for 90/10 copper-nickel. The results obtained using flow system A showed that supersaturation of dissolved copper was occurring. The copper concentration increased rapidly to about 0.82 ppm on initial exposure, which is consistent with a high initial corrosion rate. Between 1 and 40 h of exposure, the copper concentration decreased rapidly to a reasonably steady value of between 0.1 to 0.2 ppm. This suggests that supersaturation of dissolved copper tends to deposit protective films of copper salts on the metal surface. Such a build up of copper salts on the surface would be unlikely in a once through system.

To examine the effect of dissolved copper ions on the corrosion rates of copper alloys, experiments were carried out on 90/10 copper-nickel using the flow system A, in which the circulating seawater was allowed to pass through a suitably prepared ion-exchange resin. Under these conditions, the concentration of dissolved copper was substantially reduced (Fig. 13). Moreover, in the presence of the resin, when the dissolved copper and other heavy metal ions were removed from the recirculating seawater, the corrosion rates of 90/10 copper-nickel specimens remained considerably higher (Fig. 14) than in the absence of the resin, when the corrosion rates decreased steadily to a low value after ~80 h of exposure. The effect of flow velocity on the weight losses for 90/10 copper-nickel after ~200 h of exposure in the presence and absence of the resin is shown in Fig. 15. The results clearly indicate that the weight losses were significantly higher in the presence of the resin when the dissolved copper and other heavy metal ions were removed. Furthermore, in the presence of the resin, there was no evidence of an enhanced weight loss from erosion-corrosion over the range of velocities examined.

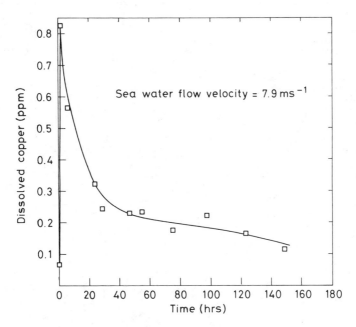

FIG. 12—*Variation of dissolved copper with time for 90/10 copper-nickel tubes in flowing artificial seawater in the absence of chelating resin (copper ions not removed).*

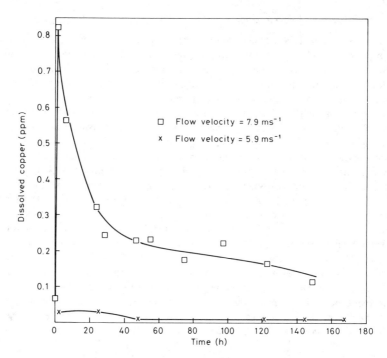

FIG. 13—*Variation of dissolved copper with time for 90/10 copper-nickel tubes in flowing artificial seawater (copper ions removed) X and (copper ions not removed)* □.

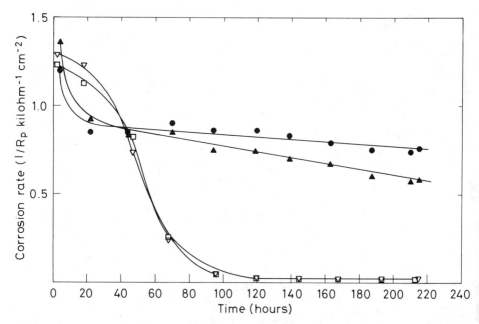

FIG. 14—*Variation with time of average corrosion rates of 90/10 copper-nickel tubes in artificial seawater flowing at* ▲ *5.5 ms⁻¹,* ● *8.2 ms⁻¹ (copper ions removed), (*▽*) 5.2 ms⁻¹,* □ *7.8 ms⁻¹ (copper ions not removed).*

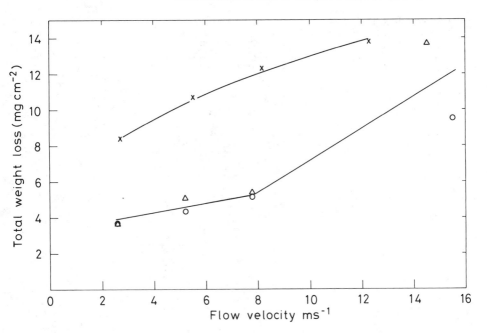

FIG. 15—*Variation of total weight losses of 90/10 copper-nickel with flow velocity of artificial seawater X (copper ions removed) ○ △ (copper ions not removed).*

Conclusions

1. The results of the present investigation have demonstrated that supersaturation of dissolved copper ions in a recirculating seawater system can significantly decrease the corrosion rate of copper-nickel alloys, and may thus affect the value of the breakaway velocity obtained in laboratory tests.

2. Further work is required to obtain breakaway velocity data for copper alloys in the presence of the ion-exchange resin, and to correlate the results with "once-through" seawater systems.

Acknowledgments

Part of this work was supported by the Heat Transfer and Fluid Flow Service at the National Engineering Laboratory. The authors wish to thank them for permission to publish this paper. The authors also wish to thank Yorkshire Imperial Alloys, who kindly supplied the copper alloy tube materials for this work.

References

[1] Mansfeld, F., *Advances in Corrosion Science and Technology*, Vol. 6, M. G. Fontana and R. W. Staehle, eds., Plenum Press, New York, 1976.
[2] Heitz, E. and Schwenk, W., *British Corrosion Journal*, Vol. 11, 1976, p. 74.
[3] Callow, L. M., Richardson, J. A., and Dawson, J. L., *British Corrosion Journal*, Vol. 11, 1976, p. 123, 132.
[4] Moreland, P. J. and Rowlands, J. C., *British Corrosion Journal*, Vol. 12, 1977, p. 72.
[5] Hines, J. G. et al., *Industrial Corrosion Monitoring*, Department of Industry, Her Majesty's Stationary Office, London, 1978.

[6] Bicicchi, R. T. et al., *NACE Corrosion/79,* Paper No. 236, Atlanta, March 1979.
[7] Syrett, B. C., *Corrosion,* Vol. 32, 1976, p. 32.
[8] Gilbert, P. T., *British Corrosion Journal,* Vol. 14, 1979, p. 20.
[9] Popplewell, J. M. and Thiele, E. A., *NACE Corrosion/80,* Paper No. 30, Chicago, March 1980.

Kenneth L. Money[1] *and Robert M. Kain*[1]

Synthetic Versus Natural Marine Environments for Corrosion Testing

REFERENCE: Money, K. L. and Kain, R. M., **"Synthetic Versus Natural Marine Environments for Corrosion Testing,"** *The Use of Synthetic Environments for Corrosion Testing, ASTM STP 970*, P. E. Francis and T. S. Lee, Eds., American Society for Testing and Materials, Philadelphia, 1988, pp. 205–216.

ABSTRACT: Comparison of synthetic and natural marine environments is the subject of this paper. The synthetic environments discussed include salt spray or fog, acetic acid salt spray and copper accelerated acetic acid salt spray. Results from corrosion tests conducted in these synthetic or simulated environments are compared to corrosion test results in natural seawater and the marine atmosphere. Laboratory and/or synthetic solutions can and often do rank or predict the relative performance or behavior of a given material, but when their results are compared to the behavior of materials in the natural environment, the investigator must use all of the available data before characterizing performance.

KEY WORDS: synthetic, natural, seawater, marine atmosphere, CASS test, salt spray (fog), stainless alloys, ferric chloride, steel, zinc, and cadmium plating, bronzes

A number of investigators have studied the use of synthetic environments for corrosion testing to find a method/environment that will yield results of predictive of or comparable to those that occur naturally. The investigator must distinguish between the simulated (synthetic) environment and the accelerated test method. The simulated or synthetic environments are selected with the hope of yielding the same results as exposures in the natural or operating environment, while the main purpose of the accelerated test is to develop reliable data in a time frame much shorter than corrosion tests in natural environments. Accelerated testing tends to seek results in minutes or hours as compared to months or even years for corrosion tests in the actual operating or natural environment. Both simulated environments and accelerated corrosion testing can, in some cases, predict the relative corrosion behavior of materials evaluated and sort out the poor performers. This paper will review data from a variety of sources on the use of synthetic (simulated) environments for corrosion testing and is based upon an earlier paper by Lee and Money [1].

The corrosion behavior of materials in marine environments can be assessed by tests either in natural environments or in synthetic environments, with each approach offering certain advantages and disadvantages. The natural environment contains the corrosive elements as they naturally occur and requires no simulation. Obviously, variations in the world's seas with respect to chlorides, oxygen, temperature, and biological influences, to name a few factors, can affect alloy or material performance. As such, even natural exposures at one location may or may not be universally applicable.

[1]Vice president and corrosion scientist, respectively, La Que Center for Corrosion Technology, Inc., P.O. Box 656, Wrightsville Beach, NC 28480.

Simulated Environment Tests

One of the first attempts to devise a simulated test for evaluating the durability of materials in atmospheric exposures was the development of a sulfuric acid test [2]. An ASTM committee reported in 1908 on the exposures of various irons and steels to a 20% sulfuric acid (H_2SO_4) solution for 1 h. Subcommittee V of ASTM Committee A-5 [3] later showed this test to be inadequate for assessing atmospheric corrosion resistance. LaQue [4] in his work of 1951 compared the differences in the behavior of steels when subjected to both the sulfuric acid test and natural atmospheres; Fig. 1 summarizes LaQue's findings. LaQue emphasized that the sulfuric acid test placed more emphasis on the corrodibility of the material and did not take into account the protective nature of the corrosion product film, which is important in atmospheric exposure. It should also be noted that the relative merits of steels in marine atmospheres may not be the same as in industrial atmospheres, and, therefore, no accelerated test can be consistent with both kinds of atmospheric exposures.

For many years, salt spray or salt fog tests have been used to assess materials' performance for atmospheric evaluations. The salt spray test was originated by Capp [5] in 1914. A number of factors that can influence the final results of the salt spray or fog tests must be considered. Among these are the size and shape of the cabinet, placement and orientation of the test specimen, concentration, temperature and pH of the salt spray, and whether or not the spray is continuous or intermittent. More recently, the importance of such factors as UV radiation, type and duration of UV rays, and effect of pollutants have been recognized and should be considered in using the salt spray humidity weatherometer type of apparatus to produce synthetic environments.

May and Alexander [6] in their work of the 1950s, demonstrated some of these factors in their studies with mild steel, iron, and zinc. Their experiments with mild steel in spray cabinets using 20% sodium chloride and synthetic seawater showed that care must be used in selecting the angle of repose of specimens. Figure 2 shows that the corrosion that occurred is nearly constant

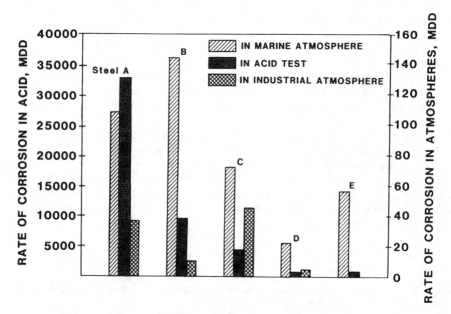

FIG. 1—*Comparison of corrosion of steels in acid tests and in natural atmospheres* [4].

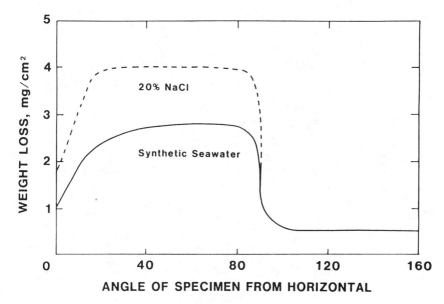

FIG. 2—*Effect of specimen angle on corrosion of iron in salt spray cabinets using synthetic seawater and 20% sodium chloride solutions* [6].

between 40 and 80 degrees to horizontal, and the effect of the angle is most notable at angles less than 15 degrees and around 90 degrees. It should be noted that specimens used in these tests were painted on one side and then exposed at the various angles with the painted side down.

May and Alexander also demonstrated that the corrosive nature of salt spray particles settling on a specimen surface in a salt spray cabinet is not necessarily comparable to the conditions occurring in a natural marine environment. Figures 3 to 5 show the corrosion behavior of ingot iron, steel, and zinc tests they performed in cabinets with sprays of 3% NaCl, 20% NaCl, and natural seawater. The weight losses in Fig. 3 to 5 represent the total from both sides of each test specimen. The environments, other than the salt spray, were the marine atmosphere at Kure Beach, NC at nominal distance of 80 ft (25 m) and 800 ft (250 m) from the ocean. While there is some similarity between the performance of iron in a 20% NaCl spray and in the 25-meter test lot, it is the natural seawater spray that most closely parallels the behavior of zinc in the natural environments. These data demonstrate the difficulties that the investigator faces in identifying a synthetic environment that will universally replicate behavior in a natural environment.

LaQue [4] has also compared results of tests on steel coated with zinc or cadmium exposed in various salt sprays and in natural atmospheres. The data summarized in Fig. 6 show that while the cadmium coating is superior to the zinc coating in all environments except seawater spray, the relative difference in behavior varies with each environment.

Because all salt sprays results cannot be universally extrapolated to anticipate service behavior, the key to using these data is to assess the *relative* behavior of closely related materials. The test can thus be useful in screening the effects of minor compositional variables on the behavior of an alloy, or in screening the effects of minor compositional variables in some plating systems.

Two accelerated salt spray tests often used in the evaluation of chromium-nickel plating systems on steel or zinc base die castings plating systems are:

(a) ANSI/ASTM Standard Method of Acetic Acid-Salt Spray (Fog) Testing B 287 [7], and

(b) ANSI/ASTM Standard Method for Copper-Accelerated Acetic Acid-Salt Spray (Fog) Testing (CASS Test) B 368 [8].

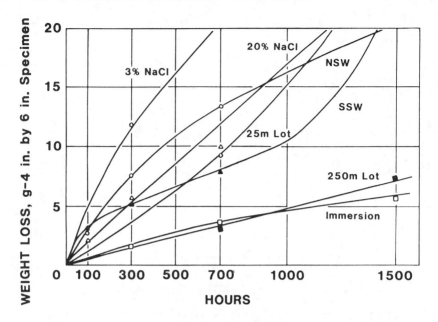

FIG. 3—*Comparative tests on ingot iron (0.04% Cu) in sprays of natural seawater (NSW), synthetic seawater (SSW), 3% sodium chloride, and 20% sodium chloride, 25 and 250 m from the ocean, and immersed in the ocean* [6].

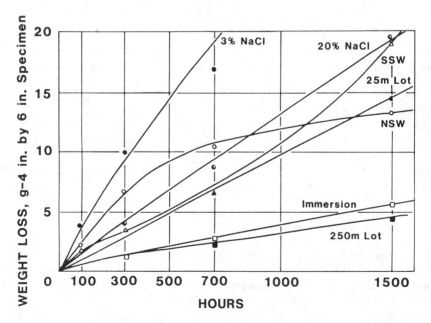

FIG. 4—*Comparative tests on steel (0.05% Cu) in sprays of natural and synthetic seawater, 3% and 20% sodium chloride, 25 and 250 m from the ocean, and immersed in the ocean* [6].

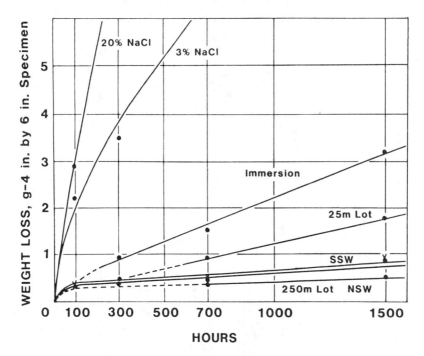

FIG. 5—*Comparative tests on zinc in sprays of natural and synthetic seawater, 3% and 20% sodium chloride, 25 and 250 m from the ocean, and immersed in the ocean* [6].

The care that should be used in extrapolating data from these types of tests to anticipated service behavior has been demonstrated by DiBari et al. [8]. These investigators compared the standard visual ratings of electroplated chromium, decorative nickel, and nickel-iron electrode-posits after the CASS test and exposure in natural marine and industrial environments. The correlation between 4 h of CASS testing and 3 months in marine or industrial atmospheres is shown in Fig. 7. The diagonal line represents perfect agreement. The greater the distance of a point from the diagonal, the less the agreement. The agreement between the 4-h CASS test ratings and the 3-month exposure data is reasonably good for the nickel coatings, but decidedly poor for the nickel-iron alloy deposits. The CASS test results predict a considerably better performance than that observed for the natural exposures.

Natural Environment Tests

While it may be difficult to develop representative and consistent synthetic environment tests, it is also difficult to clearly define a natural environment. Natural atmospheric environments are typically classified as marine, rural, industrial, or some combination of the three. Within each of these environments, a given material can exhibit a wide range of behaviors. These behaviors can be related to variations in the environmental characteristics as well as to variations in experimental techniques. Some consistency in results can be achieved through the use of the standard experimental practices outlined in ANSI/ASTM Standard Recommended Practice for Conducting Atmospheric Corrosion Tests on Metals G 50 [7]. However, variability in corrosion test results over lengthy time periods can still occur if the corrosivity of the environment is

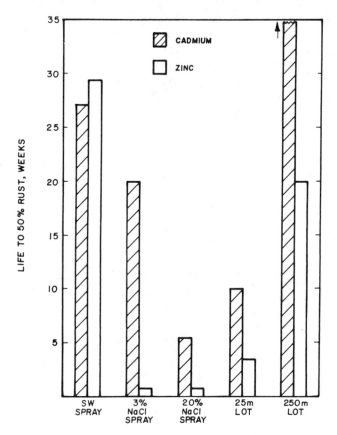

FIG. 6—*Comparison of performance of zinc and cadmium plating in salt spray tests and marine atmospheres (coatings were 0.005 μm thick on specimens in salt spray tests and only 0.001 μm thick in the marine atmospheric tests (Ref 4).*

changing. Several authors [*4,10–16*] have addressed in detail the variability of these environments and the procedures for calibration or comparison of the atmospheres.

Seawater Evaluations

Designing a synthetic test environment to simulate corrosion behavior in natural seawater can present many of the same problems encountered in atmospheric evaluations. Seawater is much more complex than a solution of sodium chloride and should be viewed as a living medium. It contains myriad biological species and organic and inorganic molecules. Compton [*17*] has reviewed the nature of some of the complexes or ligands that can form with metallic ions in seawater and notes the importance of these complexes in the metabolism and physiology of biological organisms. An additional property of seawater, as compared to many synthetic solutions, is its buffering capacity.

The most common solutions used in seawater simulations are 3.5% NaCl and synthetic seawater. Synthetic seawater (ASTM Specification for Substitute Ocean Water D 1141 [*18*]) reproduces the major inorganic constituents in natural seawater. The solution offers a convenient

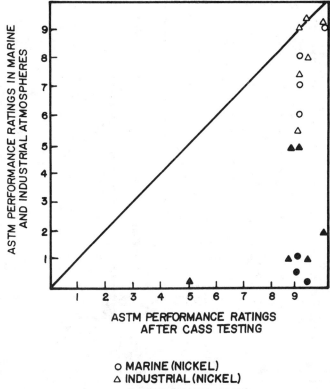

ASTM PERFORMANCE RATINGS
AFTER CASS TESTING

○ MARINE (NICKEL)
△ INDUSTRIAL (NICKEL)
● MARINE (NICKEL-IRON)
▲ INDUSTRIAL (NICKEL-IRON)

FIG. 7—*Comparison of performance of chromium electroplated nickel and nickel-iron electrodeposits in a 4-h CASS test and 3 months exposure in industrial and marine environments* [9].

method of conducting laboratory evaluations, but the results of corrosion tests can be significantly different from those conducted in natural seawater.

LaQue [19] demonstrated the differences in the corrosion behavior of bronzes when tested in natural and synthetic seawater at a velocity of 8.5 m/s as shown in Fig. 8. Stewart and LaQue [20] also reported in their 1948 work that erosion tests on bronzes in synthetic seawater corroded at a rate of 87 mg per dm^2 per day as compared with a rate of 448 mg per dm^2 per day in natural seawater. Their work shows that there are some constituents of seawater that apparently are not readily incorporated in synthetic seawater that have a profound effect on the attack of materials at high velocity.

Bogar and Crooker [21] in their work in 1979 showed that natural seawater was more aggressive and caused higher crack growth rates in high strength steels than did synthetic seawater or 3.5% sodium chloride.

Kain and Lee [22] have compared the crevice corrosion propagation behavior of Type 316 and 18Cr-2Mo stainless steels and alloy 904L (25Ni, 20Cr, 4.5Mo, 1.5Cu) in low velocity natural seawater, synthetic seawater, and 3.5% sodium chloride, and their results are shown in Fig. 9 to 11. In these experiments, the onset of crevice corrosion is illustrated by a rapid and sustained

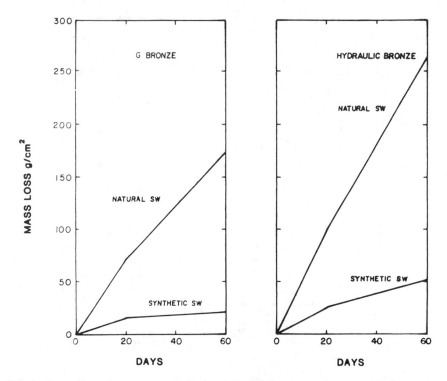

FIG. 8—*Comparison of corrosion rates of bronzes in synthetic and natural seawater (flat bar specimens were rotated to achieve a tip velocity of 8.5 m/s [19].*

increase in current as measured between physically separated but electrically connected crevice and noncrevice bearing members using remote crevice assembly tests. The initiation response for Type 316 was about the same in both the natural and synthetic environments, while alloy 18-2 and alloy 904L exhibited greater resistance in the synthetic environments. Significantly, the magnitude of current and total charge were greatest in the natural environments. The authors concluded from their studies that the unique nature of natural seawater promoted an order of magnitude more of crevice corrosion than synthetic seawater did.

Other research at the LaQue Center for Corrosion Technology, Inc. on the crevice corrosion behavior of stainless steels has also delineated differences between corrosion in synthetic and natural seawater. Table 1 summarizes both the depth of crevice corrosion incurred and the percent of crevice sites where attack was initiated. The data show significantly greater depth of crevice corrosion attack occurred in natural seawater at 25°C. No significant differences between the environments were evident at 50°C.

Nagaswami and Streicher [23] have compared the initiation of crevice corrosion in 12 stainless alloys in natural seawater, ferric chloride solution, and synthetic seawater. A summary of their findings is given in Table 2. The authors concluded that especially in initiation, the surface finish has a major effect in alloys in the medium range of resistance, and that further work is required to define laboratory tests of the initiation and growth of crevice corrosion and the interpretation of results.

The ferric chloride test (ANSI/ASTM Test Methods for Pitting and Crevice Corrosion Resistance of Stainless Steels and Related Alloys by the Use of Ferric Chloride Solution G 48) has been shown by Kain and others [24,25] to be ineffective in assessing the anticipated influence of

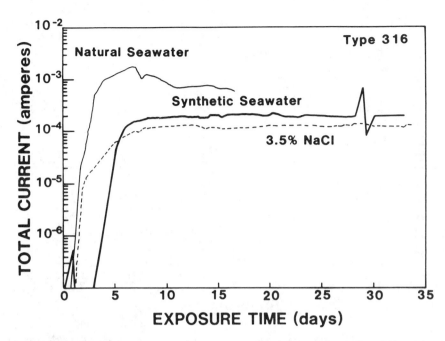

FIG. 9—*Variation in current measurements for remote crevice assemblies of Type 316 stainless steel exposed in natural and synthetic environments* [22].

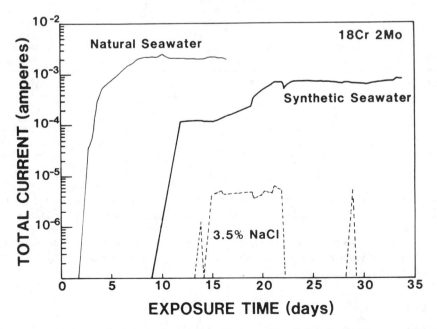

FIG. 10—*Variation in current measurements for remote crevice assemblies of 18Cr-2Mo stainless alloy exposed in natural and synthetic environments* [22].

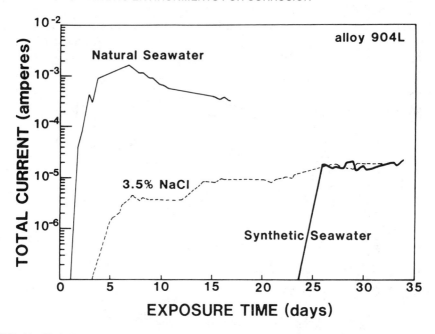

FIG. 11—*Variation in current measurements for remote crevice assemblies of alloy 904L exposed in natural and synthetic environments* [22].

TABLE 1—*Crevice corrosion of stainless alloys in natural and synthetic seawater* [24,25].

| Alloys[a] | Temperature (°C) | Synthetic Seawater | | Natural Seawater | |
		% Sites Attacked[b]	Maximum Depth of Attack, (mm)	% Sites Attacked	Maximum Depth of Attack (mm)
Type 304	25	31	0.11	42	2.91
Type 316	25	38	0.08	4	1.12
Type 304	50	31	0.15	33	0.18
Type 316	50	28	0.13	35	0.08

[a]AISI standard grades of stainless steels.
[b]A total of 40 separate crevice sites were created on each test specimen by the multiple crevice assembly.

temperature on crevice corrosion in natural seawater. He has shown the crevice corrosion of stainless alloys to be less extensive at 50°C than at 25°C in natural seawater (Table 1). However, increasing ferric chloride temperature results in a significant increase in localized corrosion. Others, however, have concluded that critical crevice temperature (CCT) tests in ferric chloride ($FeCl_3$) correlate well with the crevice corrosion resistance of fully austenitic stainless steels in ambient temperature seawater [26,27].

Summary

The use of synthetic environments or accelerated tests should be approached with caution because of the multitude of factors that can influence the outcome of the evaluations. If correla-

TABLE 2—*Initiation of crevice corrosion in immersion tests in seawater, ferric chloride, and synthetic seawater (120 grit finish)* [23].

Alloy	Filtered Seawater 30°C, 30-day tests		Ferric Chloride Tests, 10% FeCl₃ · 6H₂O, 72 h		Santron Tests, Synthetic Seawater, 20-min measuring time	
	Sites	Rank	Failure Temperature, °C	Rank	Failure Temperature, °C	Rank
A.L. 29-4C	0	1	55	1	90.0	1
Monit	0	2	47	2	67.5	2
Crucible SC-1	1	3	45	4	60.0	5
Ferralium	2	4	37	5	60.0	4
Haynes 20 Mod	6	5	28	8	47.5	7
A.L. 6X	11	6	37	6	57.5	6
254SMO	18	7	46	3	62.5	3
904L	36	8	22	10	42.5	9
Jessop 700	47	9	31	7	45.0	8
Jessop 77	60	10	14	12	30.0	12
Carpenter 329	73	11	25	9	40.0	11
Nitronic 50	112	12	15	11	40.0	10

tions are established between test results in synthetic and natural environments for a given material, care should be exercised before automatically applying the same test to other materials or alloy systems.

The natural tendency of researchers and investigators is to find a laboratory solution or test method that will yield predictable and applicable results to a given materials engineering application. Laboratory and/or synthetic solutions can and often do rank or predict the relative performance or behavior of a given material, i.e., separate good behavior from poor behavior, but when their results are compared to the behavior of materials in the natural environment, the investigator must use all of the available data before characterizing performance.

References

[1] Lee, T. S. and Money, K. L., *Materials Performance,* Vol. 23, No. 8, August 1984, p. 28.
[2] Report of Committee U on Corrosion of Iron and Steel, *Proceedings of ASTM, III,* 1907, p. 209.
[3] Report of Subcommittee V of Committee A-5, *Proceedings of ASTM,* Vol. 51, 1931, p. 176.
[4] LaQue, F. L., "Corrosion Testing," *Proceedings of ASTM,* Vol. 51, 1951, p. 495.
[5] Capp, J. A., *Proceedings of ASTM,* Vol. 14, 1914, p. 474.
[6] May, T. P. and Alexander, A. L., "Spray Testing with Natural and Synthetic Seawater, Part I—Corrosion Characteristics in the Testing of Materials," *Proceedings of ASTM,* Vol. 50, 1950, p. 1131.
[7] *1984 Annual Book of ASTM Standards,* Vol. 03.02, American Society for Testing and Materials, Philadelphia, 1984, p. 40.
[8] *1982 Annual Book of ASTM Standards,* Part 27, American Society for Testing and Materials, Philadelphia, 1982, p. 13.
[9] DiBari, G. A., Hawks, G., and Baker, E. A., "Corrosion Performance of Decorative Electrodeposited Nickel and Nickel-Iron Alloy Coatings," *Atmospheric Corrosion of Metals, STP 767,* American Society for Testing and Materials, Philadelphia, 1983, p. 186.
[10] Hudson, J. C., "Present Position of the Corrosion Committee's Field Tests on Atmospheric Corrosion (Unpainted Specimens)," *Journal British Iron and Steel Institute,* Vol. 148, 1943, p. 161.
[11] Guttman, H. and Sereda, P. J., "Measurement of Atmospheric Factors Affecting the Corrosion of Metals," *ASTM STP 435—Metal Corrosion in the Atmosphere,* American Society for Testing and Materials, Philadelphia, 1968, p. 326.
[12] "Corrosiveness of Various Atmospheric Test Sites as Measured by Specimens of Steel and Zinc,"

ASTM STP 435—Metal Corrosion in the Atmosphere, American Society for Testing and Materials, Philadelphia, 1968, p. 360.

[13] Baker, E. A. and Lee, T. S., "Calibration of Atmospheric Corrosion Test Sites," *ASTM STP 767— Atmospheric Corrosion of Metals,* American Society for Testing and Materials, Philadelphia, 1983, p. 250.

[14] Grossman, P. R., "Investigation of Atmospheric Exposure Factors That Determine Time-of-Wetness of Outdoor Structures," *ASTM STP 646—Atmospheric Factors Affecting the Corrosion of Engineering Metals,* American Society for Testing and Materials, Philadelphia, 1978, p. 5.

[15] Ellis, O. B., "Effect of Weather on the Initial Corrosion Rate of Sheet Zinc," *Proceedings of ASTM,* Vol. 49, 1949, p. 152.

[16] Dearden, J., "Climatic Effects on the Corrosion of Steel," *Journal Iron and Steel Institute,* Vol. 1959, 1948, p. 241.

[17] Compton, K. G., "The Unique Environment of Seawater and Its Effect on the Corrosion of Metals," *Corrosion,* Vol. 26, 1970, p. 448.

[18] *1980 Annual Book of ASTM Standards,* Part 31, American Society for Testing and Materials, Philadelphia, p. 1044.

[19] LaQue, F. L., "Theoretical Studies and Laboratory Techniques in Seawater Corrosion Testing Evaluation," *Corrosion,* Vol. 13, 1957, p. 303t.

[20] Stewart, W. C. and LaQue, F. L., *Metaux and Corrosion,* Vol. XXIII, No. 274, p. 147.

[21] Bogar, F. D. and Crooker, T. W., "The Influence of Bulk-Solution-Chemistry on Marine Corrosion Fatigue Crack Growth Rate," *Journal of Testing and Evaluation,* Vol. 7, 1979, p. 155.

[22] Kain, R. M. and Lee, T. S., "Recent Developments in Test Methods for Investigating Crevice Corrosion," *ASTM STP 866—Laboratory Corrosion Tests and Standards,* G. S. Haynes and R. Baboian, eds., American Society for Testing and Materials, Philadelphia, 1985, pp. 299–323.

[23] Nagaswami, N. S. and Streicher, M. A., "Accelerated Laboratory Tests for Crevice Corrosion of Stainless Alloys," *CORROSION/83,* Anaheim, CA, April 1983.

[24] Kain, R. M., "Crevice Corrosion Resistance of Austenitic Stainless Steels in Ambient and Elevated Temperature Seawater," *CORROSION/79,* Atlanta, GA, March 1979.

[25] Lee, T. S., Kain, R. M., and Oldfield, J. W., "The Effect of Environmental Variables on Crevice Corrosion of Stainless Steels in Seawater," *Materials Performance,* Vol. 23, No. 7, July 1984, pp. 9–15.

[26] Bond, A. D. and Dundas, H. J., "Resistance of Stainless Steels to Crevice Corrosion in Seawater," *Materials Performance,* Vol. 23, July 1984, p. 35.

[27] Brigham, R. O., "The Initiation of Crevice Corrosion of Stainless Steels," *Materials Performance,* Vol. 24, Dec. 1985, p. 44.

Stephen C. Dexter[1]

Laboratory Solutions for Studying Corrosion of Aluminum Alloys in Seawater

REFERENCE: Dexter, S. C., "**Laboratory Solutions for Studying Corrosion of Aluminum Alloys in Seawater,**" *The Use of Synthetic Environments for Corrosion Testing, ASTM STP 970,* P. E. Francis and T. S. Lee, Eds., American Society for Testing and Materials, Philadelphia, 1988, pp. 217–234.

ABSTRACT: The difficulties in obtaining reproducible results for the corrosion of aluminum alloys in saline waters, both in the laboratory and in the field are well known. The various factors contributing to variability in aluminum corrosion data in saline waters are briefly reviewed. This information is then used to evaluate NaCl solutions, enhanced NaCl solutions, synthetic seawaters, and stored seawater as substitutes for natural seawater in studying the corrosion of aluminum in the laboratory. The effects of such factors as dissolved oxygen concentration, pH, temperature, organic content, salinity, and stability of the solution are discussed. While no laboratory solution can replace natural seawater for long-term corrosion testing, an artificial solution composed of sodium chloride, sodium bicarbonate, and magnesium, with controlled oxygen, pH, temperature, and hydrodynamics is suitable for many laboratory purposes. This solution is simple, more stable than either natural or artificial seawater, and closely reproduces the short-term behavior of aluminum alloys in natural seawater.

KEY WORDS: seawater, aluminum, corrosion, sodium chloride, synthetic seawater, electrolyte

There has been controversy for many years over the best laboratory solution for studying the corrosion of a variety of metals and alloys in seawater. Suitable substitutes for natural seawater have been sought for several reasons, in large part because most corrosion laboratories are located some distance from the coast, making access to natural seawater difficult or impossible. Even when natural seawater is available and can be piped fresh into a running seawater system in the laboratory, it is difficult to get reproducible results because the water changes with season, daily tidal cycles, and local weather conditions [1]. Only a few laboratories in the world have consistent open ocean water available on a year-round basis.

In some cases, samples of natural seawater have been brought into the laboratory and stored for later use. In these cases, the properties of the water vary not only with the time and location of collection, but change over time as the water is stored [1–3]. Thus, there has been much interest in finding a substitute for natural seawater that would reproduce its corrosivity, but be more consistent and easier to handle in the laboratory. Important as this goal may be, it is well to remember that the near-shore seawater in which most marine structures spend their time is a variable environment [1]. Estuarine waters are particularly variable under the influence of both natural and man-made events. Thus, while it is quite possible to simulate the chemical and physical properties of any given water at a given point in time, it is nearly impossible to simulate that water by using a laboratory solution over a long period of time. For this reason, it is impor-

---

[1]Associate professor, College of Marine Studies, University of Delaware, Lewes, DE 19958.

tant to decide on the purpose of a particular experiment in which seawater will be simulated. It is relatively easy to simulate a critical physical or chemical seawater property if it is known what that critical property is for a given purpose. Simulation of the organic and biological properties of natural seawater are more difficult because less is known about these characteristics and about their influence on corrosion.

The purpose of this paper is to provide a limited review of literature pertinent to the use of synthetic solutions for laboratory studies of aluminum corrosion in seawater. The next section will present a review of factors known to affect the corrosivity of seawater toward aluminum. Following that will be an evaluation of several candidate synthetic solutions, including sodium chloride solutions, enhanced sodium chloride solutions, several synthetic seawaters, and stored seawater. Some new data on the enhanced sodium chloride solutions will also be presented.

Factors Affecting Corrosivity Toward Aluminum

There are a number of factors that may affect the variability of corrosion data for aluminum alloys in natural and artificial marine environments. The single most important factor in determining the corrosion behavior of aluminum alloys is the chloride ion activity. Bohni and Uhlig [4] have shown that the critical pitting potential is linearly related to the logarithm of the chloride ion activity. Therefore, any artificial solution that hopes to simulate the corrosive action of marine environments on aluminum alloys must correctly reproduce this factor for seawater of a given salinity.

A number of factors that are important in addition to the chloride ion activity are listed in Table 1. Four categories of factors are noted: chemical, physical, biological, and metallurgical. The metallurgical factors are outside the scope of this paper and have been covered adequately

TABLE 1—*Factors affecting corrosion potential of Al in Seawater.*

CHEMICAL

Total ionic strength
Chloride ion activity
Oxygen concentration
Other inorganic ions
pH (the CO_2 system)
Dissolved organics

PHYSICAL

Temperature
Pressure
Flow rate
Electrode surface area
and geometry
Surface films

BIOLOGICAL

Bacterial films
Macroscopic fouling

METALLURGICAL

Nominal composition
Heavy metal content
Second phase distribution
Temper and texture
Dissimilar metal coupling

elsewhere [5–7]. The physical factors are easy to control and are also well covered [5–9]. Thus, little more will be said about them in this paper, except for a short discussion of flow rate effects in a later section and a few words about pressure (below). The effect of pressure is felt through its influence on (1) the equilibrium constants of the carbon dioxide system [1] and (2) the thickness and defect structure of the oxide film and the kinetics of its interaction with chloride ions in the solution [10]. The chemical and biological effects are more difficult to control, and are the subject of this paper.

The work of Dexter [2,11] and Rowland and Dexter [3] has shown that corrosion of aluminum in marine environments depends on many subtle factors. Variations in these factors may change not only the corrosion rate, but also both the susceptibility to initiation of pitting and crevice corrosion and the rate of penetration by these types of attack. The difficulties in obtaining reproducible corrosion results from aluminum alloys in both natural and simulated seawaters under both laboratory and natural conditions are well known. Discrepancies between sets of corrosion data on aluminum alloys generated in fresh ocean water [5,12–20], stored seawater [21,22], recirculated seawater [7,23], synthetic seawater [24,25] and 3 to 3.5% sodium chloride (NaCl) [2,3,23–26] have been attributed in the literature to a number of organic and inorganic environmental factors.

The corrosivity of natural ocean water itself toward aluminum varies with depth, temperature, dissolved oxygen concentration, and pH [2,3,11]. Near-shore waters and deep ocean waters generally cause more severe attack than open ocean surface waters [1]. Variations in corrosion data taken in different natural waters have been associated with differences in dissolved organics depending on location and time of year [21,22]. The failure of synthetic seawaters to reproduce corrosion data from natural seawater has often been linked to inorganic factors, such as differences in their buffering capacities and tendencies to form complexes or calcareous deposits [7,21].

While there has been considerable argument over the degree to which various organic and inorganic factors govern the corrosivity of natural and artificial seawaters, most investigators have failed to realize that organic and inorganic factors are often interdependent in the ocean. For instance, the activities of marine organisms can influence the dissolved oxygen concentration and pH of the water [27]. It is thus important to examine the mechanisms by which these factors are related in the ocean itself before one can attempt to simulate natural seawater by means of any artificial solution in the laboratory. The relationships between organic and inorganic factors are best understood in terms of the seawater carbon dioxide system [27,28].

Relationship Between Chemical and Biological Factors

The surface waters of the ocean are usually in equilibrium with the atmosphere. Thus, the surface water concentrations of dissolved gasses like oxygen and carbon dioxide are determined by their solubilities, which are functions of salinity and temperature [29]. Since air-sea exchange of oxygen is rapid, the dissolved oxygen concentration will almost always be within 5% of air saturation at a given temperature and salinity. Exchange of carbon dioxide (CO_2), however, is less rapid and consequently, departure of CO_2 concentrations from equilibrium values are larger and more common than those for oxygen. This slower rate of CO_2 exchange is due to the buffering capacity of carbonate and bicarbonate ions formed when CO_2 is dissolved in seawater [28,29]. Upon dissolution, carbon dioxide undergoes two ionizations [28]:

$$CO_2 + H_2O \stackrel{K_1}{=} H^+ + HCO_3^-$$

$$HCO_3^- \stackrel{K_2}{=} H^+ + CO_3^{-2}$$

where the equilibrium constants, K_1 and K_2, are functions of temperature, pressure, and salinity [27,30,31].

The pH of natural ocean water is controlled by the CO_2 system described briefly above. As more carbon dioxide is dissolved, the equilibrium reactions above produce a greater concentration of hydrogen ions, thus decreasing the pH. If carbon dioxide is driven out of solution, the opposite occurs, and the pH becomes more basic. In the laboratory one can easily control the pH of seawater between about 7.0 and 8.5 by varying the CO_2 concentration in the gas mixture being bubbled through the solution [3].

In natural waters, dissolved oxygen concentration is often coupled to pH through the carbon dioxide system and the biochemical processes of photosynthesis and biodegradation [32] according to the following general reaction:

$$\text{photosynthesis} \rightarrow$$

$$CO_2 + H_2O \Leftrightarrow CH_2O + O_2$$

$$\leftarrow \text{respiration}$$

where CH_2O represents a typical organic molecule such as a carbohydrate. When the reaction proceeds to the right, photosynthesis consumes CO_2 and produces O_2. When the reaction proceeds to the left, biochemical oxidation burns up O_2 during the decay of organic matter to produce CO_2 and water. In surface waters containing an abundance of microscopic photosynthetic marine plants, CO_2 is consumed, O_2 is produced, and the waters tend to be high in both pH (about 8.2) and oxygen. In deep ocean water, however, where the supply of new oxygen and CO_2 from air-sea exchange is limited, decay of organic matter falling down through the water column from the surface layers consumes oxygen and produces CO_2. For this reason, it is common to find layers of deep ocean water having low pH and low dissolved oxygen simultaneously [1,33].

We have seen that factors in the ocean that decrease pH also decrease dissolved oxygen. This is such a strong relationship that cases in which the two factors are not closely coupled are rare [1]. Under any given set of natural seawater conditions, the total concentration of inorganic carbon (TCO_2) in the water will be partitioned between molecular CO_2; carbonate, CO_3^{-2}; and bicarbonate, HCO_3^-. The concentrations of these individual species depend on both the TCO_2 concentration and on the equilibrium constants [28].

In natural seawater of 32 parts per thousand ($^o/oo$) salinity and pH 8.2 at 25°C, the HCO_3^- ion makes up 89.3% of the total inorganic carbon, while CO_3^{-2} and molecular CO_2 make up 10.1% and 0.6%, respectively [3]. The CO_3^{-2} concentration is relatively high in surface waters, thereby causing supersaturation with respect to calcium carbonate. In deep waters, the increase in CO_2 concentration causes a decrease in both pH and the saturation state for the calcium carbonate phases calcite and aragonite [1].

All of the factors discussed above can influence the corrosivity of both natural and synthetic seawaters toward aluminum [2,3,11]. They can influence the rate of formation and composition of calcareous deposits, biological films, and passive oxide films on the metal surface. They can also influence the manner in which metallic ions are complexed as they leave the metal surface during the anodic reaction. Thus, these factors can influence both the anodic and cathodic reactions that govern the rate and mode of electrochemical corrosion. Before examining the usefulness of several substitutes for natural seawater, we consider in more detail how these factors influence the corrosion of aluminum alloys.

Effect of Chemical Factors

Dexter [2,11] varied pH, dissolved oxygen, and temperature independently in the laboratory for both 99.99% aluminum and for aluminum alloy 5052. He found that the corrosion potential for both metals was shifted in the noble (positive) direction by decreases in temperature and pH, and by an increase in dissolved oxygen concentration (Figs. 1 and 2). He then found that even though the separate pH and oxygen shifts are in opposite directions, the pH shift predominated when these two factors varied together, as they do in the open ocean. When pH and oxygen were decreased simultaneously in the laboratory test solution, the corrosion potential shifted in the noble direction as shown in Fig. 3. The noble shift was accentuated further when the temperature of the test solution was decreased by the proper amount along with pH and oxygen.

For a variety of aluminum alloys, the corrosion potential in natural seawater correlates with pitting behavior [12]. Those alloys with corrosion potentials more active than −0.9 V Ag-AgCl pitted very little if at all [12], while those with potentials more noble than −0.9 V pitted more intensely as the potential moved toward −0.7 V. Thus, the noble shifts in corrosion potential shown in Figs. 1 to 3 predict an increase in tendency to pit under those conditions.

Dexter found that the most corrosive set of conditions was high oxygen coupled with low temperature and pH [2,11]. For instance, at 4°C and an oxygen concentration of 7.2 ppm, all electrodes tested at pH 7.7 or below showed corrosion potentials right at the critical pitting potential and began to pit visibly within 1/2 h of insertion into the test cell. In agreement with

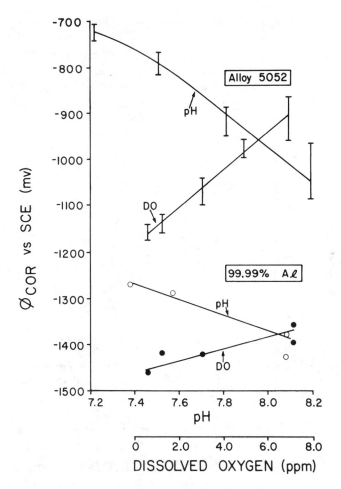

FIG. 1—*Effect of dissolved oxygen at pH 8.2 and of pH at air saturation on the corrosion potential of aluminum alloy 5052 and 99.99% aluminum in seawater* [2].

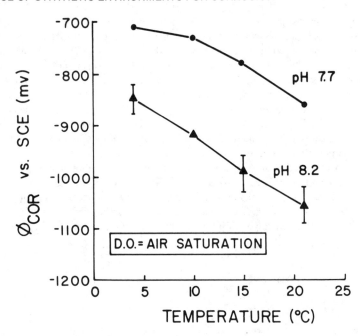

FIG. 2—*Effect of temperature at two pH values on the corrosion potential of aluminum alloy 5052 in air saturated seawater* [2].

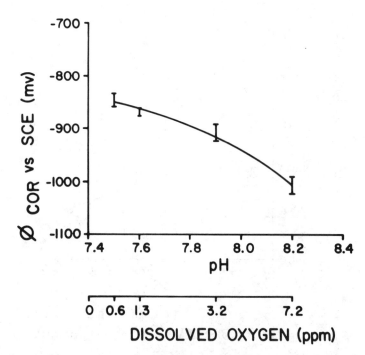

FIG. 3—*Effect of combinations of pH and dissolved oxygen found in the deep ocean on the corrosion potential of aluminum alloy 5052 in seawater* [2].

noble shift in corrosion potential upon decreasing the pH, Dexter also found that pit and crevice corrosion initiation times decreased substantially with decreasing pH in air saturated sea-water [2].

Effect of Organic and Biological Factors

Rowland and Dexter investigated the corrosion potentials and cathodic polarization behaviors of 99.99% Al and Al alloy 5052 in natural seawater and in several artificial seawaters [3]. They found that water treatment by filtration to remove microorganisms and by photooxidation to burn up all but the most refractory of dissolved organics did not appreciably change the short-term corrosion potentials or polarization curves in laboratory tests. These results indicate that the influence of microorganisms and dissolved organics on short-term laboratory corrosion behavior is not of major importance. This is in contrast to the work of Compton [21], who attributed irreproducibilities in corrosion data from different natural waters to differences in dissolved organics.

In considering the reason for this discrepancy, one must ask how organics and microorganisms interact with corrosion processes. Bacteria act as individual colloidal particles in the water, and unless the nutrient levels in the water are exceptionally high, their metabolic rates are low until they become part of a surface film where the nutrients are concentrated [34]. Thus, in open ocean seawater, bacteria must become part of a surface film to have a marked influence on corrosion.

Bacteria begin attaching to surfaces in natural seawater within the first 2 h of immersion [35,36]. It is not until 24 to 48 h after immersion, however, that a nearly continuous film containing more than 10^{10} bacteria per m^2 develops [37]. Thus, bacteria in the water are unlikely to affect most laboratory electrochemical tests, which typically last less than 2 h. On the other hand, if a bacterial film is affecting corrosion of a structure placed in the marine environment, those same laboratory tests will not detect that effect unless the film is formed on the test electrode surface before immersion in the corrosion cell.

The situation is less clear for dissolved organics, which have the potential to influence corrosion whether adsorbed or not. For example, as dissolved species, they may alter the water chemistry by changing the complexing tendencies for metallic anions entering solution, or by altering the equilibrium of the CO_2 system [1,3,27,28]. As part of an adsorbed film, however, organics will be more concentrated in the area where corrosion reactions are taking place, and their influence should be enhanced.

Adsorption of an organic film from seawater starts immediately upon immersion and is about 75% complete within the first 2 h [38]. This shorter time for adsorption, coupled with the capability of action as a dissolved species, means that organics have a potentially greater effect on corrosion in a laboratory test than do bacteria. Rowland and Dexter used offshore water, which is usually lower in organics than coastal water, for their laboratory tests [3]. This, coupled with the short duration of their tests, may explain why they saw no effect of organics, while Compton [21] did.

Because the action of organic and bacterial films is not always detected in laboratory electrochemical tests, it becomes important to know the purpose for which the experiment is being done. In finding the effect of changing a particular experimental variable on just the electrochemistry of corrosion, the effect of an organic film may not be important. If the test is being used to predict the effect of that variable on the long-term corrosion of an aluminum alloy in natural seawater, however, the effect of organic and bacterial films may need to be considered.

Effects of Electrolyte Flow Velocity

The results reported in the literature on the effect of flow velocity on corrosion are quite variable. The corrosion potential shifts in the noble (positive) direction [39], the active (negative) direction [40], or is unaffected by an increase in the flow speed of the electrolyte [41]. Recently, Rigby has found that velocities up to 0.72 m/s, characteristic of most open ocean and estuarine flows, always produce an active shift in the corrosion potential of aluminum alloys compared to that in stagnant water [42]. Jain and Larsen-Basse confirmed that low-velocity flows were conducive to the initiation of pitting on aluminum alloy tubing, particularly in deep ocean water [43].

Rigby also found that stirring the solution in a typical polarization flask introduced considerable scatter in corrosion potential data because of the variable hydrodynamics produced at the electrode surface. The above data indicate that more attention needs to be paid to electrolyte flow rate and character than has usually been the case in the past when that factor was not explicitly studied.

The factors discussed above are complex and often interdependent. This makes it difficult to specify a single laboratory solution that will be satisfactory for all applications. Nevertheless, it is often possible to find a solution that is adequate for a given purpose. Therefore, let us now evaluate the effectiveness and applicability of various substitute seawaters.

Test Solutions

A number of test solutions for studying the corrosion of aluminum alloys in saline environments have been examined [2,3]. These include: sodium chloride (NaCl) solutions, enhanced NaCl solutions, ASTM synthetic seawater prepared according to the ASTM Standard Specification for Substitute Ocean Water D 1141-75, Kester et al.'s synthetic seawater [44], and stored seawater. Let us now examine each of these in turn, attempting to list the advantages and disadvantages of each and the types of tests for which that solution is useful.

NaCl Solutions

Sodium chloride solutions have been used extensively for the corrosion testing of aluminum and many other alloy systems in chloride containing environments [2,3,23,24]. They are the simplest, most readily available, and most stable of the solutions under consideration here. Three to 3.5% reagent grade salt in distilled water is the only one of these solutions that can reasonably be expected to remain unchanged when stored longer than a few months.

The chloride ion concentration or the electrical conductivity of an NaCl solution can be matched to natural seawater of any specified salinity. There are some corrosion tests that depend almost exclusively on the chloride ion activity of the test solution. For these tests, the same results should be expected in NaCl solution as are found in natural seawater with the same chloride ion activity. An example is the measurement of the critical pitting potential [4,45]. Extensive experience with this measurement shows that all of the candidate test solutions will give the same result if they all have the same chloride ion activity and the tests are all done in the same way. Some chloride stress corrosion cracking tests that depend primarily on chloride ion activity, dissolved oxygen concentration, and temperature may also be insensitive to other characteristics of the test solution. For those tests in which NaCl solutions are suitable, they are the easiest to use and should produce the most consistent results because of their lack of variability.

These advantages are offset by the failure of NaCl solution to simulate most of the other important properties of natural seawater [3]. The buffering capacities, complexing tendencies, organic compositions, and populations of microorganisms will all be different from those in natural seawater. While the ionic strength of 3.5% NaCl solution is roughly that of 35 °/oo seawater, the lack of Mg^{+2}, Ca^{+2}, and SO_4^{-2} in NaCl solution will cause ion association in that solution to differ from that in natural seawater [1].

The ionization constants of weak acids like carbon dioxide are particularly sensitive to ion association, and the lack of these divalent ions in NaCl solution may greatly affect chemical speciation in the synthetic water [1]. For instance, Pytkowicz and Hawley [46] found that the second ionization constant of carbonic acid was four times larger in seawater than in NaCl solution. This shows that NaCl is a poor substitute for natural seawater in the study of any chemical reaction involving divalent or more highly charged ions. Indeed, Rowland and Dexter [3] found that the short-term corrosion potentials and polarization behaviors of their aluminum electrodes were quite different in 3 to 3.5% NaCl than they had been in natural seawater of the same chloride ion activity. These differences are shown for aluminum alloy 5052 in Figs. 4 and 5, and for 99.99% aluminum in Figs. 6 and 7.

It is often stated in the literature that organic or biofilm effects or both were avoided by using NaCl solution. This is a misleading statement. Sodium chloride salts are not free of organic matter, and no NaCl solution will remain free of bacteria for more than a few hours unless it is prepared and carefully maintained under sterile conditions. Organics are not absent in NaCl solutions, they are just different from those in natural seawater.

Given these differences, it is to be expected that corrosion behavior of aluminum in NaCl solutions will be different than it is in natural seawater. Indeed, most laboratory tests calling for measurement of corrosion potentials or polarization curves for aluminum in seawater cannot be successfully made in a simple NaCl solution [3]. It is possible, however, to improve the performance of NaCl by some rather simple additions, as will be shown in the next section.

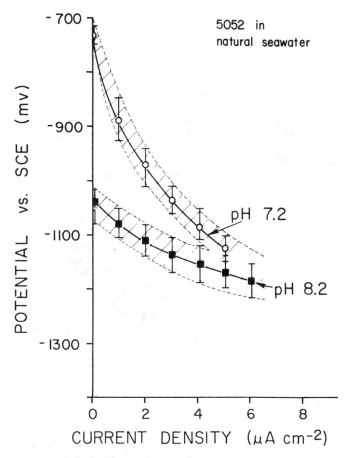

FIG. 4—*Effect of pH on the cathodic polarization of aluminum alloy 5052 in air saturated natural seawater at 21 to 25°C* [3].

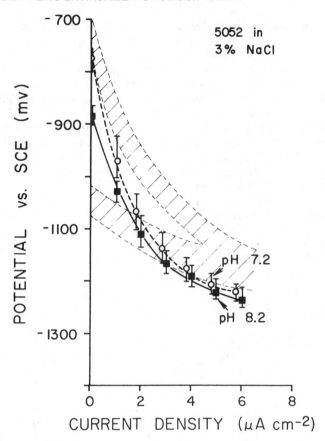

FIG. 5—*Effect of pH on cathodic polarization of aluminum alloy 5052 in air saturated sodium chloride solution at 21 to 25°C. Shaded area represents the envelope of data in Fig. 4* [3].

Enhanced NaCl Solutions

Rowland and Dexter showed that the critical ingredients missing from simple NaCl solutions are bicarbonate and magnesium [3]. In their experiments, 140 ppm bicarbonate was added as the sodium salt and 1235 ppm magnesium was added as the chloride to simulate the concentrations of those ions in full strength natural seawater. Figure 8 shows that addition of bicarbonate alone was sufficient to make the behavior of alloy 5052 in the enhanced solution match that in seawater. Such was not the case, however, for 99.99% aluminum (Fig. 9). The seawater behavior for 99.99% aluminum was not reproduced in an enhanced NaCl solution until magnesium was also added to the solution (Fig. 10).

Table 2 gives a comparison of the concentrations of the various species in the CO_2 system for a particular natural seawater and the corresponding NaCl and enhanced NaCl solutions. The carbonate alkalinities, TCO_2, molecular CO_2, and HCO_3^- values for the seawater and the enhanced solution compare quite favorably. There are still discrepancies in the CO_3^{-2} concentrations, but these are probably not important unless long-term tests involving calcareous deposition are to be done. The values for seawater will vary with temperature, pressure, and location. If these variations are known, the enhanced NaCl solution can be adjusted accordingly.

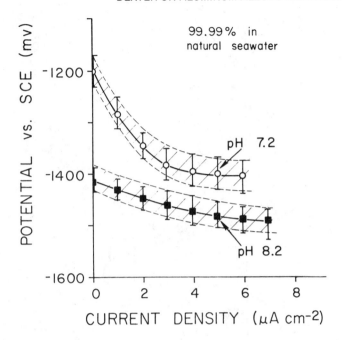

FIG. 6—*Effect of pH on the cathodic polarization of 99.99% aluminum in air saturated natural seawater at 21 to 25°C* [3].

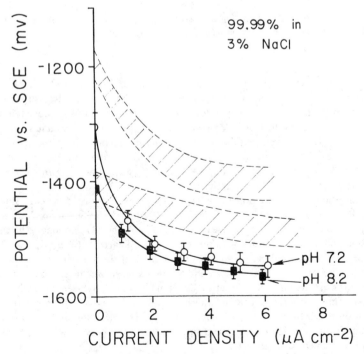

FIG. 7—*Effect of pH on cathodic polarization of 99.99% aluminum in air saturated sodium chloride solution at 21 to 25°C. Shaded area represents the envelope of data in Fig. 6* [3].

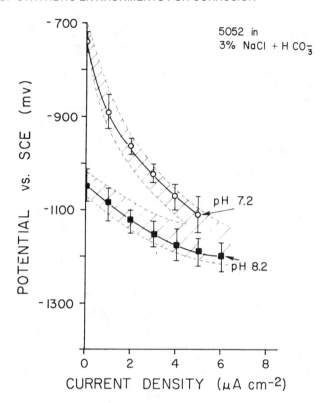

FIG. 8—*Effect of pH on cathodic polarization of aluminum alloy 5052 in 3% NaCl + HCO$_3^-$ solution at 21 to 25°C. Shaded area represents the envelope of data in Fig. 4* [3].

The CO_2 system and magnesium concentration of natural seawater of any salinity can be simulated by an enhanced solution of the type shown in Table 2. It may not be too important to match the magnesium content exactly unless the carbonate film forming capacity is important to the experiment being performed. There are over 1200 ppm of magnesium in natural seawater of 35 ^{0}/oo salinity. Dexter et al. [47], however, have shown that only about 200 ppm is needed to give the full effect of magnesium on the corrosion potential of aluminum in the enhanced solution. Figure 11 shows that additions of up to 50 ppm Mg as the chloride produced no effect on the corrosion potential of 99.99% aluminum.

At concentrations above 50 ppm, there was a gradual noble shift in the corrosion potential until the full seawater value was reached at about 200 ppm. Additions above 200 ppm gave no further change in corrosion potential. Dexter et al. [47] also showed that magnesium influenced the corrosion potential exclusively, with no effect on the polarization behavior.

Other major ions dissolved in seawater that might have an effect on either corrosion potential or polarization behavior are: Ca^{+2}, Sr^{+2}, and SO_4^{-2}. Experiments to see if calcium and strontium additions will have an effect similar to that of magnesium have been done. The results [47] indicate that there is no effect of calcium and strontium, which agrees with predictions from the mechanism proposed for the effect of magnesium [3]. The effects of adding sulfate have not been systematically investigated. However, sulfate is known to be a complexing agent for magnesium as is discussed by Dexter et al. [47].

Enhanced NaCl solutions are nearly as stable and easy to make as the simple NaCl solutions. When it comes to simulating the organic and microbiological activities of seawater, the en-

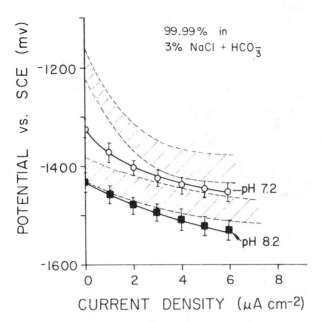

FIG. 9—*Effect of pH on cathodic polarization of 99.99% aluminum in 3% NaCl + HCO₃⁻ solution at 21 to 25°C. Shaded area represents the envelope of data in Fig. 6* [3].

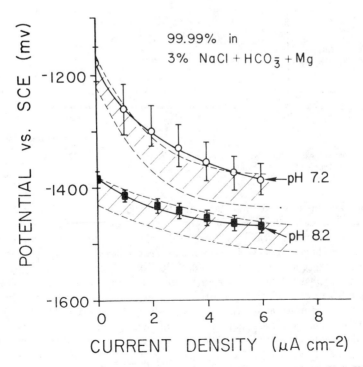

FIG. 10—*Effect of pH on cathodic polarization of 99.99% aluminum in 3% NaCl + HCO₃⁻ + Mg solution at 21 to 25°C. Shaded area represents the envelope of data in Fig. 6* [3].

TABLE 2—*Carbonate speciation at pH 8.2 and 25°C*
(after Dexter and Rowland [3]).

Species	Concentration, mol/m³		
	Seawater[a]	3% NaCl	Enhanced Solution, 3% NaCl + 140 ppm HCO$_3^-$
Alkalinity	2.2776	. . .	2.300
Total CO$_2$	2.0801	1.542	2.2304
CO$_2$	0.0122	0.0100	0.0145
HCO$_3^-$	1.858	1.475	2.132
CO$_3^{-2}$	0.2097	0.0597	0.0838

[a]For offshore water of 32 ⁰/₀₀ salinity collected as described in the section on "Stored Seawater."

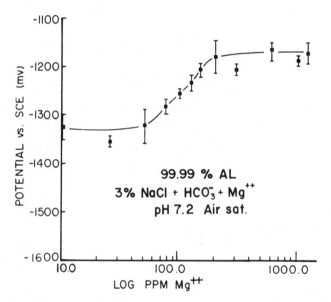

FIG. 11—*Corrosion potential of 99.99% aluminum in 3% NaCl + HCO$_3^-$ solution as a function of magnesium additions at pH 7.2 [47].*

hanced solutions will do no better than the simple NaCl solutions. But for short-term laboratory tests, especially those involving measurements of the corrosion potentials and polarization curves, they do remarkably well [3]. The enhanced solutions should also be suitable for other laboratory electrochemical tests such as potentiodynamic polarization, linear polarization, and polarization resistance measurements, although this has not been systematically documented.

Given that the enhanced solutions are unable to simulate the organic and biological properties of natural seawater, the success of experiments using these solutions in predicting the long-term corrosivity of natural seawater is somewhat surprising. This success is demonstrated by the level of agreement between predictions made from the laboratory tests and subsequent work done with aluminum tubing for Ocean Thermal Energy Conversion heat exchangers at several locations.

The laboratory work [2,3] shown in Figs. 2 and 3 suggests that warm open ocean surface waters should not cause pitting of alloy 5052 or Alclad tubing. The corrosion potential in these waters should be well below that at which pitting occurs because of the combination of the high temperature and pH. In deep water, however, one would predict that the decrease in temperature and pH should cause pitting.

In confirmation of the above predictions, Larsen-Basse [13-15] found that alloys 5052 and Alclad 3004 pitted in only 3 months in the deep water at the Seacoast Test Facility at Keohole Point, Hawaii. Pitting of the Alclad material was confined to the clad layer, and did not penetrate the base metal [20]. The same alloys remained free of pits for 1 year of exposure in surface waters at the Hawaii location. Munier and Craig [48] found similarly that alloys 5052 and 6061-T6 suffered neither pitting nor crevice corrosion during a 72-day exposure to flowing surface seawater at the St. Croix, U.S. Virgin Islands test station. Sasscer et al. [16,17] found similar results in Puerto Rican waters.

ASTM and Kester Synthetic Seawaters

It was found that corrosion potentials and potentiostatic cathodic polarization curves measured in ASTM D 1141-75 and Kester [44] synthetic seawaters did not vary significantly from their values in natural seawater of the same salinity, temperature, and pH [3]. Water made from the Kester formula was evaluated as an alternative to ASTM seawater because it is closer to the composition of natural seawater (within 1 ppm for all major constituents) [44].

The disadvantage of the Kester water is that it must be made from the individual salts, while the salts for ASTM water can be purchased premixed. The results of Rowland and Dexter [3] indicate that for corrosion studies, the small increase in accuracy of the Kester water is not worth the extra work involved in making it. They found no major difference between results from the two waters when testing both 99.99% Al and Al alloy 5052 in short-term laboratory tests, and both sets of results agreed well with those from natural seawater [3].

Neither of these artificial waters produced results any less variable than did natural water. Despite careful preparation of the synthetic solutions, carbonate precipitates were observed to form before and even during polarization runs [3]. These two artificial solutions have been found to be just as changeable in the laboratory as is natural seawater. For these reasons, it is the opinion of this author that the ASTM and Kester seawaters have no real advantage for most laboratory purposes over the simpler and more stable enhanced NaCl solutions discussed in the previous section.

Stored Seawater

For the last ten years at the University of Delaware's, College of Marine Studies, we have been evaluating the effectiveness of stored natural seawater from a variety of sources as an alternate to fresh running seawater for our laboratory experiments on aluminum alloys [2,3,11]. We have found that the results vary widely depending on how the water is collected and on what treatment the water receives between collection and use. Lower Delaware bay estuarine water is highly variable, depending on time of year, daily tidal cycle, and fresh water run-off. The organic character of the water also varies widely with weather conditions, biological activity, and agricultural activities.

Waters collected from a variety of locations within 35 km of the College facilities all gave results that were internally consistent within a single batch of water, but were inconsistent from batch to batch both at the same site and between sites. These collection locations included our own research vessel harbor at Roosevelt Inlet; the mouth of the Broadkill river, which empties into Roosevelt Inlet adjacent to the research vessel harbor; lower Delaware Bay at several locations within Lewes harbor; and at Indian River Inlet, on the Atlantic Coast 28.5 km south of the mouth of Delaware Bay. The latter of these locations was chosen to get away from the Delaware Bay plume, which follows the Atlantic Coast for 3 to 8 km south of the Bay mouth before diffusing into the open Atlantic Ocean. Waters at most of these sites have been collected at several different periods within the tidal cycle. Results for all sites were most consistent in waters collected from the incoming tide, but none of these waters was suitable as a substitute for full-strength seawater.

The only natural water suitable for laboratory tests on aluminum was full strength (32 to 35 º/oo) Atlantic Ocean surface water collected from 65 to 160 km East of the mouth of Delaware Bay. That water gave consistent results that did not vary appreciably from batch to batch or from the results obtained in the synthetic waters. Methods of collection and storage, however, were critical to the successful use of offshore water. The most successful method of collection involved using a pump on board the research vessel Cape Henlopen. The pump was mounted with its intake near the keel of the hull. This allowed collection of water from a depth of about 2.5 m below the surface while the ship was under way. Collecting surface water using a bucket from a stationary ship was unacceptable because of the extra labor involved and because the water sample often became contaminated by hydrocarbon materials floating on the sea surface.

When collecting by using the onboard pump, water was allowed to run for several minutes before collection began to flush out water that had been sitting stagnant in the system. The pump discharge was then directed immediately into the collection vessel without passing through any of the ship's various holding tanks. Water was collected in polyethylene jugs that had been "seasoned" in seawater for a minimum of 2 months to allow organic compounds to leach out of the polymer.

Water sampling was done while the ship was on the way back into port so that the sample could arrive in the laboratory within 12 h of the time it was taken from the ocean. During the trip back to the lab, water containers were kept in an air-conditioned compartment in the ship's laboratory. If the containers were allowed to sit out on the deck in direct sunlight, heat buildup would cause the water to ferment and become unusable. Upon arrival in the laboratory, the water was immediately subjected to aeration by a gas bubbler. As soon as it was convenient, the water was passed through a glass fiber filter to remove large particulates, and then through a 0.2-μm membrane filter to remove most microorganisms. Water treated in this way and kept continuously aerated gave reproducible short-term laboratory results on aluminum alloys for a period of 2 to 4 weeks. Sometime within that period, the results would begin to change, and the unused water would be discarded.

Summary and Conclusions

Aluminum alloys are quite sensitive to small changes in seawater properties. Thus, to use any of the waters discussed in this paper for the successful simulation of the seawater corrosion of aluminum alloys, one must first know the characteristics of the seawater to be simulated. First, one must know the salinity of the water. This can be simulated by adjusting the composition of a simple NaCl solution. If the corrosion phenomenon under investigation depends primarily on the chloride ion activity, a simple NaCl solution may be best. It is certainly the most stable and easiest to prepare and use.

If factors other than chloride ion activities are important, NaCl solution alone will not be adequate. The buffering and carbonate film-forming tendencies of seawater come from its carbon dioxide system. This can be simulated by adding the proper amount of NaHCO$_3$ to NaCl solution. Once the bicarbonate has been added, pH and dissolved oxygen can be controlled by bubbling the proper mixture of gaseous N$_2$, O$_2$, and CO$_2$ through the solution. Next, one needs to add at least 200 ppm magnesium as the chloride, taking care to adjust for the right Cl$^-$ ion activity if necessary. The full amount of magnesium (1230 ppm in full-strength water) should be added if one wants to simulate the magnesium component of the natural carbonate film-forming capacity. Last, one needs to control the temperature and hydrodynamics of the water, especially if corrosion potentials or polarization curves are needed. Stored ocean water treated in the proper way, and in which the above variables are controlled, can be used successfully for some applications.

None of the solutions discussed here will be able to simulate the long-term behavior of any aluminum alloy in natural seawater when the buildup of surface films on the metal is likely to have a strong influence on the corrosion rate. This is especially true when the films are expected to have substantial calcareous, organic, or biological components. Neither will any of these solutions reproduce the corrosion results from estuarine water, which is by nature highly variable.

Acknowledgements

This work was supported by Contract No. EY-76-S-02-2957 from the U.S. Dept. of Energy, Grant No. N00014-77-C-0064 from the Office of Naval Research, and Grant No. NA81AA-D-006 from the U.S. Dept. of Commerce, Office of Sea Grant.

References

[1] Dexter, S. C. and Culberson, C. H., *Materials Performance*, Vol. 19, No. 9, Sept. 1980, pp. 16-28.

[2] Dexter, S. C., *Corrosion*, Vol. 36, No. 8, Aug. 1980, pp. 423-432.

[3] Rowland, H. T. and Dexter, S. C., *Corrosion*, Vol. 36, No. 9, Sept. 1980, pp. 458-467.

[4] Bohni, H. and Uhlig, H. H., *Journal of the Electrochemical Society*, Vol. 116, 1969, p. 906.

[5] Godard, H. P., Jepson, W. B., Bothwell, M. R., and Kane, R. L., *The Corrosion of Light Metals*, Wiley, New York, 1967, p. 18.

[6] Uhlig, H. H. and Revie, R. W., *Corrosion and Corrosion Control*, 3rd ed., Wiley-Interscience, New York, 1985, pp. 341-354.

[7] LaQue, F. L., *Marine Corrosion*, Wiley-Interscience, New York, 1975.

[8] Pryor, M. J. in *Localized Corrosion*, R. W. Staehle, Ed., National Association of Corrosion Engineers, Houston, TX, 1974, p. 2.

[9] Foley, R. T., *Corrosion*, Vol. 42, 1986, p. 277.

[10] Beccaria, A. M. and Poggi, G., *Corrosion*, Vol. 42, 1986, p. 470.

[11] Dexter, S. C., *Ocean Science and Engineering*, Vol. 6, No. 1, 1981, pp. 109-148.

[12] Groover, R. E., Lennox, T. J., Jr., and Peterson, M. H., *Materials Protection*, Vol. 8, 1969, pp. 25-30.

[13] Larsen-Basse, J., *Materials Performance*, Vol. 23, 1984, p. 16.

[14] Larsen-Basse, J. and Zaidi, S. H., *Proceedings of the International Congress on Metallic Corrosion*, Toronto, June 1984, Vol. 4, p. 511.

[15] Larsen-Basse, J., *Journal of Metals*, Vol. 37, 1985, p. 24.

[16] Sasscer, D. S., Summerson, T. J., and Ernst, R., "In-Situ Seawater Corrosion of Bare, Diffusion Zinc Treated and Alclad Aluminum Heat Exchanger Materials," in *Proceedings, Oceans '82, Institute of Electronic and Electrical Engineers*, 1982, pp. 578-586.

[17] Sasscer, D. S., Ernst, R., and Summerson, T. J., "Open Ocean Corrosion Test of Candidate Aluminum Materials for Seawater Heat Exchangers," Preprint No. 67, presented at CORROSION/83, Anaheim, CA, National Association of Corrosion Engineers, April 1983.

[18] Ailor, W. H., "Ten-Year Seawater Tests on Aluminum," in *Corrosion in Natural Environments, STP 558*, American Society for Testing and Materials, Philadelphia, 1974, pp. 117-134.

[19] LaQue, F. L., "Qualifying Aluminum and Stainless Alloys for OTEC Heat Exchangers," Proceedings of the 6th Ocean Thermal Energy Conversions, Conference, G. L. Dugger, ed., Washington, D.C., June 1979, p. 12.2-1.

[20] Rynewicz, J. F. and Nagrodski, N., "Corrosion Test Results on Sea Water Heat Exchanger Materials," Preprint No. 52, presented at CORROSION/82, Houston, TX, National Association of Corrosion Engineers, March 1982.

[21] Compton, K. G., *Corrosion*, Vol. 26, 1970, p. 449.

[22] Rogers, T. H., *Journal of the Institute of Metals*, Vol. 76, 1949-50, pp. 597-611.

[23] Gilbert, P. T. and LaQue, F. L., *Journal of the Electrochemical Society*, Vol. 101, 1954, pp. 448-455.

[24] LaQue, F. L., *Corrosion*, Vol. 13, 1957, pp. 303t-314t.

[25] LaQue, F. L. and Stewart, W. C., *Metaux et Corrosion*, Vol. 23, 1948, pp. 147-164.

[26] Whitby, L., *Transactions of the Faraday Society*, Vol. 29, 1933, pp. 523-531.

[27] Riley, J. P. and Chester, R., *Introduction to Marine Chemistry*, Academic Press, London, 1971.

[28] Skirrow, G., in *Chemical Oceanography,* 2nd ed., Vol. 2, J. P. Riley and G. Skirrow, eds., Academic Press, New York, 1975.

[29] Kester, D. R., in *Chemical Oceanography,* 2nd ed., Vol. 1, J. P. Riley and G. Skirrow, eds., Academic Press, New York, 1975, p. 498.

[30] Mehrbach, C., Culberson, C. H., Hawley, J. E., and Pytkowicz, R. M. *Limnology and Oceanography,* Vol. 18, 1973, pp. 897–907.

[31] Dryssen, D. and Hansson, I., *Marine Chemistry,* Vol. 1, 1973, pp. 137–149.

[32] Fogg, G. E., in *Chemical Oceanography,* 2nd ed., Vol. 2, J. P. Riley and G. Skirrow, eds., Academic Press, New York, 1975, p. 385.

[33] Reinhart, F. M., "Corrosion of Materials in Hydrospace-Part V—Aluminum Alloys," Naval Civil Engineering Laboratory, Port Hueneme, CA, Technical Note N-1008, January 1969.

[34] Marshall, K. C., *Interfaces in Microbial Ecology,* Harvard University Press, Cambridge, MA, 1976.

[35] Dexter, S. C., *Journal of Colloid and Interface Science,* Vol. 70, No. 2, June 1979, pp. 346–354.

[36] Zobel, C. E., *Journal of Bacteriology,* Vol. 43, 1943, pp. 39–59.

[37] Dexter, S. C., *Proceedings,* Fourth Intl. Congr. on Marine Corrosion and Fouling, Juan Les Pins, Antibes, France, June, 1976, pp. 137–144.

[38] Loeb, G. I. and Neihof, R. A., in *Applied Chemistry at Protein Interfaces,* R. E. Baier, ed., Advances in Chemistry Series 145, American Chemical Society, Washington, DC, 1975, pp. 319–335.

[39] Franz, F. and Novak, P., in *Localized Corrosion,* R. W. Staehle, et al., eds., National Association of Corrosion Engineers, Houston, TX, 1971, p. 576.

[40] Davis, J. A., "Review of High Velocity Sea Water Corrosion," NACE Technical Committee T-7C-5, Task Group Report, CORROSION/77, National Association of Corrosion Engineers, Houston, TX, 1977.

[41] Mansfeld, F. and Kenkel, J. V., *Corrosion,* Vol. 35, No. 1, 1979, p. 43.

[42] Rigby, W. E., "Aluminum Alloy 5052 Corrosion in Low Velocity Seawater," Master's thesis, College of Marine Studies, University of Delaware, Dec. 1983.

[43] Jain, S. and Larsen-Basse, J., "Effect of Flow Velocity on the Corrosion of Some Aluminum Alloys in Deep Ocean Seawater," Preprint No. 224, presented at CORROSION/86, Houston, TX, National Association of Corrosion Engineers, March 1986.

[44] Kester, D. R., Duedall, I. W., Connors, D. N., and Pytkowicz, R. M., *Limnology and Oceanography,* Vol. 12, 1967, pp. 176–179.

[45] Leckie, H. and Uhlig, H. H., *Journal of the Electrochemical Society,* Vol. 113, 1966, p. 1262.

[46] Pytkowicz, R. M. and Hawley, J. E., *Limnology and Oceanography,* Vol. 19, 1974, p. 223.

[47] Dexter, S. C., Kucas, K. E., Mihm, J., and Rigby, W. E., "Effect of Water Chemistry and Velocity of Flow on Corrosion of Aluminum," Preprint No. 64, Presented at CORROSION/83, Anaheim, CA, National Association of Corrosion Engineers, April 1983.

[48] Munier, R. S. C. and Craig, H. L., "OTEC Biofouling and Corrosion Experiment, St. Croix, U.S. Virgin Islands, Part II Corrosion Studies," OTEC Report PNL-2739, Feb. 1978.

Thomas Kohley[1] *and Ewald Heitz*[1]

Particle Containing Formation Water for the Study of Erosion Corrosion

REFERENCE: Kohley, T. and Heitz, E., **"Particle Containing Formation Water for the Study of Erosion Corrosion,"** *The Use of Synthetic Environments for Corrosion Testing, ASTM STP 970,* P. E. Francis and T. S. Lee, Eds., American Society for Testing and Materials, Philadelphia, 1988. pp. 235–245.

ABSTRACT: Mass loss rates of a 13% chromium steel have been determined in particle-containing formation water using a loop with a constricted pipe as a test section. Erosion corrosion rates increase approximately with the square of the flow rate and linearly with the sand content. Under CO_2-corrosion conditions, changes of the particle parameters (size, concentration) are much more significant than medium parameters (chloride content, temperature).

KEY WORDS: chromium steel, formation water, particles, sand, mass loss rate, corrosion rate, erosion corrosion, abrasive wear, pipe, loop

Ferrous materials are increasingly being subjected to engineering conditions with high flow rates resulting in various types of corrosion. One of these is erosion corrosion, generally attributed to the combined action of chemical and mechanical processes at the metal surface [1–5]. Whereas much is known about the behavior of copper alloys, information about ferrous materials is contradictory owing to differences in the formation of surface layers [1,6]. This is especially true regarding particle containing media, for which systematic investigations have been lacking.

Erosion corrosion is a problem in engineering equipment used in oil and gas production as well as in the energy and chemical industries. Questions regarding the behavior of ferrous materials in so called formation waters of oil and gas fields are relevant. Formation waters generally contain CO_2 and high levels of salt (chlorides), but no oxygen. They arise from the geological formation of gas or oil deposits and therefore may contain small quantities of sand, which stimulates erosion corrosion processes. Another important component encountered under such conditions is H_2S, which has not been included in this study.

Experimental Procedures

Material

A 13% chromium steel (DIN specification No 1.4021; X 20 Cr 13) with the following chemical analysis (percent weight) and mechanical properties was used: Cr–12.5%, C–0.20%, Si–0.28%, Mn–0.31%; YP 500 N mm^{-2}, TS 800-900 N mm^{-2}, Hardness HB 30 230-270 HRC.

[1]Professor, Dechema-Institut, Theodor-Heuss-Allee 25, D-6000 Frankfurt/Main 97, Federal Republic of Germany.

Medium, Conditions, and Testing Methods

The following analysis of a representative formation water is the result of a comparison of various formation waters in the northern North Sea.[2] This representative formation water contains 3.8% sodium chloride (NaCl), 0.44% calcium chloride ($CaCl_2$), 0.07% magnesium chloride ($MgCl_2$), (percent weight), 3 bar carbon dioxide CO_2, no oxygen, and various amounts of sharp sand with a diameter of 0.4 ± 0.1 mm. Hydrogen sulfide (H_2S) was not present and therefore not included. A standard temperature of 60°C was chosen, resulting in a pH value of 3.6.

Corrosion rates have been obtained in the usual way by using weight loss and polarization resistance measurements. Because of the high conductivity of the medium and earlier experience with corrosion systems such as that under consideration, the linear polarization resistance method is applicable [7].

Test Apparatus and Flow Model

To investigate flow dependency, we used experimental setups with facilities appropriate for imposing particular flow conditions and system parameters.

With a liquid/solid two-phase flow, e.g., in liquids with a sand content, the problem arises of how to dose and to distribute the solid particles evenly throughout the fluid in a flow loop. In the present case, there was no need for solid dosage and separation equipment. Appropriate structural adaptations together with minimum flow rates of 0.7 m s^{-1} and particle diameters of 0.5 mm ensured an even distribution.

Figure 1 gives the flow diagram of the two-phase flow circuit made of stainless steel UNS 32550 (Ferralium Alloy 255, Deutsche Langley Alloys GMBH, Frankfurt/M). This is a ferritic/austenitic high-alloy steel with the following composition: C-0.08%; Cr-24 to 27%; Ni-4.5 to 6.5%; Mo-2.0 to 4.0%; Cu-1.3 to 4.0%; with the remainder Fe.

The circuit consists of an integral motor/transmission/pump unit (transmission type PI, manufactured by PIV, D-Bad Homburg), two heaters in front of and behind the pump (pump Type LPK 40/160, manufactured by KSB, D-Pegnitz), two parallel, vertical test channels 3.5 m in length, and a reservoir for the medium. In front of the inlet in the reservoir, which contains control equipment for cooling, gas content, liquid level, etc., is an inductive flow meter for measuring the flow rates in the test section. The temperature is controlled automatically (accuracy ±0.5°).

Ball valves are used in on or off positions only, as the flow rate can be controlled via the pump rpm. The operational data for the apparatus are as follows: rated volume 40 m^3 h^{-1}; rated pressure head 35 m.

The apparatus was designed for pressures up to 10 bar and temperatures up to 100°C. The loop had a total volume of 190 L.

The maximum mean flow rate (e.g. as measured with a usual flow meter) was, for the restricted tube section with a diameter of $d = 25$ mm, approximately 24 m s^{-1}, and for the normal tube diameter with $D = 50$ mm, approximately 6 m s^{-1}. The measuring section was designed as a segmented pipe (Fig. 2).

Following an inlet section of about $60\,D$ in length ($D = 50$ mm), the pipe suddenly constricts (diameter $d = 25$ mm, length $L = 100$ mm) and then expands, with an outlet section of $7\,D$ in length. The segments are electrically isolated from each other using Plexiglass gaskets, permitting separate mass loss measurements and electrochemical measurements along the pipe axis.

[2]German-Norwegian Joint Project, Det Norske Veritas, Oslo (private communication).

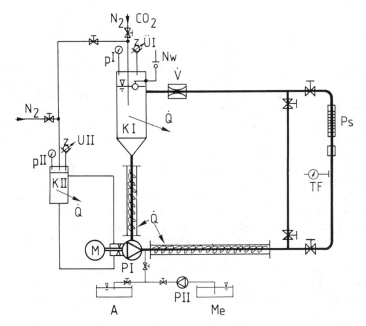

FIG. 1—*Flow diagram of a two-phase flow pipe circuit (I main circuit, II secondary circuit); Ps = test channel; A = outlet, Me = medium reservoir; PII = pump for filling circuit; Q = heat flow; V = flow velocity; Nw = level control; U = overflow valve; p = pressure gauge; K = cooler; M = motor; P = pump, TF = temperature probe.*

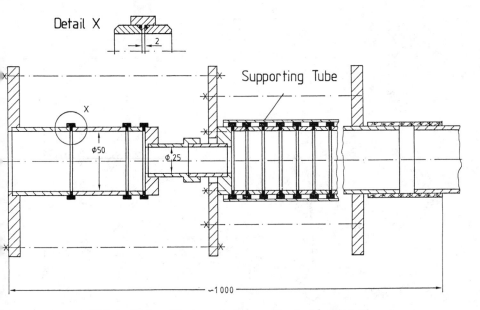

FIG. 2—*Design of the segmented pipe section (numbers in mm).*

Results

Influence of Flow Rate and Flow Geometry

Figure 3 shows the influence of flow velocity (Re number, respectively) on the corrosion rate along the pipe length in front and behind the constriction in the pipe. On the x-axis, a normalized pipe length x/D, with $D = 50$ mm, is plotted. The flow velocity is varied from 1 to 5.7 m/s in the 50 mm diameter tube. At the entrance of the pipe constriction the mass loss rates go through a maximum. For example the mass loss rate at 5.7 m/s exceeds 20 mm/year. At $x/D = 0.5$ downstream of the pipe expansion a minimum occurs, followed by a second maximum. At 1 m/s there is a finite mass loss rate but no dependence with the length of the tube. The results suggest that there exists no critical flow rate for the onset of erosion corrosion but additional experiments at low flow velocities are necessary.

A plot of the influence of the flow velocity on the mass loss rate at the maximum behind the constriction is given in Fig. 4. The corrosion rate increases to the power of 1.85 of the flow velocity, which also has been found for erosion corrosion in one-phase flow [1].

The first maximum in Fig. 3 can be mainly attributed to abrasive wear, the second to erosion corrosion. The main proof for the latter finding is given by electrochemical methods. Corrosion rates obtained by polarization resistance measurements are in accordance with the mass loss rate. At the inlet edge of the constriction, the removal rate is far higher than the electrochemically determined corrosion rate, which implies a marked mechanical abrasion of the base material.

There was an even removal over the individual pipe segments; no flow directions in the corrosion pattern were observed. There was no sign of crevice corrosion in the region of the rings separating the segments. Steady state corrosion rates were obtained soon, so that a testing time of 8 h was sufficient.

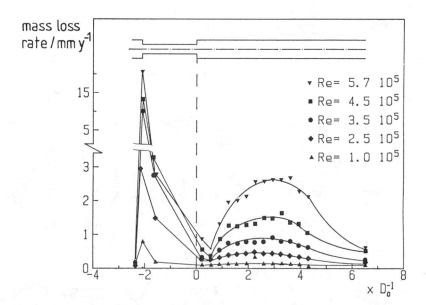

FIG. 3—Plot of the mass loss rate vs. normalized distance from the pipe expansion at different flow rates, 60°C, 3 bar CO_2 and 1000 ppm sand (Re number $5.7 \cdot 10^5$ corresponds to a flow velocity of 5.7 m/s).

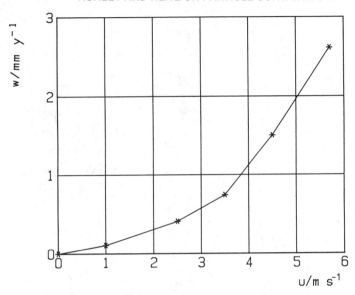

FIG. 4—*Plot of the mass loss rate* w *versus flow velocity* u *at constant sand content.*

Influence of Sand Content and Further Parameters

Figure 5 presents the dependence of the mass loss rate on the sand concentration at 3.5 m/s. At the entrance of the constriction, the influence of sand (500 to 2000 ppm) is not evident. This is the effect of the rounding off of the edge by the strong change of flow direction combined with intense abrasion. However, since steady state is soon reached, the overall mass loss is nearly constant. Similarly, after the expansion at $x/D = 0.5$, the minimum is independent of the sand concentration.

At $x/D = 2.5$, a rise in the sand concentration is responsible for a significant increase in the erosion corrosion rates. In a one-phase flow situation (0 ppm sand), no differences in removal rates along the pipe length can be observed. If the mass loss rates at the maximum of erosion corrosion are plotted against the sand concentration, a straight line is obtained (Fig. 6). This means that the erosion corrosion rate is directly proportional to the number of particles in the medium hitting the metal surface. No explanation can be given for the finite corrosion rate at zero sand content.

An important question can be raised regarding the influence of an increase in testing time. Figure 7 shows that the maximum corrosion rate at 3.5 m/s and a sand content of 1000 ppm becomes lower if the duration of the test increases. An exact interpretation of this effect may be possible based on changes in particle shape and size as well as other changes in the system.

In accordance with these findings, the change in sand geometry from sharp to rounded sand particles reduced the corrosion rate from 0.8 to 0.65 mm year^{-1}. A scanning electron microscope was used to investigate size and geometry after 20 hours under the conditions stated in Fig. 7. Particle size decreases and the edges of the particles become rounded (Figs. 8 and 9). The mean diameter is reduced from 450 μm to 380 μm, which is not very significant. Complete damage to the particles was not observed.

A number of further changes of the system at "standard" conditions (3.5 m/s, 60°C, 3 bar CO_2, and 1000 ppm sand) were made, giving the following results for the maximum of erosion corrosion rate:

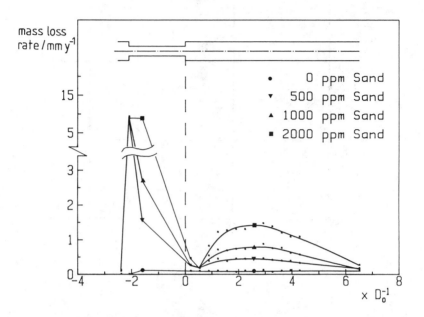

FIG. 5—*Influence of various sand concentrations on the mass loss rate* w *along the pipe length at a flo* *rate of 3.5 m/s.*

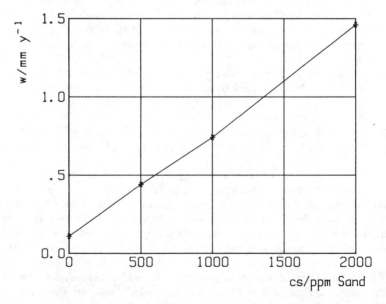

FIG. 6—*Plot of the mass loss rate* w *versus sand concentration.*

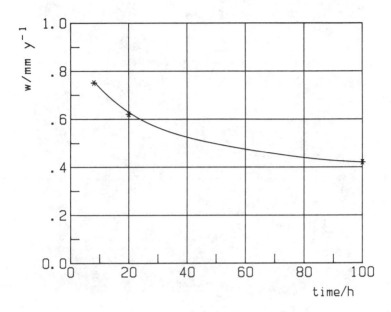

FIG. 7—*Plot of mass loss rates w versus test duration at 3.5 m/s and with a sand content of 1000 ppm.*

FIG. 8—*Sand particles before the tests (SEM 90:1), mean diameter $\phi = 450$ μm (from sieve analysis).*

FIG. 9—*Sand particles after test (SEM 90:1), mean diameter $\phi = 380$ μm (sieve analysis).*

1. The reduction of the chloride content from 26 to 10 g/L has no influence on the corrosion rate.

2. The decrease of the temperature from 60 to 25°C has no effect on the corrosion rate.

3. The consequence of a temperature rise from 60 to 80°C is a corrosion rate increase of 25%.

If the diameter of the sand particles is reduced from 0.4 mm to 0.09 ± 0.01 mm at constant particle concentrations, the corrosion rate decreases from 0.8 to 0.3 mm/year.

In conclusion, changes in the flow conditions and particle parameters are much more significant than changes in the medium parameters in CO_2-corrosion.

Possible Mechanistic Interpretations

The findings can be explained by the breakdown in passivity of 13% chromium steel in sand containing formation water, which is a consequence of particle impacts on the metal surface. High kinetic energy particles not only destroy the passive layer, but also the base material, and mechanical wear is the result. This holds for the entrance region of the constriction of the segmented tube. Medium energy particles only destroy the passive layer, and if the kinetics of the healing process is slower than the particle impact kinetics, erosion corrosion arises. The intensity of the corrosion is therefore a function of the hydrodynamics of the flow. As a result the area of maximum erosion corrosion behind the orifice can be explained as the region in which reattachment of the flow with high particle impact frequency occurs.

Hydrodynamic investigations using laser-doppler anemometry support these findings. A plot of the kinetic energy of the particles along the pipe axis gives a curve of the same shape as the mass loss rate.[3] However, as long as single events are not completely understood, mechanistic interpretations can only be made on a qualitative base.

Acknowledgments

The research has been sponsored by the Bundesministerium für Forschung und Technology within the German-Norwegian joint project on erosion corrosion in multiphase systems. Their excellent cooperation with the partners of this project is gratefully acknowledged.

We thank the following firms that contributed with equipment or financial support to the project: KSB, Frankenthal, Hoechst AG, Frankfurt, Mannesmann Röhrenwerke AG, Düsseldorf (FRG).

References

[1] Lotz, U. and Heitz, E., *Werkstoffe und Korrosion,* Vol. 34, pp. 454–461, 1983.
[2] Ellison, B. T. and Wen, C. J.; AICHE Symposium Series No. 204, Vol. 77, 1981, pp. 161–169.
[3] Heitz, E., Kreysa, G., and Loss, C., *Journal of Applied Electrochemistry,* Vol. 9, pp. 243–253, 1979.
[4] Syrett, B. C., *Corrosion,* Vol. 32, pp. 242–252, 1976.
[5] Efird, K. D., *Corrosion,* Vol. 33, 1977, pp. 3–8.
[6] Lotz, U., Schollmaier, M., and Heitz, E., *Werkstoffe und Korrosion,* Vol. 36, 1985, pp. 163–173.
[7] Grauer, R., Moreland, R., and Pini, G., *A Literature Review of Polarisation Resistance Constant (B) Values,* NACE Publications, Houston, TX, 1982.

[3]Kohley, T., Blatt, W., Lotz, U., and Heitz, E., *Corrosion,* National Association of Corrosion Engineers, in preparation.

DISCUSSION

Turnbull[1] *(discusser's question)*—Is it possible to quantify the frequency of particle impact upon the surface?

T. Kohley and E. Heitz (authors' response)—Possible methods are (a) high-speed kinematography, (b) laser-doppler-anemometry and (c) acoustic methods. Experiments with methods (a) and (b) are under way. Results will be published in Corrosion NACE.

D. R. John[2] *(discusser's question)*—Did you find that the relative ranking of the materials tested was the same under all conditions applied?

T. Kohley and E. Heitz (authors' response)—The relative ranking of 13% chromium steel and ferritic-austenitic stainless steel (results are not part of the paper) are the same both in the erosion corrosion and abrasive wear region. In general, ferritic austenitic steels showed half of the mass loss rate as compared to 13% chromium steel.

T. S. Lee[3] *(discusser's question)*—In comparing the polarization resistance and mass loss data, is it possible to identify what percent of the metal removal is attributable to electrochemical versus mechanical processes in both the abrasion and erosion-corrosion regions? What is the effect of the particle angularity on this distribution?

T. Kohley and E. Heitz (authors' response)—The distribution is approximately as follows:

abrasive wear region:
mechanical 90 $\pm$ 5%
electrochemical 10 $\pm$ 5%

erosion corrosion region:
electrochemical 70 $\pm$ 10%
mechanical 30 $\pm$ 5%

The scatter of data is due to restrictions in accuracy of the polarization resistance method. The effect of particle angularity on the distribution is not known. Further results on particle dynamics will be published in *Corrosion NACE*.

Y. F. Van Baar[4] *(discusser's questions)*—(a) Can you specify the role of oxygen? and (b) Can you quantify the influence of CO_2 partial pressure?

T. Kohley and E. Heitz (authors' response)—(a) Oxygen content was below 20 ppb throughout the experiments. Increasing oxygen content leads to pitting on 13% chromium steel in particle-free formation water. It is not clear if the "residual" corrosion rate at zero sand content in Fig. 6 of the paper is due to traces of oxygen. In general, the influence of oxygen has not been investigated. (b) In particle-free formation water, nearly no difference between corrosion rates of carbon steel at 3 and 5 bar CO_2 at various flow rates have been found in the present investigation. However, this finding disagrees with practical experience. The influence of CO_2 partial pressure under conditions of erosion corrosion of 13% chromium steel is just now being investigated.

J. M. Sykes[5] *(discusser's question)*—You have attributed the corrosion losses in pure water to the removal and reformation of the passive oxide film, yet the addition of inhibitor was able to reduce this loss very significantly. Do you have any ideas on how the inhibitor is able to do this?

---

[1]National Physical Laboratory, Teddington, Middlesex TW110LW, United Kingdom.
[2]BSC Swinder Labs., Moongate, Rotherham S60 3AR, United Kingdom.
[3]National Association of Corrosion Engineers, P.O. Box 218340, Houston, TX 77218.
[4]Shell Amsterdam, Shell Research B.V., Badhusiwegs, 1031 GM, Amsterdam, Netherlands.
[5]University of Oxford, Department of Metallurgy and Science of Materials, Parks Road, Oxford OX1 3PH.

Can an inhibitor reduce the amount of charge required to form the passive oxide? Is it possible that passive oxide growth is accompanied by the formation of soluble species?

T. Kohley and E. Heitz (authors' response)—Your question refers to results already presented at the 6th European Symposium on Corrosion Inhibitors, Ferrara, Italy, Sept. 1985. Erosion corrosion as referred to in this paper can be effectively inhibited according to those results. Even if particle-containing distilled water with 3 bars nitrogen is used, a finite but small corrosion can be observed, which also can be inhibited. There is no definite answer to this problem. Possibly, the proposal that passive layer growth is accompanied by a parallel formation of soluble species is correct. This process should be susceptible to inhibition.

Dale R. McIntyre[1]

Pitting of AISI 410 Stainless Steel in CO₂-Saturated Synthetic Seawater and Condensate

REFERENCE: McIntyre, D. R., **"Pitting of AISI 410 Stainless Steel in CO₂-Saturated Synthetic Seawater and Condensate,"** *The Use of Synthetic Environments for Corrosion Testing, ASTM STP 970,* P. E. Francis and T. S. Lee, Eds., American Society for Testing and Materials, Philadelphia, 1988, pp. 246–254.

ABSTRACT: Specimens of American Iron and Steel Institute (AISI) Type 410 stainless steel were exposed to aerated and deaerated solutions of artificial seawater and distilled water, with and without quaternary amine additions, with a CO₂ atmosphere. Pitting rates were determined from autoclave exposures, and electrochemical parameters were obtained from potentiodynamic polarization curves.

The studies showed that in deaerated solutions, the amine additions had a mild inhibiting effect on liquid phase pitting in both seawater and distilled water. The amine additions increased the rate of attack in aerated solutions of distilled water. Correlation between electrochemical test methods, laboratory autoclave tests, and field experience is good.

KEY WORDS: stainless steels, corrosion tests, seawater, condensate

Increased use of stainless steels in hydrocarbon production wells, especially offshore, is forcing many oil and gas companies to re-examine traditional corrosion inhibition practices. Most commercial inhibitor mixtures have been formulated for carbon steel; their effectiveness on active passive alloys, such as stainless steels, has not always been demonstrated. Indeed, there is some field experience that suggests that some compounds that inhibit general corrosion of carbon steels may actually accelerate localized corrosion of stainless steels.

This paper describes an investigation into the possible effects a particular class of common inhibitor might have on localized corrosion of American Iron and Steel Institute (AISI) 410 stainless steel. The "inhibitor" studied was a commercial mixture specifically compounded to inhibit packer fluids in the annular space between casing and tubing in hydrocarbon wells. Its active ingredient is a complex blend of quaternary amines.

The inhibitor mixture could be present in wells under a wide variety of situations other than its intended use in the packer fluid. The packer fluid may be based on solutions ranging from seawater to freshwater, with or without oxygen control. Initial installation of the packer fluid requires displacement of the completion fluid into the annulus, with subsequent production up the tubing. Shut-in wells will have gas saturated condensed water forming during shutdown or at pressure drops during normal production, creating the possibility of a mixture of condensate with diluted packer fluid and inhibitor. The possibility of contamination caused by packer leaks, particularly of carbon dioxide (CO₂) from formation gases, cannot be ignored. Thus the conservative approach was to test the proposed inhibitor compound in a variety of environ-

[1]Cortest Laboratories, Inc., 11115 Mills Rds., Suite 102, Cypress, TX 77429.

ments, both high and low in chlorides, with and without oxygen control, with a substantial CO_2 partial pressure.

Laboratory simulation of such environments requires test methods which realistically reflect possible service conditions, yet sensitively detect the onset of any localized attack in a reasonable amount of time. Pitting is a notoriously erratic phenomenon, involving initiation time with the possibility of wide scatter. The experimental method is detailed below.

Experimental Procedure

Two experimental methods were used: evaluation of coupons exposed for four weeks in autoclaves under simulated shut-in conditions and electrochemical potentiodynamic polarization studies on electrodes exposed under the same conditions.

The CO_2 partial pressure used was 3.75 bar (375 kPa). Therefore, the autoclave experiments and potentiodynamic polarization tests were conducted at 3.75 bar (375 kPa) with 100% CO_2. Solutions were maintained at 65°C during the tests.

Autoclave Tests

In the autoclave experiments, 25- by 13- by 1.6-mm 410 stainless steel coupons were cleaned, weighed, and then exposed in 1-L 316 stainless steel pressure vessels. Approximately 500 mL of solution was used. Duplicate specimens were exposed in both the vapor space and the liquid in each vessel. Teflon® washers, serrated as described in ASTM Guide for Crevice Corrosion Testing of Iron Base and Nickel Base Stainless Alloys in Seawater and Other Chloride-Containing Aqueous Environments (G 78), were bolted onto each coupon to create an artificial crevice and insulate the coupons from electric contact with the specimen rack and autoclave.

Deaerated solutions were sparged with nitrogen overnight before being poured into the autoclaves. After the autoclaves were sealed, the deaerated solutions were evacuated to 740 mm of mercury with a mechanical vacuum pump before the CO_2 was introduced into the vessel. Experience indicates that this procedure (nitrogen purging followed by vacuum) reduces oxygen in solution to levels comparable to downhole environments. Aerated solutions were not sparged or evacuated before testing.

Inhibited solutions had a commercial mixture of the quaternary amine inhibitor added in the proportion 3 to 500 mL of solutions. This is equivalent to the recommended field dosage level of 25 gal/100 bbl H_2O.

Seawater solutions were mixed by adding 41.97-g/L artificial sea salt (as described in ASTM Specification for Substitute Ocean Water [D 1141]) to distilled water at room temperature.

After four week's test time, the coupons were removed, cleaned, and reweighed. Maximum pit depth was determined on a metallograph with a micrometer stage.

Electrochemical Tests

The potentiodynamic polarization experiments were conducted in a 250-mL Hastelloy C autoclave. A test electrode of 6.35-mm-diameter 410 stainless steel rod was inserted into the autoclave through packing glands at the top and bottom. The exposed area of each test electrode was 7.9 cm² (1.22 in.²). The test electrodes were insulated from electrical contact with the vessel using Teflon® bushings and Ryton (polyphenoline sulfide) spacers.

A Hastelloy C indicator electrode was introduced into the autoclave through a 1/4-in. (6.35-mm) NPT fitting. This indicator electrode was electrically insulated from the body of the vessel with Teflon® and polyethylene fittings. The stability of the Hastelloy C indicator electrode was checked against a commercial silver/silver chloride (Ag/AgCl) reference electrode. The body of the autoclave itself was used as the counter, or auxiliary, electrode. The indicator electrode was used in place of a true reference electrode to eliminate the introduction of chloride ion into the solutions from the reference electrode's electrolyte.

Each test was conducted in 200 mL of solution. Solution mixing and aeration or deaeration practice were as described above for the autoclave tests. The commercial quaternary amine inhibitor was added to the inhibited solutions in the proportion of 1.19 to 200 mL of solution. Two tests in aerated freshwater were conducted in Houston tap water.

After the autoclave was sealed and charged to 3.75 bar (375 kPa) with CO_2, the vessel was brought to 65°C with electric resistance heaters controlled through a Type J thermocouple. When the temperature had stabilized, the open circuit potential of the test electrode was measured with a digital voltmeter.

Starting with the test electrode at open circuit, applied current was measured as the potential of the test electrode was dynamically polarized in the cathodic direction at the rate of 5 V/h. When either the current limit of the instrument or the onset of concentration polarization effects was reached, the scan direction was reversed and continued in the anodic direction. The test electrode was polarized back through the open circuit potential and through the anodic passive current zone.

When the test electrode was well into the anodic transpassive zone, or when the current limit of the instrument was reached, the scan direction was reversed once more and continued in the cathodic direction until the appearance of the minimum on the applied current (anodic and cathodic currents are equal) signaling that the repassivation potential was reached.

Results and Discussion

"General Corrosion" rates calculated from weight loss can be highly misleading on active-passive alloys in which the primary attack is by pitting. Observed weight losses are given in Tables 1 and 2 for comparison. Maximum pitting rates were obtained by dividing the maximum pit depth observed on each coupon by the exposure time and converting this number to millimetres per year.

Tables 1 and 2 present weight loss and pitting corrosion rates calculated for vapor space and liquid exposure in each autoclave. The values given are the average of the duplicate coupons from each condition. Photographs of coupons after exposure are presented in Figs. 1 through 4.

The results of the electrochemical experiments are presented in Table 3: Typical polarization curves are presented in Figs. 5 through 8. These figures show applied current as a function of

TABLE 1—*Autoclave test results deaerated solutions.*

Solution	Inhibitor	Weight Loss, mg	Maximum Pitting Rate, mm/year
	LIQUID PHASE		
Seawater	no	21.3	0.15
	yes	15.9	0.091
Distilled water	no	15.9	0.53
	yes	15.9	0.38
	VAPOR PHASE		
Seawater	no	10.5	1.04
	yes	10.5	0.36
Distilled water	no	5.4	0.22
	yes	10.5	0.31

TABLE 2—*Autoclave results aerated solutions.*

Solution	Inhibitor	Weight Loss, mg	Maximum Pitting Rate, mm/year
LIQUID PHASE			
Seawater	no	37.2	1.73
	yes	10.5	0.44
Distilled water	no	138.2	0.48
	yes	222.0	5.02
VAPOR PHASE			
Seawater	no	201.0	3.22
	yes	339.2	2.03
Distilled water	no	10.5	0.58
	yes	15.9	1.07

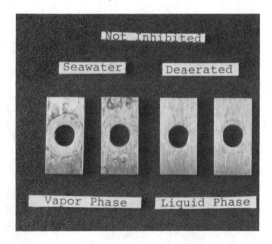

FIG. 1—*Coupons after exposure to deaerated seawater.*

potential for the different environmental conditions tested. Direction of the scan at any given instant (cathodic or anodic) is given by the arrows on the potential current traces.

Type 410 stainless steel pitted at rates high enough to imperil the sealing surface of a valve in all the autoclave tests except deaerated seawater, both with and without inhibitor.

The maximum pitting rate in the "inhibited" solution of aerated, distilled water was approximately ten times the rate observed in the same solution without inhibitor. The pitting rate in the vapor phase was approximately doubled above the inhibited, aerated, distilled water compared to the uninhibited solution. Therefore, it seems clear that the commercial inhibitor dosage accelerates pitting in aerated distilled water.

The presence of the amine in the liquid phase of all other test environments resulted in a slight reduction in pitting rate. The pitting rate in the deaerated distilled water vapor phase exposure was slightly, although not significantly, higher in the inhibited solution as compared to the uninhibited solution. Vapor phase exposures above deaerated and aerated seawater showed lower pitting rates above the inhibited solutions when compared to the corresponding

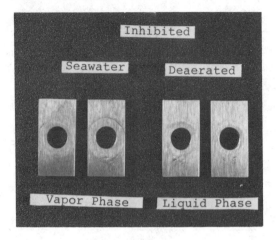

FIG. 2—*Coupons after exposure to deaerated inhibited seawater.*

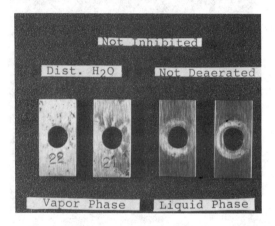

FIG. 3—*Coupons after exposure to aerated distilled water.*

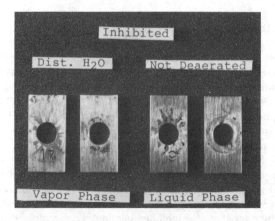

FIG. 4—*Coupons after exposure to aerated distilled water with amine inhibitor.*

TABLE 3—*Summary of electrochemical potentiodynamic polarization studies (all tests conducted at 65°C with 3.75 bar (37.5 kPa) CO$_2$ gas phase).*

Solution	Deaerated	Amine Addition	OCP Volts Versus Ag/AgCl	Transpassive Potential, Volts Versus OCP	Repassivation Potential, Volts Versus OCP	Anodic Hysteresis Loop
Synthetic sea-H$_2$O	yes	no	−0.633	+0.200	+0.131	yes
Synthetic sea-H$_2$O	yes	yes	−0.695	+0.263	+0.096	yes
Distilled H$_2$O	yes	no	−0.198	+0.375	0	yes
Distilled H$_2$O	yes	yes	−0.686	+0.700	+0.331	yes
Synthetic sea-H$_2$O	no	no	−0.523	0	−0.081	yes
Synthetic sea-H$_2$O	no	yes	−0.435	0	−0.062	yes
Distilled H$_2$O	no	no	−0.044	+0.600	−0.062	yes
Distilled H$_2$O	no	yes	−0.253	+0.400	−0.163	yes
Tap water	no	no	−0.395	+0.400	0	yes
Tap water	no	yes	−0.364	+0.350	−0.056	yes

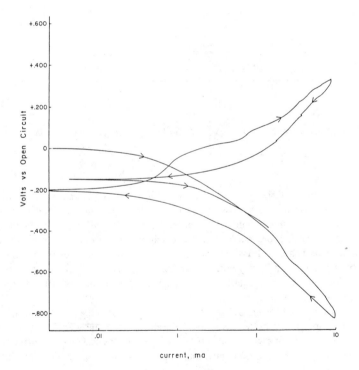

FIG. 5—*Aerated seawater with 3.75-bar (375-kPa) CO$_2$.*

inhibited solution. This is somewhat surprising, since the inhibitor mixture was never meant to be a vapor phase inhibitor.

The electrochemical tests supported the results of the liquid phase exposures in the autoclave tests. (Electrochemical tests of this sort are not useful for predicting vapor phase corrosion, since the tests require a continuous electrolyte).

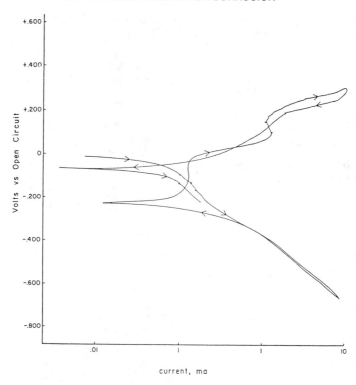

FIG. 6—*Aerated seawater with 3.75-bar (375-kPa) CO_2 with quaternary amine addition.*

On potentiodynamic polarization curves, such as Figs. 5 through 8, the presence of measured minima in applied current, which are cathodic to the open circuit potential (OCP), and a hysteresis loop in the anodic current potential trace, indicate potential pitting problems [2]. If the repassivation potential is more noble than the original open circuit potential, then pits that initiate should repassivate. If, however, the repassivation potential is less noble than the original open circuit potential, pits may both initiate and propagate. In general, the more cathodic the repassivation potential is in relation to the original open circuit potential, the more severe the pitting rate will be [3].

As shown in Table 3, all the environments tested showed applied current minima cathodic to OCP and significant anodic hysteresis loops. Therefore, all of these environments could produce pitting under conditions where the potential of the 410 stainless steel is shifted in the noble direction (because of scale formation, galvanic coupling, or variations in local solution composition).

However, only the aerated solutions showed a repassivation potential less noble than open circuit. Therefore, severe and active pitting should take place in any of the aerated solutions. This supports the observations from the autoclave tests and much field experience with chromium stainless steels.

The inhibited, aerated distilled water solution showed the largest shift of the repassivation potential to a value less noble than the open circuit potential, and also the largest anodic current hysteresis loop. Therefore, the potentiodynamic polarization experiments would predict that the inhibited, aerated distilled water solution would be the most severe pitting environment. This is confirmed by the autoclave experiments.

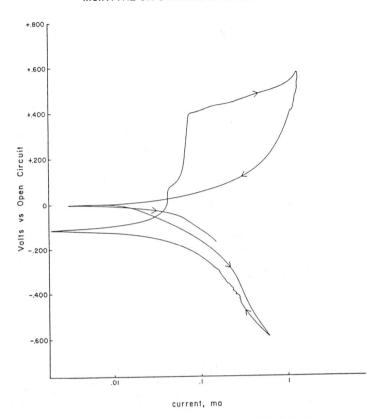

FIG. 7—*Aerated tap water with 3.75-bar (375-kPa) CO_2.*

Autoclave tests were not conducted on the aerated freshwater solutions. However, the electrochemical polarization tests predict that such solutions would be intermediate in severity between the distilled water and seawater solutions. In the aerated, inhibited freshwater solution, the repassivation potential is somewhat less noble than the open circuit potential, whereas in the uninhibited solution the repassivation potential and the open circuit potential were identical. Thus, these tests would predict that in aerated freshwater, the presence of the commercial inhibitor in the recommended amounts would slightly accelerate pitting over an uninhibited solution.

Although formulated to prevent general corrosion on carbon steels, the commercial inhibitor reduced pitting corrosion on 410 stainless in the brine solutions for which it was intended whether these solutions were aerated or deaerated.

Conclusions

Based on the experimental program summarized in this report, the following conclusions are made:

1. The commercial mixture of quaternary amine reduces pitting corrosion on 410 stainless steel in brine solutions, whether these solutions are aerated or deaerated.
2. The commercially prepared quaternary amine accelerates pitting in aerated solutions of distilled or freshwater. The inhibitor would not accelerate pitting in deaerated condensate water.

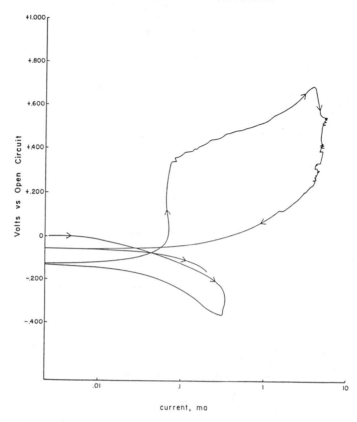

FIG. 8—*Aerated tap water with amine additions with 3.75-bar (375-kPa) CO₂.*

References

[1] Robinson, J. N., et al., "Estimation of the Water Content of Sour Natural Gas," *Society of Petroleum Engineers' Journal,* Vol. 17, No. 4, Aug. 1977, p. 281.
[2] Chamberlain, T. E., "Method Can Improve Corrosion Evaluation," *0.1 & Gas Journal,* 12 Nov. 1984, pp. 144–150.
[3] Efird, K. D. and Moller, G. E., "Electrochemical Characteristics of 304 and 316 Stainless Steels in Fresh Water as Functions of Chloride Concentration and Temperature," Paper 87, *CORROSION/78.*

John M. Sykes[1] and Peter H. Balkwill[1]

Simulating the Pitting Corrosion of Steel Reinforcement in Concrete

REFERENCE: Sykes, J. M. and Balkwill, P. H., **"Simulating the Pitting Corrosion of Steel Reinforcement in Concrete,"** *The Use of Synthetic Environments for Corrosion Testing, ASTM STP 970,* P. E. Francis and T. S. Lee, Eds., American Society for Testing and Materials, Philadelphia, 1988, pp. 255–263.

ABSTRACT: The pitting behavior of steel in solid gels made from alkaline solutions containing chloride ions is shown to be markedly different from the behavior that occurs in the solutions themselves. After an induction period during which composition changes take place at the metal-gel interface, pitting can start even at bulk chloride levels at which pitting would not occur in solution. Addition of calcium hydroxide to the gel delays the onset of pitting. Portland cement mortar is a less aggressive environment then the alkaline gels, probably because of the formation of a portlandite layer on the steel surface.

KEY WORDS: simulation, steel reinforcement, corrosion in concrete, gels, alkaline environments, pitting corrosion, chloride

Steel reinforcing bars in damp concrete are embedded in a porous solid matrix permeated by an alkaline aqueous solution. If this solution contains sufficiently high levels of chloride ions, or its pH is not sufficiently high, pitting corrosion can result. Electrochemical studies of steel in concrete seem to suggest that we can understand behavior in concrete largely in terms of the composition of the aqueous phase within the pores [1]. However, the importance of the solid phase as a barrier to aggressive species and of the portlandite it contains as a reserve of alkalinity that prevents pH shifts has also been emphasized [2]. It is clear, though, that this solid environment is not really like an aqueous solution; diffusion constants will be smaller and, we believe more importantly, solution convection will be suppressed such that diffusion phenomena are not limited to a thin boundary layer at the surface of the metal. To explore this proposition, experiments have been carried out on hot-rolled mild steel in solid alkaline aqueous gels containing sodium chloride.

Experimental Procedure

In the first instance, polarization curves were determined in aqueous calcium hydroxide or sodium hydroxide solutions of pH 12.6 or 13.5 containing different levels of chloride as a basis for comparison. The steel specimens were cylinders 20 mm in height by 15 mm in diameter. A typical analysis of this steel is given in Table 1. These cylinders were polished with 1200-grade silicon carbide paper, degreased, then sealed into plastic holders with epoxy resin. The counter electrode was of platinized titanium, the reference electrode was a standard calomel electrode, and the cell a 1-L Kilner jar (Fig. 1). The solution was deaerated with "white-spot" nitrogen.

---

[1]Senior research fellow and research student, Department of Metallurgy and Science of Materials, University of Oxford, Parks Road, Oxford OX1 3PH, U.K.

TABLE 1—*Composition of hot-rolled mild steel bar.*

Element	C	Si	S	P	Mn	Cu	Ni	Cr	Mo	Sn
Weight%	0.16	0.18	0.032	0.039	0.65	0.30	0.17	0.14	0.02	0.037

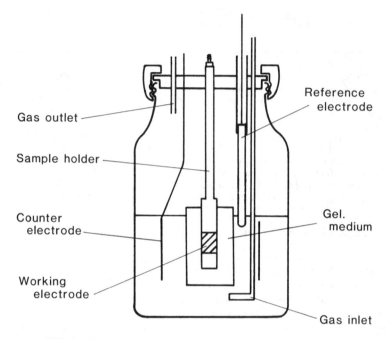

FIG. 1—*Cross-section of electrochemical test cell with gel specimen.*

Further experiments were carried out for similar specimens embedded in ordinary Portland cement mortar containing fine sand as the aggregate (sand:cement 3:1 by weight) and having a water-cement ratio of 0.5 by weight. The specimen was centrally placed in a cylindrical block of mortar 80 mm in height and 55 mm in diameter. After casting, the samples were demoulded and cured for 30 days at 100% relative humidity. To facilitate the ingress of chloride to the surface of the steel, the samples were partially dried at a temperature of 40°C to constant weight to remove capillary pore-water. When the specimen is immersed in salt solution it soaks up the solution at once, taking the chlorides with it, so that pitting can be observed in tests carried out after a few hours. This technique also permits experiments to be conducted in deaerated conditions, revealing the whole of the anodic behavior of the steel. To do this, the dried specimen was placed in the cell without solution; the cell was deaerated with nitrogen for 48 h then deaerated solution was run into the cell. A highly negative rest potential is achieved immediately after reduction of the air-formed oxide. Similar samples of concrete were given the same treatment, then crushed to express pore solution using a technique similar to that of Diamond [3] to ascertain the pH and level of chloride to which the steel had been exposed. The tests were carried out in synthetic seawater (BDH seawater corrosion test mixture) or in 3% NaCl by weight, with the level of the solution kept below the top of the mortar. Both test solutions contain about 0.5 M chloride.

For the gel tests, sodium hydroxide solutions containing 0.05, 0.1, 0.25, or 0.5 M NaCl were thickened with 1.5% by weight of agarose (No. A-6877 Type II Medium EEO). Two different concentrations of alkali were used to give pH 12.6 or 13.5. The warm gel was cast around the steel specimens in plastic molds (35 mm in diameter, 80 mm in height), which were removed before immersing the specimens in a test solution of the same composition (without agarose) in the cell. In one case, a thin layer of gel (<1 mm) was applied to a steel specimen to limit the thickness of the diffusion layer. These tests were not deaerated. Further tests were carried out in which up to 1.0% by weight of solid calcium hydroxide ($Ca(OH)_2$) was dispersed in the gel using an ultrasonic bath to ensure even dispersion. All solutions were made with "Analar" grade chemicals and singly distilled water.

Polarization curves were determined in all three types of media using normal potentiostatic equipment by sweeping the potential at 0.33 mV s^{-1} with a linear scan generator. No facility for IR compensation existed, but as the resistance of the concrete was determined to be ~100 Ω by ac impedance, this will not be significant except at the highest currents. Errors are negligible in the gel. Most of the pitting studies in gel used the galvanostatic method. That is, a constant current was applied and the potential was recorded as a function of time on a chart recorder.

Results

The polarization curves in saturated $Ca(OH)_2$ solution were similar in form to those described elsewhere [1,4]. They showed a single active peak and pitted at very low potentials (Fig. 2). In pH 13.5 sodium hydroxide (NaOH), the active region shows twin peaks at more negative potentials and a wide passive range. The current does not rise significantly until the onset of oxygen evolution, though it is evident from a reverse scan (which does not fall back to the passive level at first) or from examination of the specimen that pitting has started at this point. Comparison

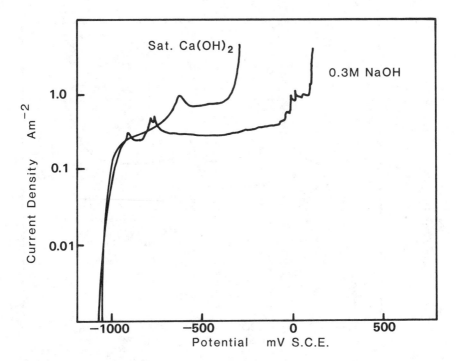

FIG. 2—*Polarization curves for mild steel in alkaline solutions with 0.5 M sodium chloride (room temperature).*

of these two results shows a clear influence of pH. The behavior in cement mortars is more complex; on the first day of immersion these specimens clearly exhibit pitting (Fig. 3), but on subsequent days no pitting is observed until eventually, after some weeks, pitting breakdown can again be seen. It has previously been shown that similar mortars containing higher contents of sand aggregate show pitting whenever tested.

Preliminary results of the pore-water expression tests show that the pH is in excess of 13 and that the chloride concentration is about 0.25 M, that is, about half the value in the solution, but changes with time of immersion.

Preliminary experiments in gels made with $Ca(OH)_2$ solution containing sodium chloride (NaCl) gave polarization curves in which pitting was evident, even with 0.05% Cl$^-$, (0.0086 M), but if the gel contained solid particles of $Ca(OH)_2$, pitting was not seen. To characterize this in more detail, galvanostatic tests were used. These permit extended experiments that allow time for pits to initiate and local concentration changes to occur. In the galvanostatic tests, the potential jumps suddenly when the current is applied, then rises linearly with time to the oxygen evolution potential as the passive film grows (Fig. 4). It then remains stationary until pitting commences, when it falls in an irregular manner. When these tests are made in 0.3 M NaOH without agarose, pitting is observed with solutions containing 0.25 M Cl$^-$, but tests with 0.1 M Cl$^-$ showed no evidence of pitting after 22 h. When similar tests are carried out in gels containing chloride levels of 0.1 M or below, pitting is eventually observed after a delay time, which is a function of the applied current (Fig. 5). No threshold concentration below which pitting does not occur has so far been detected. When tests were carried out with gels containing suspended fine particles of $Ca(OH)_2$, pitting still took place eventually, but only after a delay, which is a function of the $Ca(OH)_2$ content of the gel (Fig. 6). A further potentiostatic test was carried out with a specimen embedded in pH 13.5 gel containing 0.05 M Cl$^-$; this was held at a potential

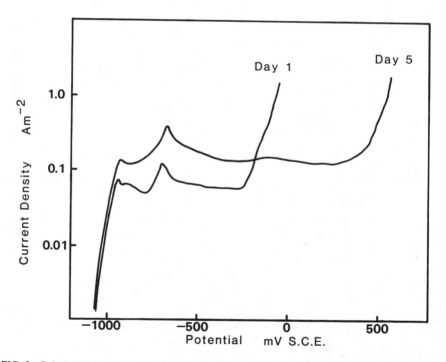

FIG. 3—*Polarization curves for mild steel in Portland cement mortar in 3% sodium chloride (room temperature).*

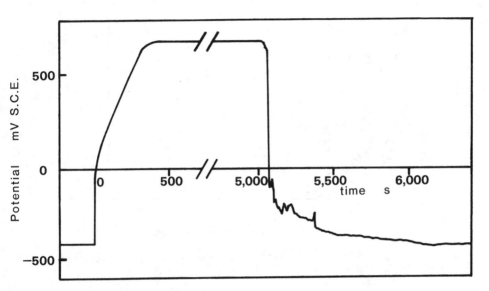

FIG. 4—*Potential variation in typical galvanostatic experiment (pH 13.5 gel with 0.1 M sodium chloride, current = 0.1 A m⁻²).*

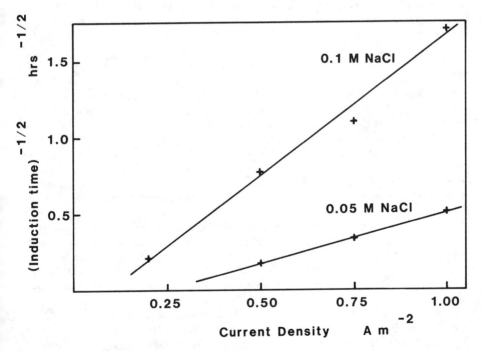

FIG. 5—*Effect of applied current density on initiation time in galvanostatic experiments in pH 13.5 gel.*

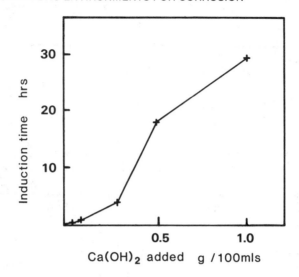

FIG. 6—*Effect of calcium hydroxide content on initiation time in galvanostatic experiments in pH 12.6 gel with 0.05% sodium chloride, current = 0.1 A m^{-2}.*

just below the oxygen evolution potential (+550 mV SCE), at which the anodic current was small (0.25 μA cm^{-2}). Even though the potential was high, no pitting had occurred, even after 5 days of testing. Galvanostatic tests with a very thin layer of pH 13.5 gel containing 0.1 M NaCl showed no pitting after 24 h, even though thick gel samples of the same composition pitted after about 800 s.

Discussion

Many workers have described the pitting behavior of iron or steel in alkaline chloride solutions [1,4,5-7]. Most of these workers report a similar conclusion—pitting is only seen above a certain threshold concentration of chloride, which depends upon the pH. Hausmann's idea [6] of a critical ratio of chloride to hydroxide, with pitting occurring only if Cl$^-$/OH$^-$ is greater than 0.6, is now widely accepted, though Gouda [5] has advanced a more complicated relationship. Henrikson [7], using scratch tests, was able to produce pits at much lower chloride levels, which suggests that these criteria are concerned with pit initiation rather than propagation. The results in calcium hydroxide and sodium hydroxide presented here do not disagree with these conclusions.

There have also been many studies of mild steel in chloride-containing concrete or mortar [1,8-11], sometimes with samples immersed in chloride solution, but more usually in those containing admixtures of CaCl$_2$ or NaCl. In this latter case, and probably in the former, part of the chloride becomes bound by reaction with the portland cement paste such that the actual chloride content in the pore solution at the steel-concrete interface is uncertain. Typically, admixtures of a few percent calcium chloride (by weight of cement) will cause pitting. Holden, Page, and Short [12] have analyzed pore solution for cement paste containing 0.4% Cl$^-$ and found between 0.041 and 0.083 M Cl$^-$, giving a Cl$^-$/OH$^-$ ratio of between 0.06 and 0.11, which would not be expected to cause corrosion. Dehghanian and Lock [10] found breakdown of passivity in concrete immersed in a 0.5% NaCl solution, but not in a 0.1% solution.

With the experiments presented here, the mortar is freshly cured and the experiments are short, so the pH in the mortar will be high, with a hydroxyl concentration of the same order as that of chloride. Because of this, it is not certain whether the Cl^-/OH^- ratio is greater or less than the critical 0.6. It is not easy to explain why pitting occurs in some cases, but not in others, simply on the basis of solution chemistry.

The results in gel do not easily fit into this picture unless we consider the effects of ion migration; pitting is seen in gels with Cl^-/OH^- ratios as low as 0.166. This, we believe, is because in a solid medium there will be no convection and the passage of an anodic current through the gel causes migration of anions (OH^- and Cl^-) towards the metal and of cations away from the metal. The electrode reactions at the metal anode (oxygen evolution or metal dissolution followed by cation hydrolysis) will both tend to acidify the environment near the steel more than compensating for the flux of OH^- ion towards the anode. Thus, with an anodic current imposed, the chloride level at the interface will rise and the OH^- level will fall; in principle, no matter what the original levels of chloride and hydroxide, the critical ratio can eventually be reached.

It is possible to quantify these changes by considering the balance between migration of ions in the electric field (or their rate of generation at the electrode) with diffusion into the bulk environment. Because the characteristic diffusion layer thickness, δ, increases as the square root of time up to the thickness of the gel layer, the concentration at the interface continues to rise until δ equals this thickness, or salt precipitation or some similar process intervenes. In a galvanostatic experiment, the increase in chloride concentration at the interface should be proportional to the applied current and the square root of the time (Appendix 1). Therefore, there should be a linear relationship between the applied current and the root of the time to the onset of pitting if we assume that this delay represents the time taken for the surface concentrations to reach the appropriate value. This relationship is seen for two different chloride concentrations in Fig. 5.

We see then that if cement paste were simply an inert solid, like the gel containing an alkaline solution, the corrosion behavior of steel in it would be quite different, with pitting taking place quite readily. It is possible to begin to understand the actual behavior in concrete if we take into account Page's ideas [2] on the role of Portlandite, which acts as a reserve of alkalinity, hindering pH changes at the interface and in incipient pits. A dispersion of $Ca(OH)_2$ in the gels, even in the absence of a compact portlandite layer on the steel as described by Page, has a dramatic effect. Yet though it can interfere with the downwards shift in pH, it is not likely to affect chloride accumulation, so that in these experiments, pitting eventually results. It is not clear why this eventual breakdown does not appear to occur with cement pastes containing low chloride levels.

Finally, slower diffusion in cement paste or portlandite will not prevent accumulation of chloride at the metal surface. Far from it; it is diffusion that allows the excess to escape back to the bulk.

Conclusions

It can be misleading to consider the corrosion of metals in solid media such as concrete solely in terms of the aqueous solutions that permeate them. In particular, the lack of convection can bring about large concentration changes at the surface of the metal, in this instance enhancing the local levels of chloride ion. Concrete is more protective than we might expect.

Acknowledgments

We are grateful to the Science and Engineering Research Council (SERC) and Harwell Laboratory for financial support, to Professor Sir Peter Hirsch for provision of laboratory facilities, and to Dr. N. J. M. Wilkins for advice and encouragement.

APPENDIX

To ascertain the concentration of ionic species at the metal-electrolyte interface, it is necessary to solve Fick's 2nd Law under nonsteady state conditions; in the case of a planar electrode for transport in one dimension:

$$\frac{\partial c}{\partial t} = D \times \frac{\partial^2 c}{\partial x^2} \tag{1}$$

where

D = the diffusion coefficient of the ions under consideration, and
C = the concentration of ions at a distance x from the electrode at time t.

If a constant current is applied and the ions concerned (say chloride) are not consumed at the electrode, then migration in the field towards the metal must be balanced by diffusion, giving a boundary condition at the metal surface:

$$D\left(\frac{\partial c}{\partial x}\right)_{x=0} = t^- \times \frac{i}{zF} \tag{2}$$

where

t = transport number.

This leads to a solution for the concentration of chloride at the metal surface [15]:

$$C_{x=0} = C_{x=\infty} + \frac{2t^- it^{1/2}}{zFD^{1/2}\pi^{1/2}} \tag{3}$$

where

$c_{x=\infty}$ = concentration at $x = \infty$

A similar relationship can be derived more simply but less rigorously by treating this as a quasi-steady state situation and linearizing the concentration gradient [16]. The mass balance at the interface becomes:

$$\frac{t^- \times i}{zF} = D \times \frac{c_{x=0} - c_{x=\infty}}{\delta} \tag{4}$$

where

δ is the diffusion layer thickness given approximately as $D^{1/2}t^{1/2}$ as a function of time, which yields:

$$c_{x=0} - c_{x=\infty} = t^- \times \frac{it^{1/2}}{zFD^{1/2}} \tag{5}$$

References

[1] Page, C. L. and Treadaway, K. W. J., *Nature*, Vol. 297, 1982, p. 297.
[2] Page, C. L., *Nature*, Vol. 258, 1975, p. 514.
[3] Barneyback, R. S., Jr. and Diamond, S., *Cement and Concrete Research*, Vol. 11, 1981, p. 229.
[4] Westcott, C., Lunn, F. C., and Sykes, J. M., *Proceedings of the 8th International Congress of Metallic Corrosion*, Vol. II, 1981, p. 1047.

[5] Gouda, V. K. and Halaka, W. Y., *British Corrosion Journal*, Vol. 5, 1970, pp. 198, 204.
[6] Hausmann, D. A., *Materials Protection*, Nov. 1967, p. 19.
[7] Henriksen, J. F., *Corrosion Science*, Vol. 20, 1980, p. 1241.
[8] Baumel, A. and Engell, H. J., *Archiv fur das Eisenuhuttenwesen*, Vol. 30, 1959, p. 417.
[9] Treadaway, K. W. J. and Russell, A. D., *Highways and Public Works*, Aug. 1968, p. 19.
[10] Dehghanian, C. and Locke, C. E., *Corrosion*, Vol. 38, 1982, p. 494.
[11] Wilkins, N. J. M. and Lawrence, P. F., *Corrosion of Reinforcement in Concrete Construction*, ed., A. P. Crane, Ellis-Horwood, London, 1983, p. 119.
[12] Holden, W. R., Page, C. L., and Short, N. R., *Corrosion of Reinforcement in Concrete Construction*, ed., A. P. Crane, Ellis-Horwood, London, 1983, p. 143.
[13] Lambert, P. and Page, C. L., *26th Corrosion Science Symposium*, UMIST, 1985.
[14] Bard, A. and Faulkner, L. R., *Electrochemical Methods*, John Wiley and Sons, New York, 1980, p. 252.
[15] Bockris, J. O. M. and Reddy, A. K. N., *Modern Electrochemistry*, Macdonald, 1970, p. 105.

J. L. Dawson[1] and P. E. Langford[2]

The Electrochemistry of Steel Corrosion in Concrete Compared to Its Response in Pore Solution

REFERENCE: Dawson, J. L. and Langford, P. E., "**The Electrochemistry of Steel Corrosion in Concrete Compared to Its Response in Pore Solution,**" *The Use of Synthetic Environments for Corrosion Testing, ASTM STP 970,* P. E. Francis and T. S. Lee, Eds., American Society for Testing and Materials, Philadelphia, 1988, pp. 264–273.

ABSTRACT: Corrosion control of steel reinforcement relies on the maintenance of the high alkalinity of the hydrated cement especially in the presence of chloride ions. The present paper also indicates the adverse effect that chloride contamination of the initial mix can have on the permeability of the matrix to oxygen diffusion.

Data obtained on steel specimens embedded in ordinary portland cement paste prepared with varying additions of sodium chloride are presented. Potentiodynamic polarization curves obtained after 400-days exposure in a controlled humidity environment are compared to curves from steel immersed in extracted pore solutions. Pore solution measurements are shown to provide analogous information to embedded steel samples on the anodic and passivation processes. However, pore solution experiments do not simulate the influence of the matrix in controlling diffusion and resistivity, which influence the long-term durability and macro-cell development.

KEY WORDS: cement/concrete corrosion, pore water, electrochemical measurements, AC impedance, chloride contamination, permeability, porosity

Reinforced concrete is a widely used composite material whose civil engineering design parameters are based on the excellent compressive strength of the concrete in combination with the tensile properties of the steel. Selection of the appropriate starting materials in conjunction with adequate codes of practice should typically ensure lives in excess of 50 years. However the durability of concrete can be adversely affected by physical damage, cracking, shrinkage, deformation and frost, and from chemical attack including acids, sulfates, alkali-aggregate reactions, soft water leaching as well as corrosion of the reinforcement [1,2].

Corrosion of steel reinforcement is controlled by physical diffusion processes and chemical/electrochemical reactions; they are linked in a complex manner, and it is these aspects that have been made the subject of considerable research and practical interest worldwide. The present paper reports on electrochemical measurements made on steel in pore water and compares these to the response from steel embedded in a typical test matrix. The results obtained formed part of a program of investigations into the effect of cement chemistry and chloride additions on the electrochemistry of the corrosion process [3]. The corrosion information from pore waters must be considered in terms of the known chemical and diffusion processes operating within a concrete matrix, and outlined below, in order to provide a greater understanding of steel reinforcement corrosion.

[1]Chemist, Corrosion and Protection Centre, University of Manchester Institute of Science and Technology, P.O. Box 88, Manchester M60 1QD, United Kingdom.
[2]Chemist, Taylor Woodrow Construction, Taywood House, Ruislip, Middlesex, United Kingdom.

The protection of steel reinforcement relies on the passivity produced by the high alkalinity of the concrete matrix, the result of calcium hydroxide in combination with low levels of sodium and potassium ions, a typical pH being greater than 12. Essentially calcium hydroxide is precipitated from the saturated solutions produced during the hydration process [4]; a fully hydrated cement paste would contain approximately 30% calcium hydroxide with the rest as calcium silicate hydrate (C-S-H), an amorphous gel containing varying amounts of aluminum, iron and sulfate ions. The solid calcium hydroxide is largely responsible both for the residual alkalinity of the matrix and the interfacial bond between the steel and the bulk matrix. The calcium hydroxide nucleates on the steel to form a "discontinuous polycrystalline layer of portlandite of variable thickness and coarse grain size, containing inclusions of C-S-H gel" [5]; this zone could act as a local source of hydroxyl ions and provide a diffusion barrier.

The development of the calcium hydroxide and C-S-H gel during hydration controls the strength of the matrix and its permeability, a higher strength being associated with a lower pore size. Micropores less than 2.5 nm diameter form part of the C-S-H structure while it is the larger pores, and in particular the capillary or macropores >50 nm, often the remnants of water filled space, that control the permeability. Permeability is not only important in restricting the access of atmospheric oxygen but also in controlling the ingress of moisture and the diffusion of chloride ions as well as the acidic gases carbon dioxide and sulfur dioxide, which may affect the long-term durability against corrosion. The diffusion and chemical reaction of these gases in the concrete matrix causes a progressive decrease in the pH to ≈ 9. The development of such a "carbonation" front typically follows a diffusion or square root time law [6]. In the presence of chloride ions there is the increased probability of corrosion as demonstrated by E-pH diagrams.

The tensile stresses produced in the concrete by the formation of the corrosion products are the major durability problem, the relative increase in volume of the oxide to original metal resulting in cracking and spalling of the concrete with subsequent loss of strength. Pre-stressed or post-stressed systems tend to contain higher strength steels, compared to conventional reinforced structures, and are more susceptible to stress corrosion cracking and corrosion fatigue. It is the processes associated with the corrosion of conventional systems that is of concern in the present paper.

Chloride, present in the original mix in the form of additives or contaminated aggregates, or both, or the result of diffusion from marine environments or deicing salts, or both, is probably responsible for the majority of the reported corrosion problems, particularly where the carbonation has proceeded to the depth of the reinforcement. The ratio of chloride ions to hydroxyl ions is an important consideration although the concentration of sulfate ions and the tricalcium aluminate (C_3A) content of the original cement will influence the reactions [7]. The sulfate releases bound chloride by the preferential formation of calcium sulpho-aluminate hydrates, while C_3A may form an insoluble complex calcium chloro-aluminate hydrate (Friedel's salt). The lime-rich zone adjacent to the steel may exclude the chloro-aluminate hydrates that could liberate chloride ions.

The use of electrochemical measurements for the study of reinforcement corrosion is a subject of continuing research and scientific discussion, particularly with regard to selection of appropriate techniques and data interpretation. The practical difficulties include electrode design, the positioning and attachment of reference electrodes, and counter electrodes as well as the development of macro-corrosion cells. Nevertheless IR compensation can allow for the high resistance of the concrete/pore solution [8] while the restricted oxygen diffusion control processes are amenable to investigation using AC impedance techniques [9,10]. An alternative approach has been to use electrodes immersed in simulated pore solutions, generally calcium hydroxide solution with additions of sodium and potassium hydroxide [11,12]. The effect of chloride additions to such solutions has been used in mechanistic studies [11-15] to show an inter-dependency of chloride concentrations and pH. This gave rise to consideration of the chloride to hydroxyl ratio required to produce steel depassivation [13]. There are obvious differences between concrete and bulk solution experiments that are related to the effect of solid calcium hydroxide, the lime rich layer, and the critical effect of oxygen, particularly the restricted diffusion process, which will be considered below.

Experimental Procedure

A series of cement paste specimens were prepared using a water to cement ratio of 0.5 and a range of sodium chloride additions. The sodium chloride was dissolved in the water before mixing to give 0, 0.4, 1.5, and 3.0% by weight of cement. The cement analysis is given in Table 1.

The cell used for the investigations comprised a plastic bottle, 150 cm^3 in volume and 5 cm in diameter, in which were positioned two mild steel electrodes and an agar/potassium nitrate salt bridge. The mild steel electrodes were prepared from 10 cm lengths of 3-mm-diameter bar which were degreased in acetone and polished using 200-, 400-, and 600-grit silica carbide papers before use. Copper connecting wires were soldered onto one end, which was then encapsulated or masked in a polymer modified cement paste of low water/cement ratio, leaving an exposed area of 5 cm^2. The cement coat was cured for seven days and then coated with epoxy resin to seal the masking. Two of these electrodes were then positioned in the bottle test cell on each side of a 5-mm-diameter saltbridge filled with saturated potassium nitrate agar gel. The hand mixed cement paste was poured into the inverted bottle and mechanically vibrated for 5 min to remove entrapped air. The bottles were then sealed and cured for ten days, after which they were opened and placed in an exposure tank maintained at 95% relative humidity.

The corrosion potential was monitored versus a saturated calomel reference electrode, first daily during the curing period, then weekly over the first six months and finally periodically over the final six months. AC impedance measurements were made at intervals throughout this exposure using the two electrodes and a Solartron 1250" frequency response analyser controlled by an HP-85 microcomputer. The sine-wave amplitude was 20-mV RMS and the frequency range was 65 kHz to 1 mHz, the interface to the cell was via a "Thompson 251 Ministat" with phase compensation in this frequency range. Potentiostatic DC polarization measurements were made using a three electrode arrangement, the external platinum auxiliary electrode being attached by a potassium chloride gel soaked into a tissue. The potentiostatic polarization curves obtained at 18-mV/min sweep rate were collected via a digital voltmeter and the HP-85 microprocessor. The two electrode method was also used to obtain linear polarization data.

At the end of the exposure period the samples were broken up; scanning electron microscope examinations, mercury penetration measurements and pore water extractions were then made.

TABLE 1—*Chemical analysis.*

Component	Percentage
CaO	63.34
SiO_2	20.94
Al_2O_3	5.26
Fe_2O_3	2.35
MgO	2.41
K_2O	0.89
SO_3	2.85
LOI (loss on ignition)	0.68
Tri calcium silicate (C_3S)	48.0
Di calcium silicate (C_2S)	22.0
Tri calcium aluminate (C_3A)	10.3
Tetracalcium alumino-ferrite (C_4AF)	7.5

The pore water was extracted using a specially manufactured press fit double cylinder vessel having a central sample volume of 43 mm diameter by 140 mm high. The load was applied by a Amsler 3000-kN concrete press at a linear rate up to 350 kN and then cyclic, dropping the load approximately 100 kN each time until 400 kN was applied. The pore solutions were collected in a syringe for analysis and transferred to a small test cell containing small mild steel electrodes, saltbridge, and a counter electrode. This aqueous cell steel electrode was allowed to stabilize for seven days before the electrochemical measurements.

Results

The corrosion potential for all the specimens was initially low in all cases, -0.3 to -0.4 V versus saturated calomel electrode (SCE), but quickly rose to the passive state. The 0% sodium chloride (NaCl) additions maintained a high potential throughout the exposure period, -0.15 to -0.20 V versus SCE, but increasing the chloride content decreases the time the specimens remained passive. The 0.4% NaCl maintained noble potentials until day 105 then a rapid drop in potential occurred at around 200 days to -0.6 to -0.7 V versus SCE. The 1.5% NaCl showed an even earlier drop around 20 days while the 3% NaCl gave very early unstable behavior and then base steel potentials of about -0.6 V versus SCE. Larger additions of sodium chloride cause an increasing rate of depassivation of the steel.

Impedance measurements typically gave a low-frequency response comprising either part of a very large semi-circle or a plot indicative of a diffusion controlled process (Fig. 1); the low frequency point is 100 mHz. At day one a charge transfer semi-circle could be inferred, but thereafter a diffusion process predominates. A precise analysis is not possible and the corrosion rate can not be extrapolated. The response also moves along the real axis with time suggesting a progressive increase in the solution resistance. The insert shows the development of a high frequency response, first seen as a small inflexion at day 3, which progressively increases and becomes less depressed. The capacitance values of this inflexion are of the order of 1 to 50-nF

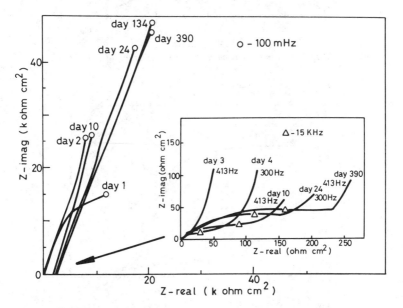

FIG. 1—*AC impedance for a cement paste with 3% NaCl addition.*

cm^{-2} and are assumed to be associated with the dielectric properties of the hydrated cement matrix.

Typical polarization curves obtained on the aerated specimens after 440-days exposure are shown in Figs. 2 and 3 for the 0 and 3% NaCl mixes, respectively. Values for the passivation current, that is, above the transition at −0.1 V versus SCE, and for the cathodic limiting current are listed in TAble 2; all data are IR corrected. Addition of chloride increases the passive current density, but there is also a decrease in the limiting current density. This is an indication of the change of pore structure since the moisture content and the exposure environment is constant. Mercury penetration data confirm that although the total pore volume increases with increase of chloride in the initial mix, the pore size between 60 and 100 nm decreases, and it is presumably these larger pores which are controlling the oxygen diffusion processes. These two factors, an increased passivation current and decreased diffusion or increased polarization, means that active conditions are recorded.

Analysis of the pore water extracted from the samples is shown in Table 3. Addition of sodium chloride increases the calcium concentration or presence of soluble calcium chloride; the

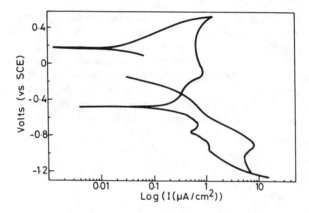

FIG. 2—*Polarization in cement paste, 0% NaCl addition, after 440 days.*

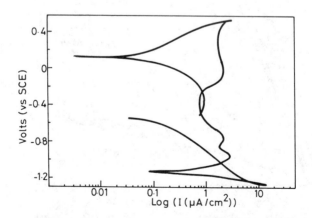

FIG. 3—*Polarization in cement paste, 3% NaCl addition, after 440 days.*

TABLE 2—*Passive and limiting current densities.*

Paste	i_p, $\mu A/cm^2$	i_L, $\mu A/cm^2$
Cement paste + 0% NaCl	0.7	9
Cement paste + 0.4% NaCl	1.2	2
Cement paste + 1.5% NaCl	2.0	3
Cement paste + 3% NaCl	2.0	6

TABLE 3—*Cement composition.*

Cement	pH	Ca^{++}, ppm	Na^+, ppm	K^+, ppm	Cl^-, ppm	Cl^-/OH^-
+0% NaCl	13.4	55	6 250	24 375	1 750	0.19
+0.4% NaCl	13.4	77	7 190	21 875	3 750	0.42
+1.5% NaCl	13.5	183	15 940	20 375	12 190	1.14
+3% NaCl	13.3	199	21 560	15 580	22 190	3.2

pH is constant hence the chloride/hydroxide ratio increases. The 3% NaCl additions decrease the pH slightly, but the ratio increases.

Polarization curves in the extracted solutions are shown in Figs. 4, 5, and 6, for the 0, 0.4, and 3% NaCl additions, respectively. The chloride free pore solution protected the steel, $E_{corr} = -0.2075$ to -0.2250 V versus SCE. The passive currents are about twice as large as those from embedded electrodes, which suggests that the lime-rich layer is either increasing the anodic polarization or shielding the steel. Increasing the chloride content decreased the corrosion potential, that is, -0.547 V versus SCE for the 3% NaCl mix. The polarization data tends to show higher passive currents than in the cement paste, which again suggests that the buffering capacity of the lime rich layer is influencing the process.

The effect of deaeration and a slow sweep rate can also be seen; the results from the slow sweep rate of 3.6 mV/min were essentially the same as the standard procedure of 18 mV/min adopted in the present study. The deaerated solution results indicate the influence of the cathodic kinetics. Note that the pore water was protected from reaction with atmospheric carbon dioxide.

Discussion

The data presented above are in general agreement with results obtained by previous workers in terms of anodic polarization in cement and concrete samples as well as the lower passivation current densities compared to measurements in the extracted or simulated pore solutions [8]. However, in the present work a more direct comparison of the anodic processes can be made since the same potential transitions appear in both the cement paste samples and the pore solutions, aerated and deaerated.

The hydrated cement system can be compared directly to other alkaline environments and some transients observed on the polarization curves can be identified from the E-pH diagram [16]. Typically hydrogen evolution was observed at a reversible potential of about -1.00 V versus SCE and oxygen evolution at about 0.23 V versus SCE in the present samples. Formation of Fe_3O_4 could be identified at -0.95 V versus SCE with conversion to Fe_2O_3 at -0.80 V. Further oxidation of Fe_2O_3 to FeO_4^{2-} appeared to take place at -0.20 V versus SCE [16]. This anionic species is considered as being present in alkaline batteries [17] and appears to be stabilized by sodium and potassium ions [18]; it has also been reported in the concrete system [19].

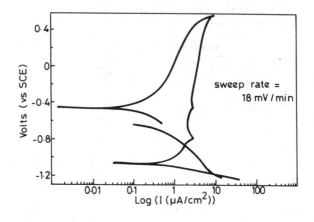

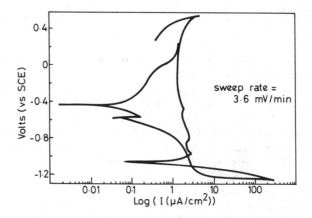

FIG. 4—*Polarization in deaerated pore solution, 0.4% NaCl solution, showing effect of the potentiodynamic sweep rate.*

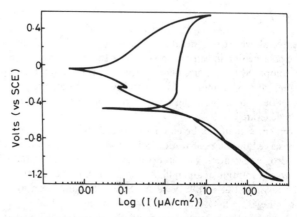

FIG. 5—*Polarization in aerated pore solution, 0% NaCl addition.*

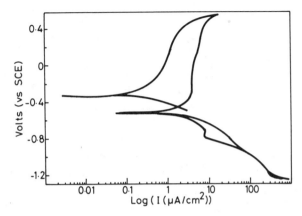

FIG. 6—*Polarization in aerated pore solution, 3% NaCl addition.*

This higher oxidation species formation suggests that the lower potential transient of -0.5 V versus SCE is probably due to the formation of a Fe(IV) species, although this requires further confirmation. The passive region typically extended from -0.7 to -0.4 V versus SCE and agrees with Preece et al. [8].

The importance of the pore solution polarization, compared to the cement based samples, is that they demonstrate that the anodic kinetics are essentially the same. Any difference in the current magnitude arises from the apparent effect of the lime rich layer, which is bonding the steel to the hydrated cement and thereby decreasing the area available for reaction. Pore solution studies would therefore appear to provide a basis for further investigations of the anodic processes. Indeed in the case of the present cement paste samples the results suggest that corrosion may not be a problem provided the chloride to hydroxyl ratio is less than 1.0. This ratio could however vary in other types of sample since the cathodic kinetics will also influence the overall rate and determine whether the steel is in the passive or active region.

The chloride additions increase the anodic polarization and the passive current density, however, the corrosion potentials tend towards the active region rather than the higher pitting region. The observed corrosion is therefore general corrosion, which tends to be localized at specific areas of the specimens, presumably where there is less adhesion of the lime rich layer. The chloride additions also influence the initial hydration processes in that the number of pores greater than 50 nm has decreased; this in turn decreases the oxygen diffusion and cathodic kinetics. There is therefore a balance operating in the matrix, a high chloride level, and high oxygen diffusion should give pitting corrosion, but the cement chemistry has resulted in a set paste which has a low oxygen permeability, and hence active corrosion is observed.

The impedance data on the cement pastes are typical of results obtained on concrete samples [10] and can be characterized by an equivalent circuit having a solution resistance, a hydrated cement dielectric component, and the electrochemical double layer in parallel with a charge transfer and diffusion process (Fig. 7). The solution resistance and dielectric component develops during the exposure period (Fig. 1). The low frequency data are electrochemical and are dominated by the diffusion process. Although diffusion processes can be observed within passive films, it is more probable that the diffusion is associated with the cathodic kinetics in the paste samples. The predominance of the diffusion process makes it impossible to obtain a true linear polarization resistance measurement as the real or resistive value is frequency dependent even down to 1 mHz. Although the DC, or zero frequency, value of the resistance is not observed, it may be possible to measure a frequency dependent (apparent) polarization resistance value, which could be used to compare samples prepared in a similar manner.

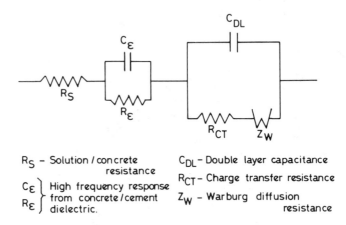

R_S – Solution / concrete resistance

C_ε
R_ε } High frequency response from concrete / cement dielectric.

C_{DL} – Double layer capacitance

R_{CT} – Charge transfer resistance

Z_W – Warburg diffusion resistance

FIG. 7—*Equivalent circuit representation.*

The impedance data and the cathodic polarization results clearly show the important influence of the hydrated cement matrix in determining the corrosion processes. Thus although th pore-water measurements are useful in assessing the anodic kinetics, such an approach ha limited application in long-term concrete durability studies.

Acknowledgments

The authors wish to recognize the financial support provided by the Building Research Establishment and also to the Science and Engineering Research Council for the student grant t P. E. Langford. Our appreciation is also expressed to K. W. J. Treadaway of BRE for invalu able technical assistance and to C. Page of Aston University for pore-size analysis measure ments.

References

[1] Page, C. L. and Treadaway, K. W. J., *Nature*, Vol. 297, 1982, p. 109.
[2] Slater, J. E., *Corrosion of Metals in Association with Concrete, STP 818*, American Society for Testin and Materials, Philadelphia, 1983.
[3] Langford, P. E., Ph.D. thesis, Manchester, United Kingdom, 1986.
[4] Lea, F. M., *The Chemistry of Cement and Concrete*, 3rd ed., Arnold, Garden Grove, CA, 1970.
[5] Al-Khalaf, M. N. and Page, C. L., *Cement and Concrete Research*, Vol. 9, 1979, p. 197.
[6] *Durability of Concrete Structures*, Beton Information, Bulletin 148, Paris, 1982.
[7] Holden, W. R., Page, C. L., and Short, N. R., *Corrosion of Reinforcement in Concrete Construction* SCI, London, 1983, pp. 143–150.
[8] Preece, C. M., Gronvold, F. O., and Frolund, T., *Corrosion of Reinforcement in Concrete Construc tion*, SCI, London, 1983, pp. 143–150.
[9] Dawson, J. L., *Corrosion of Reinforcement in Concrete Construction*, SCI, London, 1983, pp. 173 191.
[10] John, D. G., Searson, P. C., and Dawson, J. L., *British Corrosion Journal*, Vol. 16, 1981, p. 103.
[11] Berman, H. A., "The Effect of Sodium Chloride on the Corrosion of Concrete Reinforcing Steel an the pH of Calcium Hydroxide Solution," FHWA-RD-74-1, Federal Highway Administration, Wash ington, DC, 1974.
[12] Ftikos, C. L. and Parissakis, G., *Cement and Concrete Research*, Vol. 15, 1985, p. 593.
[13] Hausmann, D. A., *Materials Protection*, Vol. 6, Nov. 1976, p. 19.
[14] Shalon, R. and Raphael, M., *Journal of the American Concrete Institute*, Vol. 55, 1959, p. 1252.
[15] Gouda, V. K., *British Corrosion Journal*, Vol. 5, 1970, p. 198.

[6] Pourbaix, M., *Atlas d'Equilibres Electrochimiques*, Pub Gauthies-Villars, 1963.
[7] Uno Falk, S. and Salkind, A. J., *Alkaline Storage Batteries*, Wiley, New York, 1969.
[8] Cotton, F. A. and Wilkinson, G., *Advanced Inorganic Chemistry*, Wiley, New York, 1980.
[9] Silverman, D. C., *Corrosion*, Vol. 38, No. 8, 1982.

DISCUSSION

J. E. Castle[1] *(written discussion)*—There seemed to be a disparity between the scales used for the inset diagram and the main diagram on its AC impedance plot. Can you comment further on its high frequency and low frequency and of the Z'-Z plot?

J. L. Dawson and P. E. Langford (authors' closure)—Figure 1 shows how the impedance changes with time, the concrete resistance R_s increasing from a few tens of Ω cm^2 to between 2 to kΩ cm^2. The insert shows the development of the high frequency spur, the dielectric effect; these data have been normalized by removal of the R_s value to aid the visual comparison.

T. B. Turnbull[2] *(written discussion)*—What is the effective diffusion co-efficient for oxygen in the concrete layer?

J. L. Dawson and P. E. Langford (authors' closure)—We do not have a value; it would require determination over a range of well characterized concretes.

N. J. M. Wilkens[3] *(written discussion)*—Can you comment on the way in which partial replacement of OPC with BFS or PFA influences some of the observations you have described?

J. L. Dawson and P. E. Langford (authors' closure)—Addition of the blending agents (PFA and BFS) gives a finer pore size and a decreased porosity, which is important in controlling diffusion. There is also a possible decrease in pH and subtle changes in the chemistry, which may with certain cements, influence the passivation and the upper limit of chloride to hydroxyl ion ratio criterion.

[1]University of Surrey, Guildford, Surrey GU 2XH, United Kingdom.
[2]National Physical Laboratories, Teddington, Middlesex TW11 OLW, United Kingdom.
[3]AERE Harwell.

Barry Hindin[1] and Arun K. Agrawal[1]

Materials Performance for Residential High Efficiency Condensing Furnaces

REFERENCE: Hindin, B. and Agrawal, A. K., **"Materials Performance for Residential High Efficiency Condensing Furnaces,"** *The Use of Synthetic Environments for Corrosion Testing,* *ASTM STP 970,* P. E. Francis and T. S. Lee, Eds., American Society for Testing and Materials, Philadelphia, 1988, pp. 274–286.

ABSTRACT: In high efficiency furnaces, the condensate that is produced from the flue gas contains trace amounts of such contaminants as $SO_x^=$, Cl^- (chloride ions), NO_3^- (nitrate ion), NO_2^- (nitrate ion), F^- (fluoride ions), and organic acids. The condensate after concentration by re-evaporation becomes very corrosive to the heat exchanger materials. An electrochemical investigation has been conducted using 316L as a reference alloy and synthetic condensates to identify the environmental factors most significant in corrosion. The condensates were formulated using a statistical experimental design approach. Through a regression analysis of the results, chloride and temperature were identified as the most damaging factors. Investigation of several other alloys in an aggressive condensate showed that their corrosion performance was dependent upon their Cr and Mo contents. Subsequently, a large number of alloys were evaluated in a gas-fired furnace by exposing them to chloride-spiked condensates under alternate wetting and drying conditions for 90 days. The alloys included austenitic, ferritic, and duplex stainless steels, and also a few aluminum, copper, and nickel-based alloys. Most of the alloys performed well in the condensate having 26 ppm Cl^-; a few of the alloys did well in 200 ppm Cl^-, but at 1000 ppm Cl^- only those alloys which had high Cr and Mo contents survived.

KEY WORDS: condensing furnaces, pitting, chloride attack, screening of alloys, stainless steels, corrosion

Typical efficiencies of standard residential gas furnaces range from 60 to 70%. In recent years, gas furnaces having efficiencies of up to 90% have been introduced. Higher efficiency realized by extracting the heat of vaporization of water from the flue gas, which otherwise wasted. The flue gas is cooled below its dew point (54 to 60°C) before it leaves the furnace. The cooling condenses the water vapor from the gas, and the released latent heat is used to warm the house air. The condensate produced is drained from the furnace.

Water vapor present during condensation traps some of the contaminants contained in the flue gas. These contaminants include $SO_x^=$, Cl^-, NO_2^-, NO_3^-, F^-, and some organic acids. Sources of the contaminants are both the combustion air (N, Cl, and F) and the natural gas (S). A recent survey [1] of 572 houses in the United States found that the nominal (or geometric mean) composition of the condensate was as follows: NO_3^-, 8.5 ppm; $SO_4^=$, 5.5 ppm; NO_2^-, 1. ppm; Cl^-, 1.1 ppm; and F^-, 0.21 ppm, with a pH of 2.9. These houses used indoor air for combustion. In the survey, analyses were not made for organic acids, and all the S compounds were oxidized to $SO_4^=$ to facilitate analyses.

Though the initial condensate is very dilute, it becomes extremely concentrated in certain parts of the heat exchanger through re-evaporation processes. The re-evaporation is caused by

[1]Principal research scientist and senior research scientist, respectively, Battelle Columbus Division, 505 King Ave., Columbus, OH 43201.

temperature fluctuations in the heat exchanger as the furnace cycles on and off in response to the heating load. The condensate at times is almost completely evaporated in certain localized areas of the heat exchanger, depending upon the design of the furnace. Concentrated condensate poses serious corrosion problems for the heat exchanger material(s). Pitting has been reported as a major corrosion concern in a previous study [1].

The criteria used by industry in setting the maximum acceptable general corrosion rate for an alloy used in a furnace depend upon several parameters, including expected life, cost, and wall thickness. In Europe, rates less than 0.1 mm/year for uniform corrosion and 0.2 mm/year for pitting are required for a 15-year life [2]. In the United States, a maximum acceptable uniform corrosion rate of 0.05 mm/year is required for a 20-year service life.

The present study was concerned with identifying the environmental factors which most affect the pitting resistance of the heat exchanger alloys, and evaluating the corrosion performance of a number of candidate alloys in a condensing furnace. An electrochemical anodic polarization technique was used to identify the critical environmental factors and the role of alloy composition in the pitting susceptibility of the heat exchanger materials. An alternate immersion and drying technique was used to evaluate the long-term performance of the candidate alloys exposed to chloride-spiked condensates generated by an actual furnace.

Experimental Procedure

Electrochemical Studies

The ranges of the environmental factors considered for the electrochemical investigation are given in Table 1. An overall concentration factor of 10^4 was assumed for the dilute condensate in determining the ranges in Table 1. Because some of the more volatile components of the condensate are lost to the flue gas at a concentration factor of about 10^3, their concentration factors were considered to be less than 10^4, as given in Table 1. The test matrix for the investigation was developed using a statistical experimental design approach, which was based on the "2^{8-4} partial factorial design" of Box and Hunter [3]. The test matrix is shown in Table 2. The main advantage of the Box and Hunter method is that the significance of each factor is identified without interference from the other factors. Two factor interactions also are identified, but these are confounded.

The chemical solutions listed in the test matrix were prepared using reagent grade chemicals and demineralized water. The chemicals used were H_2SO_4, HCl (hydrochloric acid), HNO_3, HF (hydrogen fluoride), Na_2SO_4, NaCl (sodium chloride), $NaNO_3$, $NaNO_2$, NaF (sodium fluoride), and $Na_2S_2O_3$. The latter was used to simulate the presence of reduced sulfur oxyanions (S^{+x}; $x < 6$) in the condensate. Condensate in a furnace is always in contact with the flue gas. There-

TABLE 1—Condensate composition.

Factors	Nominal Value	Concentration Factor	Possible Range
$SO_4^=$, ppm	5.5	10^4	5 to 5×10^4
Cl^-, ppm	1.1	10^4	1 to 1×10^4
F^-, ppm	0.2	2.5×10^3	1 to 5×10^2
NO_2^-, ppm	1.9	10^3	2 to 2×10^3
NO_3^-, ppm	8.5	10^3	10 to 1×10^4
S^{+x}, ppm	trace	10^4	1 to 1×10^3
pH	2.9	...	0.4 to 4.0
Temperature, °C	54	...	49 to 82

TABLE 2—*Partial factorial design—test matrix and results for alloy 316L.*

Solution No.	pH	Temperature, °F[a]	Concentration, log ppm						ΔE_{pit}
			$SO_4^=$	Cl^-	F^-	NO_3^-	$S_2O_3^=$		
1	0.4	120	4.7	0	0	4	3		560
2	4.0	180	0.7	0	0	1	3		630
3	4.0	120	4.7	4	0	1	0		390
4	0.4	180	0.7	4	0	4	0		320
5	4.0	120	0.7	0	2.7	4	0		470
6	0.4	180	4.7	0	2.7	1	0		700
7	0.4	120	0.7	4	2.7	1	3		55
8	4.0	180	4.7	4	2.7	4	3		500
9	4.0	180	0.7	4	2.7	1	3		110
10	0.4	120	4.7	4	2.7	4	0		760
11	0.4	180	0.7	0	2.7	4	3		700
12	4.0	120	4.7	0	2.7	1	3		860
13	0.4	180	4.7	4	0	1	3		70
14	4.0	120	0.7	4	0	4	3		440
15	4.0	180	4.7	0	0	4	0		490
16	2.9	120	0.7	0	0	1	0		1010
17	2.2	150	2.7	2	1.3	2.5	1.5		610

[a]$1°C = 33.8°F$.

fore, each solution was saturated at the time of use with a gas mixture comprised of N (nitrogen) (86%), carbon dioxide (8.8%), and O (oxygen) (5.2%). This gas mixture simulated the combustion product of methane using 30% excess air.

The polarization experiments were conducted using cylindrical specimens 1.25 cm long × 0.625 cm in diameter, which were given a 600 grit surface finish. The 12 alloys used were 304L, 316L, Nitronic®-50, 904L, 44LN, Ferralium® 255, Monit®, AL-6XN®, Sea-Cure®, 254 SMO®, 29-4-2, and Hastelloy® C-276.[2] Their nominal compositions are listed in Table 3. The counter electrode was made of platinum, and the reference electrode was a standard calomel electrode (SCE). Specimens were allowed to equilibrate with their test solution for 3 to 4 h in the presence of the gas mixture prior to starting their polarization scans. The anodic scan was started from approximately 10 mV negative of the specimen's corrosion potential, and continued until pitting occurred and the current reached 5 mA/cm². The potential scan rate used in the polarization experiments was 600 mV/h. Values of the passivation potential, E_{pas}, and the pitting potential, E_{pit}, were determined for each specimen from its polarization curve. These potentials are identified on the idealized polarization curve shown in Fig. 1.

Alternate Immersion Tests

A schematic of the alternate immersion test apparatus is shown in Fig. 2. Twenty-nine alloys were tested; their compositions are listed in Table 3. The specimens were suspended from racks that immersed them in the condensate for 4 min, and then lifted them for drying in warm flue-gas for 11 min. The 15-min cycle was repeated continuously for 90 days. The temperature of the condensate was kept at 49°C (120°F). The temperature of the flue gas over the specimens during the drying cycle ranged from 77 to 99°C (170 to 210°F). Two sets of tests were conducted. In the first set, the chloride concentration of the condensate was raised to 26 ppm by adding 0.1 M

[2]Nitronic is a registered trademark of Armco Steel. Ferralium is a registered trademark of Bonar Langley Alloys, Ltd. Monit is a registered trademark of Nyby Uddeholm AB. Sea-Cure is a registered trademark of Colt Industries. Hastelloy is a registered trademark of Cabot Corporation. 254 SMO is a registered trademark of Auesta Jernverks AB. AL-6XN is a registered trademark of Allegheny Ludlum Steel Corporation.

TABLE 3—*Nominal composition of alloys used in tests.*

Alloy	Class[a]	Cr	Ni	Mo
		Weight Percent		
304	A	19	9	0
304L	A	19	10	0
309	A	23	14	0
316	A	17	12	2.5
316L	A	17	12	2.5
Alloy 20Cb3	A	20	35	2.5
Nitronic 50	A	22	12	2.5
317	A	19	12	3.5
904L	A	21	25	4.5
254 SMO	A	20	18	6.2
AL-6XN	A	21	24	6.5
44LN	D	25	6	1.6
2205LCN	D	22	5.5	3
Ferralium 255	D	25	5.5	3
409	F	11	0.8	0
430	F	17	0.8	0
439	F	18	0.5	0
Ebrite-26-1	F	26	0.5	1.3
ELI-T 18-2	F	17	0.3	2
444	F	18	1	2
Sea-Cure	F	27	1.2	3.5
Monit	F	25	4	4
29-4C	F	29	0.2	4
C-276	Ni	15	57	16
Al-1100	Al	99% Al		
Al-3003	Al	1.3% Mn, 0.7% Fe, balance Al		
Finned-Tube	Al	same as Al-3003		
Al-Bronze	Cu	7% Al, 0.2% Fe, balance Cu		
70Cu/30Ni	Cu	70% Cu, 30% Ni		

[a]A = austenitic, D= duplex, F = ferritic, Ni = nickel-based, Al= aluminum-based, Cu = copper-based.

HCl, and in the second set it was raised to 1000 ppm. Data points from previous studies on the corrosion behavior of several alloys tested at 100 and 200 ppm chloride levels were included for purposes of comparison. (As will be shown later, the chloride concentration was identified by the electrochemical investigation as the key environmental factor responsible for pitting.)

The test specimens included weight-loss coupons, ASTM (Practice for Making and Using U-Bend Stress Corrosion Test Specimens [G 30]) U-bends, and weld-crevice, finned-tube, and expanded-tube specimens. Most of the austenitic and ferritic alloys were tested using weight-loss, U-bends, and weld-crevice, and expanded-tube specimens. Duplex alloys were tested using only U-bends and weld-crevice specimens. The nickel-based and copper-based alloys were tested using only the weld-crevice specimen. The aluminum alloys were tested using only the weld-crevice and finned-tube specimens. The weld-crevice specimen, shown in Fig. 3, is unique because it allows a metal's resistance to weld deterioration, crevice corrosion, and pitting to be evaluated simultaneously on a single specimen. The expanded-tube specimen shown in Fig. 4a was made from as-received tube and had one end roll-expanded by approximately 5%. This

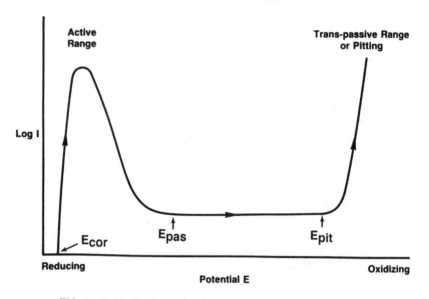

FIG. 1—*An idealized polarization curve, identifying various transitions.*

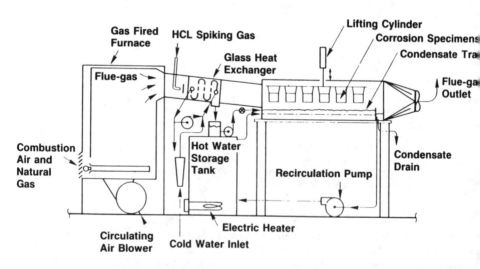

FIG. 2—*Schematic of alternate immersion test apparatus.*

expansion was done to simulate the stresses created in rolled-tubes that are used in certain heat exchangers.

The U-bends and expanded-tube specimens were used to evaluate the susceptibility of alloys to stress-corrosion cracking (SCC). U-bend specimens were tested in two conditions, namely, mill-annealed (as-received) and sensitized (650°C/4 h/furnace cooled) heat treatments.

Aluminum finned-tubing has been used in some European heat exchangers. Sections of these tubes were cut to form specimens having either one or three fins; an illustration of the latter is

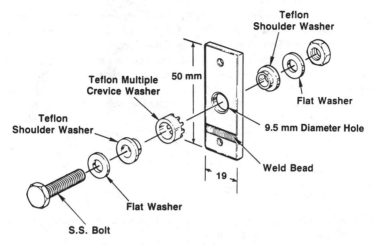

Weld-Crevice Specimen

FIG. 3—*Exploded view of weld-crevice specimen.*

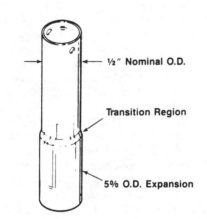

Expanded Tube Specimen
(a)

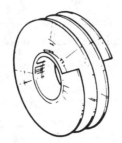

Finned Tube Specimen
(b)

FIG. 4—*Illustration of expanded-tube specimen (a), and finned-tube specimen (b).*

shown in Fig. 4b. A minimum of three samples of each specimen type were tested. Galvanic coupling between specimens was prevented by suspending the specimens from their racks using chromel wire covered with Teflon® shrink-tubing. Cross contamination of the condensate was also prevented by testing dissimilar alloys in their own condensate trays.

The specimens were evaluated for general corrosion rate, maximum pit depth, and the presence of stress corrosion cracks. The corrosion rate for an alloy was calculated by using the areas exhibiting deepest attack on the multiple specimens and averaging their values. The area having the deepest attack was always found at the bottom section of a specimen. The bottom section most likely remained wet for a longer time in each cycle than the upper section.

Results and Discussion

Electrochemical Studies

The values of E_{pas} and E_{pit} measured from each polarization curve were used to calculate a "pitting margin" parameter, that is defined as:

$$\Delta E_{pit} = E_{pit} - E_{pas}$$

The pitting margin represents the potential jump that is required to initiate pits on the passive surface of the specimen. Thus, the larger the value of ΔE_{pit}, the greater the resistance of the alloy to pitting in the test environment. The values of ΔE_{pit} calculated for 316L in various synthetic condensates are given in Table 2. A multiple regression analysis was conducted for the values of ΔE_{pit} and the corresponding environmental factors from the test matrix. (See also Koch et al. in this publication.) The probable significance of each factor and its regression coefficient are summarized in Table 4. A significance value of 80% or more indicates a definite effect of the factor on the measured parameter, i.e., ΔE_{pit}; the effect is not certain for a significance less than 80%.

The results for 316L given in Table 4 indicate that chloride and temperature are the most significant factors that affect the pitting margin. The other environmental factors also affect the pitting margin, but these are of little significance in the ranges investigated; their effects are overwhelmed by those of chloride and temperature. The latter two reduce the pitting margin as indicated by their negative regression coefficients. The results are in agreement with those reported in the literature for 316 stainless steel [4].

One of the most aggressive solutions found in the test matrix for 316L was solution No. 13. This solution was used to determine the pitting margins of some additional alloys containing Cr and Mo. The results are plotted against the Cr and Mo contents of the respective alloys in Fig. 5.

TABLE 4—*Results of multiple regression analysis of ΔE_{pit} for 316L stainless steel.*

Main Factor	% Significance	Regression Coefficient
Cl^-	100	−173
Temp.	93.7	−64
$SO_4^=$	74.8	...
S^{+x}	59.1	...
NO_2^-/NO_3^-	56.8	...
pH	45.3	...
F^-	35.5	...

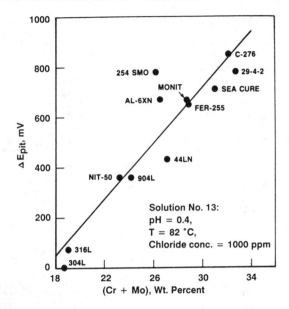

FIG. 5—*Correlation between pitting margin,* ΔE_{pit}, *of selected alloys and their Cr and Mo content.*

The trend in Fig. 5 clearly indicates that the pitting resistance of stainless alloys in aggressive condensates increases with increasing Cr and Mo contents. This again was to be expected in an acidic chloride environment from data in the literature [5].

Alternate Immersion Tests

Over 200 specimens of 29 alloys, including those screened in the electrochemical investigation, were evaluated by alternate immersion tests (Table 3). Table 5 lists the average corrosion rates and maximum pit depth under crevices measured for alloys tested for 90 days in spiked condensate containing 26 and 1000 ppm chloride. Corrosion rates averaged over the specimens' entire surface area and the heavily attacked areas only have been included. The results are summarized below according to alloy class, i.e., austenitic, ferritic, etc. Corrosion rate data for chloride levels of 100 and 200 ppm, available from other similar investigations [1,6] using weld-crevice and weight-loss coupons, have been included in the corrosion rate plots to more clearly define the effect of chloride concentration. All the corrosion rates reported in the text and figures of this paper refer to average rates from the multiple specimens' surface areas exhibiting the deepest attack.

Austenitic Alloys

Corrosion rates of several alloys are plotted in Fig. 6 as functions of Cl^- concentration in the condensates. As expected, the rates increased with the Cl^- concentration. The rate was < 0.04 mm/year up to 200 ppm Cl^- for all the alloys, but at 1000 ppm Cl^- the rates ranged from 0.02 to 0.1 mm/year depending upon the alloy (Fig. 6). There was no significant pitting or crevice corrosion observed in specimens at 26 ppm Cl^-. At 200 ppm Cl^-, the pitting and crevice corrosion was moderate in the 300 series stainless steels, but at 1000 ppm Cl^-, all the austenitic alloys had severe crevice corrosion and pitting attack of approximately 0.2 mm/year, except for the Nitronic® 50, 254 SMO, and Al-6XN alloys.

TABLE 5—*Corrosion rates and maximum pit depth measurements of weld-crevice specimens after exposure to 26 and 1000 ppm chloride for 90 days[a] in the alternate immersion test.*

Alloy	26 ppm Chloride Level			1000 ppm Chloride Level		
	Corrosion Rate, mm/year		Maximum Pit Depth, mm	Corrosion Rate, mm/year		Maximum Pit Depth, mm
	Entire Surface[b]	Heavily Corroded[c]		Entire Surface	Heavily Corroded	
304	<0.003	0.01	0.01	nt(d)	nt	nt
304L	<0.003	0.02	0.02	0.10	0.11	0.03
309	<0.003	0.01	<0.01	nt	nt	nt
316	<0.003	0.01	<0.01	0.06	0.07	0.04
316L	<0.003	0.01	0.02	0.08	0.09	0.01
Alloy 20Cb3	<0.003	...	<0.01	0.03	0.10	0.03
Nitronic 50	<0.003	...	<0.01	0.04	0.05	0.03
317	<0.003	...	<0.01	0.02	0.03	0.06
904L	<0.003	...	<0.01	0.01	0.03	0.06
254 SMO	<0.003	...	<0.01	<.003	0.04	0.08
AL-6XN	<0.003	...	<0.01	<.003	0.04	0.02
44LN	<0.003	...	<0.01	0.02	0.16	0.03
2205LCN	<0.003	...	<0.01	0.04	0.06	0.00
Ferralium 255	<0.003	...	<0.01	0.02	0.14	0.02
409	0.01	0.01	0.16	nt	nt	nt
430	0.005	...	0.11	0.19	0.20	0.61
439	<0.003	0.01	0.04	0.15	0.20	0.03
Ebrite-26-1	0.01	...	0.05	0.05	0.10	0.03
ELI-T 18-2	<0.003	0.01	0.04	nt	nt	nt
444	<0.003	0.01	<0.01	0.10	0.13	0.20
Sea-Cure	<0.003	...	<0.01	<.003	0.04	0.10
Monit	<0.003	...	<0.01	<.005	0.06	0.03
29-4C	<0.003	...	<0.01	<.003	0.05	0.05
C-276	<0.003	...	<0.01	<.003	<.003	0.01
Al-1100	0.03	...	0.15	1.22	1.22	1.60
Al-3003	0.03	...	0.07	1.14	1.14	2.21
Al-Bronze	0.04	0.05	<0.01	nt	nt	¯nt
70Cu/30Ni	0.07	0.09	0.04	nt	nt	nt

[a]Copper-based alloys exposed for 17 days only.

[b]Corrosion rate averaged over entire exposed surface area.

[c]Corrosion rate of heavily corroded surface area (dashed line indicates heavily attacked region did not exist).

[d]nt = alloy not tested at the 1000 ppm chloride level.

Stress-corrosion cracks (SCC) were observed in the welded areas of the weld-crevice of specimens of 304L, 316, and 316L at the 1000 ppm chloride level. None of the U-bends tested in the mill-annealed condition exhibited SCC. Only sensitized U-bends made from alloys 304L, 309, and 316 exhibited SCC. Metallographic examination of these specimens indicated that the cracking was predominantly intergranular.

Duplex Stainless Steels

Corrosion rates of duplex stainless steels are shown in Fig. 7. There was essentially no corrosion attack of any kind at 26 and 100 ppm Cl⁻. The rates were <0.01 mm/year at 200 ppm

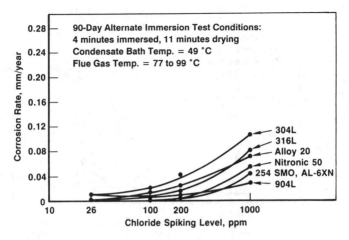

FIG. 6—*Corrosion rate (from areas of deepest attack) versus chloride concentration of condensate for austenitic stainless steels. Data points at 100 and 200 ppm chloride were taken from Ref 1.*

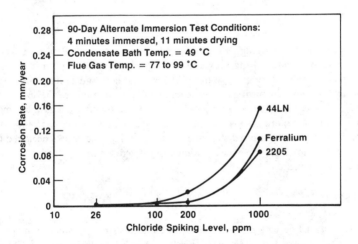

FIG. 7—*Corrosion rate (from areas of deepest attack) versus chloride concentration of condensate for duplex stainless steels. Data points at 100 and 200 ppm chloride were taken from Ref 1.*

Cl^-, but were considerably higher (0.05 to 0.15 mm/year) at 1000 ppm Cl^-. The alloys suffered detectable crevice corrosion and pitting attack at 200 ppm Cl^-, and were severely attacked at 1000 ppm Cl^-. No other forms of attack were observed in the weld-crevice specimens up to the highest Cl^- level. The U-bend specimens made from the duplex alloys did not exhibit SCC, but shallow intergranular attack was found on mill-annealed 2205 LCN and Ferralium 255 at the 1000 ppm Cl^- level.

Ferritic Alloys

Wide ranges of uniform corrosion rates were observed in ferritic alloys; several examples are shown in Fig. 8. The alloys having the lowest rates were Monit®, 29-4C, and Sea-Cure. Their rates ranged between 0.03 to 0.06 mm/year at the highest Cl^- level. The other alloys, 409, 430,

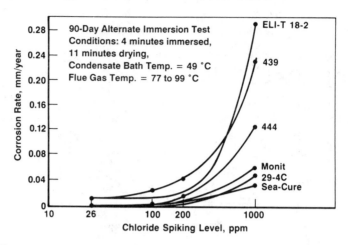

FIG. 8—*Corrosion rate (from areas of deepest attack) versus chloride concentration of condensate for ferritic stainless steels. Data points at 100 and 200 ppm chloride were taken from Ref 1.*

439, 444, and Ebrite 26-1, had higher rates (0.1 to 0.2 mm/year at 1000 ppm Cl⁻) and suffered from crevice corrosion and pitting attack even at the lowest Cl⁻ concentration of 26 ppm. Monit® had no detectable pitting at 200 ppm Cl⁻, but had deep pits in its welded areas at 1000 ppm. Sea-Cure had sporadic pitting in crevice and weld areas at all three Cl⁻ levels. At the 1000 ppm Cl⁻ level, 29-4C exhibited no pitting in its welded areas, and the maximum measured pit depth was 0.05 mm compared to 0.10 mm for Sea-Cure and 0.03 mm for Monit®. Monit® was the only ferritic alloy that exhibited stress-corrosion cracking; the cracks occurred at the weld area on a Monit® weld-crevice specimen and in a sensitized Monit® U-bend specimen tested in 1000 ppm Cl⁻ condensate.

Nickel Alloy

Alloy C-276 was the only nickel-based alloy tested in this study. It showed the greatest resistance to all forms of corrosion at all the Cl⁻ levels. The only attack observed on C-276 was some surface staining in the crevice areas of its weld-crevice specimens exposed to 1000 ppm Cl⁻.

Aluminum Alloys

The general corrosion rate (Fig. 9) of alloys Al-1100 and Al-3003 was 0.03 mm/year at 26 ppm Cl⁻, and approximately 1.2 mm/year in 1000 ppm Cl⁻. Severe pitting occurred in the alloys at 1000 ppm Cl⁻.

Copper Alloys

Aluminum bronze and 70Cu-30Ni were tested at only 26 ppm Cl⁻. Their corrosion rates were 0.06 and 0.09 mm/year, respectively. No significant crevice corrosion or pitting was observed in these alloys. An earlier study [1] at 200 ppm Cl⁻ had found corrosion rates of approximately 0.2 mm/year in the copper alloys, and therefore they were not tested at 1000 ppm Cl⁻.

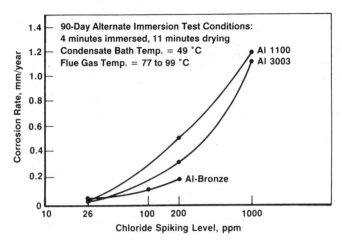

FIG. 9—*Corrosion rate (from areas of deepest attack) versus chloride concentration of condensate for copper and aluminum alloys. Data points at 100 and 200 ppm chloride were taken from Ref 1.*

Conclusions

Flue gas condensate became increasingly corrosive to heat exchanger materials after it is concentrated by repeated condensing and evaporation on heat exchanger surfaces. The most significant environmental factors responsible for the condensate's corrosivity were chloride ions and elevated temperature. A wide range of corrosion behavior was exhibited by the various classes of tested alloys. Resistance of the stainless steel and nickel-based alloys to pitting attack in the condensate environment was strongly dependent upon their Cr and Mo content; these elements are known to improve the pitting and crevice corrosion resistance of alloys in the presence of chloride ions. The alloys that showed the best overall corrosion performance were, in decreasing order, C-276, 29-4C, 254-SMO, and Al-6XN. Sea-Cure exhibited corrosion behavior similar to that of 29-4C, except that Sea-Cure exhibited slightly lower pitting resistance. Aluminum and copper alloys had considerably higher uniform corrosion rates than did the other alloys tested.

Acknowledgment

This work was supported by the Gas Research Institute (GRI) of Chicago, IL under contract number 5084-24-1022. The support and technical input of John Lockwood and Bob Hemphill of GRI and E. White, S. Talbert, G. H. Stickford, M. Spangler, and J. H. Payer of the Battelle Columbus Division are greatly appreciated.

References

[1] Razgaitis, R., Payer, J. H., Stickford, G. H., Talbert, S. G., Farnsworth, C., and Locklin, D. W., "Research on Heat Exchangers Corrosion," Interim Report (May 1983 to Aug. 1984), GRI Report 84/0157, Gas Research Institute, Chicago, IL (Sept. 1984).
[2] Britton, R. N. and Stevens, R. L., "Materials Requirements for Condensing Boilers," *Dewpoint Corrosion*, D. R. Holmes, Ed., Ellis Horwood Ltd., London, 1985.
[3] Box, B. E. D., Hunter, K. W. G., and Hunter, J. S., *Statistics for Experiments*, Wiley, New York, 1978.

[4] Effird, K. D. and Moller, G. E., Paper No. 87, presented at Corrosion/78, Houston, TX, March 1978.

[5] Streicher, M. A., *Metal Progress*, Oct. 1985, p. 29.

[6] Stickford, G. H., Hindin, B., Talbert, S. G., Agrawal, A. K., Murphy, M. J., Razgaitis, R., Payer, J. H., Cudnik, R. A., and Locklin, D. W., "Technology Development for Corrosion-Resistant Condensing Heat Exchangers," Final Report (Sept. 1984 to Oct. 1985), GRI Report 85/0282, Gas Research Institute, Chicago, IL, Oct. 1985.

Summary .

In recent years researchers have realized that natural environments are far more complex than, for example, a 3.5% sodium chloride solution often used to simulate seawater, and that trace elements present in real environments can have considerable effects on the corrosion performance of metals. Investigations in complex environments can lead to a large number of experiments. Koch et al. have described a method to handle the effects of a large number of variables on the corrosion behavior of metals and to enable better simulation of chemical species within these complex environments. With a suitable experimental design a mathematical expression can be developed that describes the effect of several independent variables on corrosion behavior. Several examples are given in this paper to demonstrate the technique. Two of these examples are covered in greater detail by other authors. Mansfeld and Jeanjaquet used synthetic environments to study the corrosion of materials used in flue-gas desulfurization plants. Their study on Ferralium 255, Hastelloy G3, stainless steel (Type 317L), Monit and titanium (Grade 2) highlighted the importance of the composition of the synthetic environment used for these tests. Relatively minor variations in the chemistry of the synthetic test solutions had a marked effect on the corrosion performance of the alloys tested. For a given solution composition, high corrosion rates could occur for some alloys while others remained virtually unattacked. Hindin and Agrawal used synthetic environments to study the corrosion performance of candidate materials for use in residential high-efficiency condensing furnaces.

Haynes and Baboian described experience with synthetic environments for testing automobile body trim. In addition to road salts, atmospheric pollutants, such as SO_2 and NO_x have an important bearing on automobile corrosion. These pollutants can react on the metal surface to form corrosive acids. The industry has not been able to simulate this complex environment accurately in the laboratory and relies heavily on proving ground tests.

Fiaud supports the above findings, and in his paper comments that despite extensive work over a period of 70 years in the field of atmospheric corrosion many questions remain and no satisfactory method for predicting the performance of materials in the atmosphere exists. Fiaud lists the requirements of indoor and outdoor environments for corrosion testing and highlights the importance of SO_2, NO_x, and Cl_2 in the atmospheric corrosion of metals.

Murphy and Pape described the role of synthetic media in the study of metal food containers. Even very low corrosion rates can be harmful to the contents of the containers, for example, dissolution of tin or iron from tinplate or of aluminum from the body or end stock. Slight stains on the metal surface might not be harmful, but are undesirable. Synthetic environments have to be carefully formulated to simulate complex natural food materials. They concluded that the use of synthetic media is useful for studying basic principles like the complexing action of fruit acids, but there is always a need to support tests using synthetic environments with studies on actual products to be packed.

Kuhn et al. reviewed the use of synthetic environments for testing metallic biomaterials. They suggest that the best experimental techniques have not always been applied to measurements of corrosion in body fluids, and that the correlation between in-vivo and in-vitro corrosion rates is far from satisfactory in many cases.

Stott et al. described the evaluation of materials for resistance to sulfate reducing bacteria (SRB). Traditionally, the testing of materials against anaerobic corrosion caused by SRB has been carried out in small-scale "batch cultures" using filled and stoppered vessels or small cells contained in "anaerobic jars." These authors point out that such tests give much lower corrosion rates than those frequently experienced in the field. The reasons for this are outlined in the

287

paper. Some workers have used semi-continuous cultures to overcome some of the shortcomings of the "batch culture" test. The authors have outlined the features required for a satisfactory test environment for measurement of the resistance of materials to SRB. They describe an experimental rig design that produced an environment containing SRB, which generated without artificial acceleration high metallic corrosion rates comparable to those in naturally occurring corrosion processes.

The microbial theme was continued by Hill et al. who described the use of synthetic environments to simulate the microbial environment in the metal-working fluid in a machine tool coolant system.

These authors stressed the difficulty in simulating metal-working fluid spoilage in the laboratory using simple shake flasks. For organisms to develop and an environment to become established which relates closely to those in machine tool coolant systems, the test procedure must include an aeration phase, a piped phase, an adequately sized sump, and appropriate metal swarf. A synthetic environment has been used by these authors to rank biocides for use in machine tool coolant systems.

Kolts et al. investigated standard tests and environments for ranking the pitting and general corrosion resistance of stainless steels and nickel-base alloys. They found excellent correlation between field tests and pitting resistance determined from measurements of critical pitting temperatures in chloride solutions containing ferric ions. Correlations for general corrosion were found to be poor.

Parkins pointed out that evidence for the cracking domain for ferritic steels and α-brasses suggests that SCC in a range of environments occurs at potentials and pH values where the lower oxide of the relevant metal form. This suggests that laboratory tests for SCC should involve more than a single standardized environment and that variation from test to test of potential and pH, guided by the potential pH diagram may provide a more reliable guide to SCC propensity than that obtained from tests in a single solution. Various alloy systems were considered in the paper.

Johns described a method for assessing the sensitization to intergranular corrosion of austenitic stainless steel based on potentiokinetic polarization in a perchloric acid/sodium chloride solution. The data obtained correlated well with penetration rate values recorded in the boiling nitric acid test, ASTM C262-75 Practice C.

Mattsson et al. have described an ammonia test for the evaluation of the stress corrosion resistance of copper alloy products. The test procedure involves a 24-h exposure of the test material at room temperature to an ammoniacal atmosphere in equilibrium with an ammonium chloride solution. The vapor pressure of ammonia in equilibrium with the test solution depends on the pH value. Hence the severity of the test can be regulated by changing the pH value. With a pH of 9.5 the test was shown to correlate well with four-year field trials. The test has been proposed for (ISO) standardization.

Goodman et al. have described a test cell that provides the conditions that lead to Type I pitting of copper in potable waters.

A detailed explanation of the mechanism of pit initiation in the test cell is given, and a synthetic pitting solution, which may be used to check the quality of copper tubes, was described.

Several contributors discussed the use of natural and synthetic seawater for corrosion testing. Castle et al. described their experience in holding natural seawater in the laboratory. They made a detailed study of the microorganisms that developed in the holding tanks over a period of 78 days. They also separated the organic component from seawater by ultrafiltration and added it to synthetic mixes. Using these techniques they described the influence of organic natural products on the corrosion of Kunife 10.

Thomas and Cheung draw attention to the inaccuracies that occur in measurement of corrosion and erosion-corrosion of copper-nickel alloys in synthetic seawater contained in a recirculating loop. Buildup of small quantities of heavy metals in the loop environment leads to spurious results. A method of overcoming this problem was described.

Money and Kain reviewed relative merits of synthetic and natural marine environments (atmospheric and immersed) for corrosion testing. They drew attention to the difference between the use of synthetic environments to simulate natural environments and accelerated test methods that are used to reduce the timescale of the tests. Synthetic environments established for one material must be applied to other materials with great care.

Dexter reviewed the factors contributing to the variability of aluminum corrosion data in saline waters. He has evaluated the requirements of synthetic seawater for studies on the corrosion of aluminum which can reproduce closely the short-term behavior of aluminum alloys, and described an enhanced sodium chloride solution, more simple and more stable than either natural or standard synthetic seawater, which can closely reproduce the short-term behavior of aluminum alloys in seawater.

Kohley and Heitz described a suitable test medium and test method for examining the erosion-corrosion of 13% chromium steel in formation waters containing particulate matter (sand particles). Continuing the theme of hydrocarbon production, McIntyre described suitable methods of assessing the pitting propensity of seawater environments and condensates containing quaternary amines as corrosion inhibitors at 65°C in a CO_2 atmosphere.

Two papers addressed the problem of simulation of synthetic environments for steel in concrete. Sykes and Balkwill compared results from tests on specimens embedded in ordinary portland cement mortar, calcium hydroxide solutions, and a gel made from alkaline solution and chloride ions. They demonstrated that the pitting behavior of steel in gels was markedly different from that in alkaline solution themselves, and that portland cement mortar was less aggressive than the gel.

Dawson and Langford presented data on steel embedded in portland cement mortar and extracted pore solutions. They concluded that although measurements in pore solutions are useful, such an approach has limited application to longer term concrete durability studies.

P. E. Francis

National Physical Laboratory, Teddington, Middlesex, England; symposium chairman and editor.

T. S. Lee

National Association of Corrosion Engineers, Houston, TX 77218; symposium chairman and editor.

Author Index

Subject Index

293